Textile Microplastics in the Environment

Textile Microplastics in the Environment

Sources, Impacts, and Solutions

Aravin Prince Periyasamy
VTT Technical Research Centre of Finland
Espoo, Finland

WILEY

Published by John Wiley & Sons, Inc., Hoboken, New Jersey.

Library of Congress Cataloging-in-Publication Data applied for:

Hardback ISBN: 9781394273140
ePDF ISBN: 9781394273164
ePub ISBN: 9781394273157
oBook ISBN: 9781394273171

Cover Design: Wiley
Cover Image: © Kinga Krzeminska/Getty Images

Set in 9.5/12.5pt STIXTwoText by Straive, Chennai, India

Contents

Preface

The proliferation of microplastics in our environment has emerged as one of the most pressing ecological and public health challenges of the 21st century. Among the myriad sources of these persistent pollutants, textiles and garments have gained increasing attention for their significant contribution to microplastic contamination across air, water, and soil ecosystems.

This book, *Textile Microplastics in the Environment: Sources, Impacts, and Solutions*, is a comprehensive exploration of the multifaceted relationship between textiles and microplastic pollution. It is designed to serve as a foundational resource for researchers, environmental scientists, textile engineers, policymakers, and sustainability advocates who are seeking to understand and mitigate the environmental footprint of textile-derived microplastics.

The chapters are structured to provide a coherent advancement from basic principles to sophisticated solutions. We start with a comprehensive introduction of microplastics and their categorization, followed by a detailed examination of how textiles, via their composition, application, and lifetime, contribute to microplastic emissions. Emphasis is placed on domestic laundering and drying, which are sometimes neglected but substantial contributors to microfiber discharge. The next chapters explore the health risks associated with microplastics, techniques for their detection and quantification, and the increasing significance of machine learning and machine vision in microplastic analysis. The book emphasizes proactive solutions for reduction, including sustainable design, creative business models, and regulatory frameworks. Ultimately, we investigate innovative techniques for the extraction of microplastics from aquatic ecosystems and wastewater, alongside the significance of textile recycling in mitigating pollution.

This work is the result of interdisciplinary collaboration and reflects the latest scientific insights and technological advancements. It is our hope that this book not only informs but also inspires action toward a cleaner, safer, and more sustainable future.

1

Microplastics: General Prospectives

1.1 Introduction

Microplastic is the most significant and widespread change that has been made to the surface of the Earth in the contemporary era [1]. As of 2024, global plastic production has continued to rise, reaching approximately 400 million metric tons (MMT). This marks a significant increase from the 359 MMT produced in 2018 [2]. Plastic is an essential material in our daily lives because of its versatility, durability, and cost-effectiveness [3]. It is utilized in several sectors, including healthcare, transportation, construction, electronics, apparel and textile, home goods, energy, communication, and personal hygiene items. In healthcare, plastics are utilized in sterile and flexible materials such as syringes, intravenous bags, surgical gloves, catheters, and prostheses. Pharmaceutical packaging consists of lightweight plastic containers designed to store and safeguard drugs against infection. Food preservation and packaging utilize plastic wraps, containers, and bottles, which diminish transportation expenses and energy consumption. In transportation, plastics are used in automotive interiors, bumpers, fuel tanks, and lightweight components [4]. Plastic foams offer thermal insulation in structures, enhancing energy efficiency. Plastics are utilized in plumbing, insulation, flooring, and roofing applications. Because of their adaptability, durability, and versatility, synthetic polymers are an indispensable component of the textile and apparel industry. Because they are not only lightweight but also comfortable and resistant to wear and tears, they are an excellent choice for technical and clothing applications. Additionally, they are hydrophobic, which makes them appropriate for clothing that is water-resistant or that dries quickly. Polymers such as spandex have a high degree of flexibility, which makes them suited for use in athletic wear and clothing that are tailored to the body. Moreover, they may be created to imitate natural fibers like wool or silk, which provides additional advantages such as resistance to wrinkles and ease of maintenance

Textile Microplastics in the Environment: Sources, Impacts, and Solutions,
First Edition. Aravin Prince Periyasamy.

[5, 6]. The production of synthetic polymers is typically less expensive than that of natural fibers, which results in textiles being more accessible and inexpensive to a wider audience. Chemical engineering allows for the modification of these materials so that they can display desirable features such as resistance to ultraviolet light, flame retardancy, or antibacterial effects.

Microplastics are tiny plastic particles that are smaller than 5 mm. They can stem from a range of resources consisting of synthetic textiles, tires, roadway markings, aquatic coverings, personal care/cosmetic products, and crafted plastic pellets. They also originate from the fragmentation of bigger plastics gradually overtime [7–9]. Among these sources, synthetic textiles have the highest possibilities for the generation of microplastics. Microplastics pose an escalating environmental issue owing to their ubiquity in ecosystems, possible detrimental effects on animals, and implications for human health. Textiles composed of synthetic polymers such as polyester, nylon, acrylic, and spandex substantially contribute to microplastic pollution [10–12]. These fibers decompose into microplastics due to abrasion, laundering, and landfill disposal. A single washing cycle can discharge hundreds of thousands of microplastic strands. Most wastewater treatment plants are not equipped to filter out microplastics effectively, leading to their discharge into rivers, lakes, and oceans. Abandoned garments and textiles in landfills deteriorate over time, emitting microplastics into soil and aquatic systems. In addition to textile origins, particles from various plastic uses have amassed in terrestrial atmospheres, marine ecosystems, coastal areas, remote islands, the deep ocean, and even on Mount Everest [13]. A significant majority of plastics are manufactured for single-use applications, leading to considerable waste accumulation. Terrestrial regions are the principal origin of plastic pollution in marine ecosystems [14].

Beginning in 2007, there has been an increase in the number of scientific publications on microplastics. These publications include reviews, research papers, and book chapters. Beginning in 2015, there has been a significant exponential surge in the number of publications (see Fig. 1.1). The purpose of these studies was to address the need to understand the link between the environmental repercussions of exposure to microplastics in the marine ecosystem. There have been a total of 6715 articles published throughout this year as of 9 December 2024.

It is possible to divide the publication of microplastics into different subject areas, as shown in Fig. 1.2. By doing so, we are able to comprehend the relationship between microplastics and various subjects, such as environmental science, agricultural sciences, chemistry, earth and plant sciences, and other fields that are related to microplastics.

The textile industry's contribution to microfiber pollution is intensified by fast fashion, which emphasizes the mass manufacture of inexpensive clothes crafted from synthetic fibers [15, 16]. These clothing are more prone to microfiber shedding and disposal, increasing the environmental impact. Moreover, most customers remain oblivious to the environmental contamination linked to

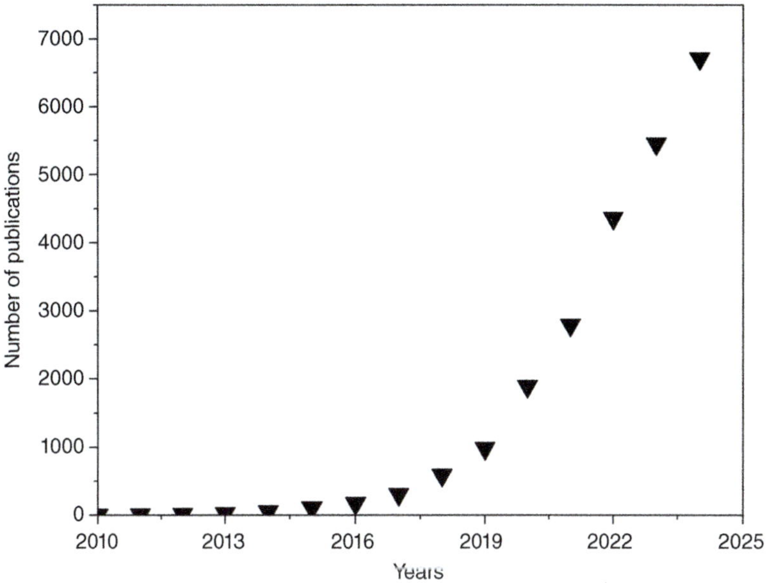

Figure 1.1 Number of publications in the field of microplastics. (*Source:* Scopus; keyword: microplastic on 09.12.2024.)

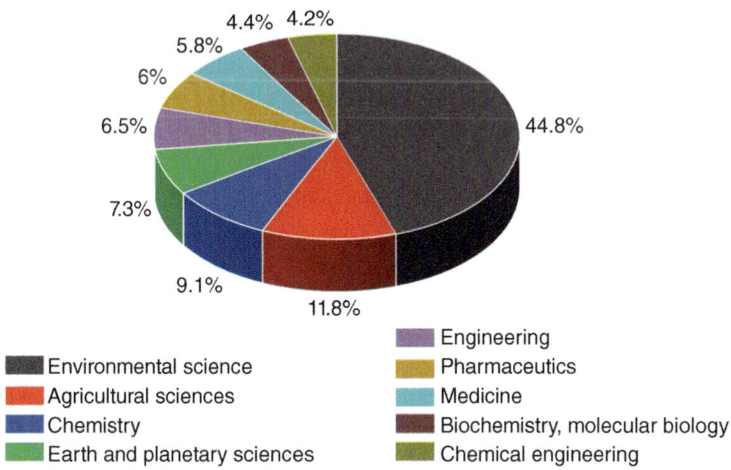

Figure 1.2 Publication by subject area. (*Source:* Scopus; keyword: microplastic on 01.04.2025.)

synthetic and regenerated fibers [17]. In general, microplastics can enter the marine ecosystem through a variety of different pathways; this can be better understood by looking at Fig. 1.3 [18]. Further breakdown of microplastics into nanosized plastics, which are referred to as nanoplastics, occurs in the marine environment. This degradation occurs when the size of the plastic particles is less than 100 nm [19, 20]. Due to the fact that microfibers are so small, fish and other aquatic creatures are the only marine organisms that consume them. After that, these little pieces of microfiber are transmitted into the digestive tracts of aquatic species. Microfiber particles are mistakenly consumed as food by a variety of marine species, including fish, shellfish, and other large marine animals. Both marine creatures and humans consume microfibers as a component of their diets. Microfibers are absorbed by the cells of marine animals. Additionally, several birds that are dependent on small fish are mistakenly swallowing these microfibers [21]. This is a similar situation to the last one. There is a possibility that microfibers in bodies of water might be the cause of a variety of problems, including cosmetic problems, complications, and so on. It has been estimated that mussels consume roughly 11,000 microplastic objects on an annual basis [22].

1.1.1 Issues of Microplastics

Microplastics represent a considerable environmental hazard, polluting ecosystems worldwide. Notwithstanding their little size, they provide considerable dangers to the environment, human health, and marine ecosystems. Their persistence, bioaccumulative potential, and widespread presence render them a considerable threat. Aquatic ecosystems are significantly impacted, with millions of tons of microplastics permeating marine areas each year. These particles are ingested by marine organisms, accumulating along the food chain, jeopardizing marine biodiversity and impacting fisheries and livelihoods [15, 16]. Microplastics provide considerable health hazards to people, as they are ingested via seafood, drinking water, and airborne particulates. They frequently convey dangerous substances such as phthalates and bisphenol A (BPA), which can interfere with biological systems and induce hormonal imbalances, developmental issues, and chronic diseases [23, 24]. The resilience of microplastics enables them to transport pollutants such as heavy metals and hydrophobic organic compounds, endangering fauna, soil integrity, water quality, agricultural methods, and drinking water sources. Addressing microplastics necessitates a multifaceted approach that encompasses policy, technology, and public awareness. Governments and international organizations must implement regulations to curtail plastic manufacturing and waste, while innovations in materials science might mitigate microplastic dissemination. Public education initiatives are essential for fostering sustainable practices and behavioral modifications.

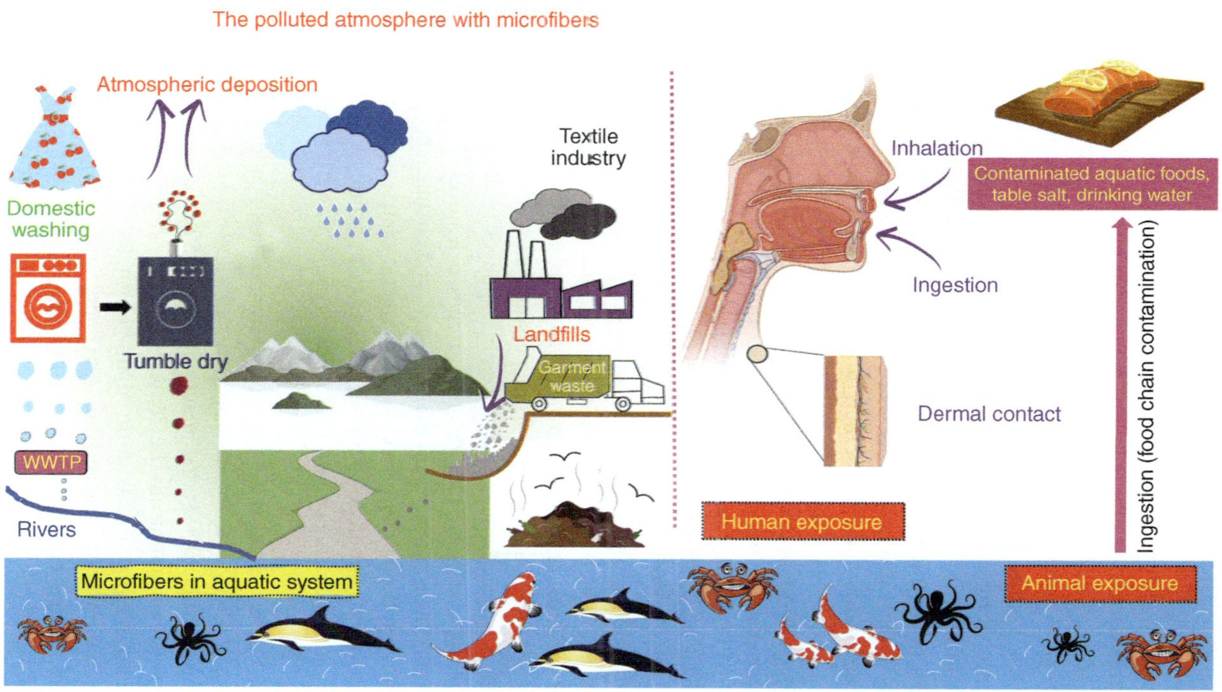

Figure 1.3 Textile and apparel as a source of microplastic migration into the aquatic environment, and their possible routes to humans. (*Source:* Periyasamy [17] / with permission of MDPI / CC BY-NC 4.0.)

1.1.2 Definition of Microplastics and Nanoplastics

The term "microplastic" was coined by Thompson et al. [25] to describe the accumulation of minute plastic particulates in aquatic detritus throughout Europe. Arthur and coworkers [26] suggested a restriction on the top dimension for the first term, as well as microplastics, which are defined as "plastic particles smaller than 5 mm" [27]. In another definition, microplastics are synthetic solid particles or polymeric matrices characterized by regular or irregular forms and dimensions ranging from 1 μm to 5 mm [28]. The origin of microplastics was further refined by Cole et al. [29] by distinguishing between primary and secondary sources. Synthetic fibers, plastic granules, and personal care items are all sources of primary microplastics, which are particulates that are intentionally manufactured to be extremely small. The degradation of larger plastic objects is a result of environmental factors, including sunlight, UV radiation, and other light sources, as well as wind and wave action. Secondary microplastics are generated as a consequence of this process. For example, the primary sources of secondary microplastics are textile materials, such as fibers, filaments, fabrics, and garments. Additionally, recyclable materials include fishing gear constructed from abandoned nets and lines, as well as bottles and containers. Some of the microplastics that may be present in the environment include pearls, fibers, foams, and fragments. The degradation of larger plastics leads to the formation of fragments, which are irregularly shaped pieces; fibers, which are thin particles that resemble thread and are derived from textiles; beads, which are spherical particles that are derived from personal care products; and foams, which are microscopic particles that resemble sponges and are derived from polystyrene. Frias and Nash [30] recently proposed a novel interpretation of microplastics that takes into account their physical and chemical properties, size, and origin: Any synthetic solid particle or polymeric matrix, with a regular or irregular shape and a dimension ranging from 1 μm to 5 mm, of either primary or secondary manufacturing origin, that is insoluble in water is referred to as a "microplastic."

Plastics are further categorized into five primary classifications for a better understanding of their size, for instance, nanoplastic (<0.3 mm), microplastic (<5 mm), mesoplastic (0.5–5 cm), macroplastic (5–50 cm), and megaplastic (>50 cm), which can graphically be represented in Fig. 1.4. Microplastics are plastic particles having a maximum diameter under 5 mm, encompassing nanoplastics. Nanoplastics, measuring less than 1 μm, are generated by the decomposition of bigger plastic pieces or industrial operations, enabling them to infiltrate ecosystems more profoundly and exhibit varied distribution patterns. Both particle kinds are becoming a concern owing to their potential

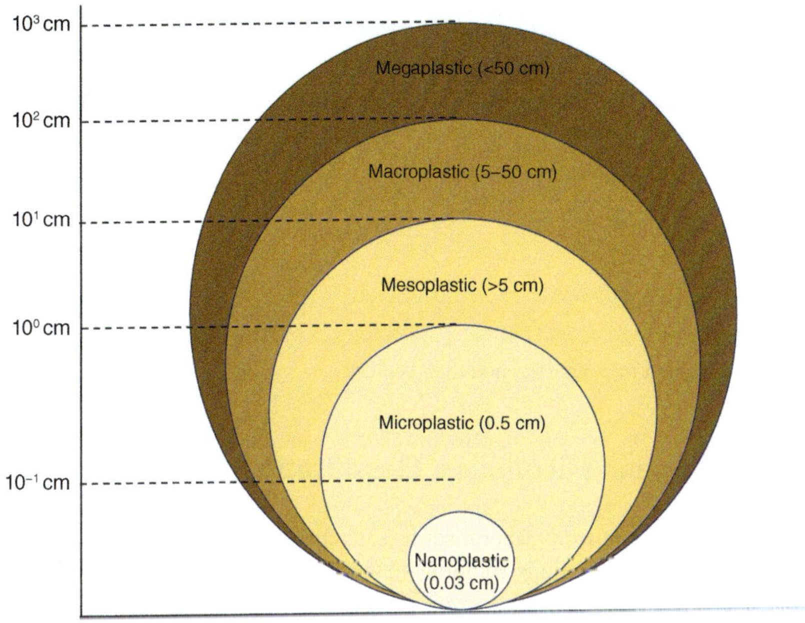

Figure 1.4 Categories of plastic classified by size: nanoplastic (<0.03 cm), microplastic (0.05–0.5 cm), mesoplastic (0.5–5 cm), macroplastic (5–50 cm), and megaplastic (>50 cm) (according to prior research).

environmental and health hazards. This book solely focuses on micro-level plastic particles (fibrous fragments) that are emitted from textile and apparel origin.

Various authors offer distinct definitions; hence, the decisions made by the European Union (EU), American Association of Textile Chemists and Colorists (AATCC), and The Microfibre Consortium (TMC) have yielded a more precise definition that alleviates the general confusion around the current nomenclature. Microplastics are water-insoluble macromolecular plastic particles measuring less than 5 mm. AATCC defines a fiber fragment as a small segment (usually $<5 \times 10^{-3}$ m in length) of textile fiber that has detached from a textile structure. For TMC, a microfiber or fiber fragment denotes a brief segment of textile fiber that has detached from the primary textile structure or has subsequently fractured in the natural environment. Microplastics denote tiny fragments of plastic garbage of 5 mm or less, present in the environment due to the dumping or degradation of consumer goods and industrial waste. Figure 1.5 illustrates the many morphologies of microplastics, specifically emphasizing fibrous fragments detached from textiles, characterized by a greater length-to-diameter ratio.

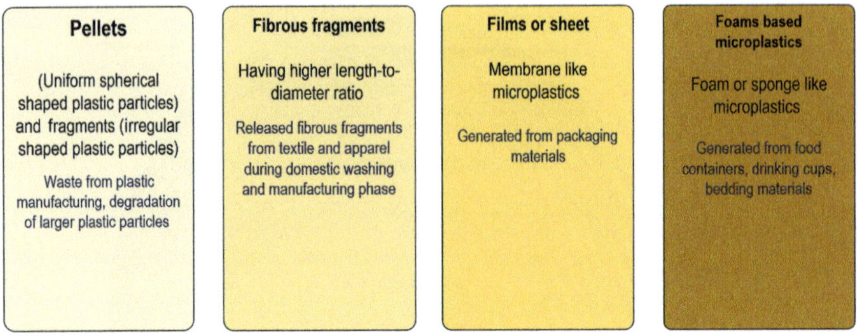

Figure 1.5 Different microplastics and their sources.

1.2 Textiles and Microfibers, Classification

Microplastics originate from several sources, such as personal care items, tires, road markings, marine coatings, building materials, agricultural plastics, landfills, urban dust, and textiles, which is well studied in the previous section of this chapter. Textiles contribute to microfibers through the utilization of man-made cellulosic fibers (MMCFs); natural fibers such as cotton, wool, and silk; and synthetic fibers like polyethylene terephthalate (PET), polyamide (PA), and polyacrylonitrile (PAN). Clothing accessories such as buttons and zippers are made from different polymers, which contribute to environmental pollution by detaching micro-level particles.

Global fiber production has increased fourfold in the past twenty years, attaining 124 MMT in 2023, with projections of reaching 160 MMT by 2030 [31]. Synthetic fibers comprised 54% of the worldwide textile industry's production in 2023. Nevertheless, customers continue to select synthetic materials because of their cost-effectiveness and availability. The durability, accessibility, and cost-effectiveness of synthetic-based polymers contribute to environmental pollution (see Fig. 1.6) [32]. Nonetheless, consumers choose clothes composed of synthetic materials due to their affordability and accessibility. The advancement of fashion would undoubtedly be impractical without synthetic fibers [33]. The demand for and utilization of synthetic fibers has markedly escalated over the past twenty years, prompting apprehensions about soil pollution and the possible discharge of microplastic particles at both consumer and post-consumer stages. China constitutes around 65% of global synthetic fiber output, positioning the Asia-Pacific region as the leader in both production and consumption metrics [31].

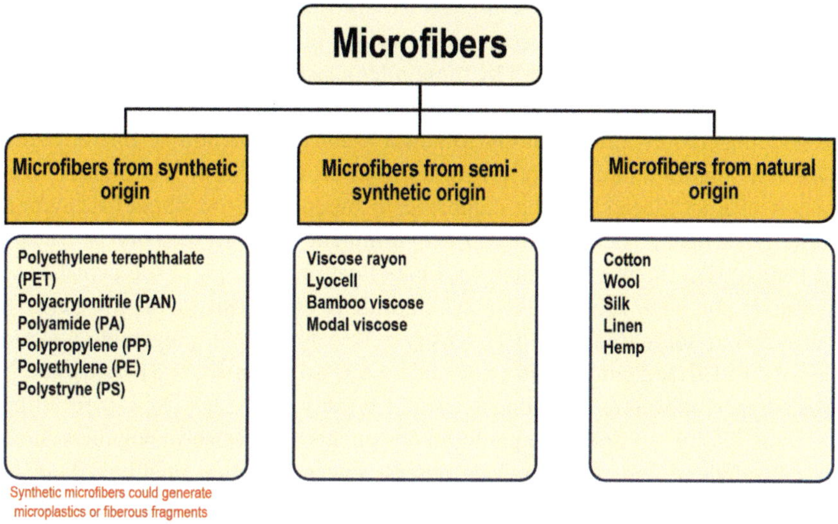

Figure 1.6 Different types of microfibers originated from textiles. *Note: Microfibers generated from synthetic origins are called microplastics.*

1.2.1 Synthetic Textiles and Emission of Microfibers

Synthetic fibers provide several benefits; nonetheless, they are non-biodegradable and need substantial fossil-based ingredients for production, which are classified as non-renewable resources [34–36]. Currently, synthetic fibers comprise over 60% of total fiber demand and consumption in the market. Furthermore, the recycling rate of synthetic fiber or whole textiles is around 10–15%, while the remainder is sent to landfills, energy generation, and other negligible uses. Conversely, the manufacture of synthetic fibers necessitates substantial energy consumption and results in the emission of numerous greenhouse gases, including N_2O and CO_2 [37, 38]. For example, N_2O is emitted during the synthesis of polyamide, which is regarded as particularly severe compared to the emissions of other greenhouse gases, including CO_2. Despite several downsides, the shedding of microfibers and particles from synthetic fibers during home washing remains a significant concern [39]. Microplastics are accumulating in numerous environmental sectors and are currently seen as a worldwide concern. The principal causes of microfiber pollution include household laundering of synthetic fiber clothes, the synthetic textile industry, textile waste, and landfills located near rivers and residential drainage systems. These microfibers may be more detrimental than the more spherical microplastics due to their increased surface area. The ocean is the largest biome on Earth, encompassing approximately 71% of the planet's surface area. Each biome possesses a unique variety that encompasses the smallest microbes and the largest species, such as the blue whale. Nevertheless, the deep ocean has been infiltrated by these microplastics, potentially leading to an exponential increase in pollution. Observations indicate that both developed and developing nations are primarily responsible for the creation of microplastics [40]. The predominant use of polyester in clothes results in the greatest likelihood of microfiber shedding compared to other synthetic fibers [41]. When these microfibers reach the marine environment, polymer density is crucial, as it directly influences whether the fibers float or settle [42]. The densities of synthetic polymers such as polyvinyl chloride (PVC, density 1.14–1.56 $g.cm^{-3}$), polyethylene terephthalate (PET, density 1.32–1.46 $g.cm^{-3}$), polyamide 6 (PA6, density 1.31–1.36 $g.cm^{-3}$), and polyamide 6,6 (PA6,6, density 1.31–1.36 $g.cm^{-3}$) exceed that of water (including seawater, density 1.02 $g.cm^{-3}$) [43], resulting in their ability to sink in aquatic environments. Conversely, polymers such as polyethylene and polypropylene possess lower densities, enabling them to float on water surfaces. Nevertheless, the aquatic biota augments the polymers' weight through biofouling and the colonization process [44]. The International Union for Conservation of Nature (IUCN) assessment indicates that synthetic clothes contribute 34.8% of microfiber emissions to the ocean [40]. Nonetheless, emissions might differ according to clothing material, with fleece exhibiting

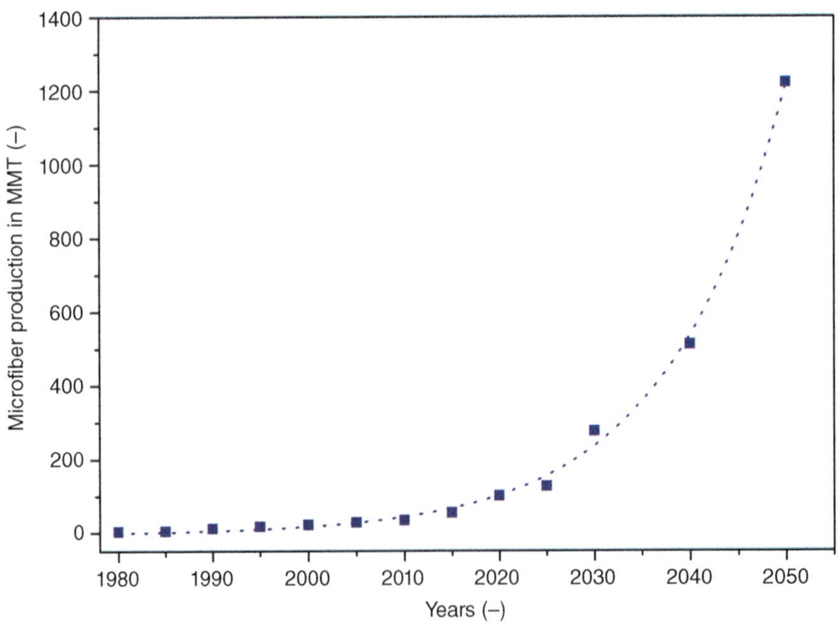

Figure 1.7 Historical and expected rate of microfiber production.

the highest propensity for microfiber release. In addition to standard apparel, home-furnishing textiles also generate elevated levels of microfibers, including curtains, rugs, beds, and tablecloths. Furthermore, in addition to textile items, plastic bottles, bottle caps, food packaging materials, everyday bags composed of polyolefins, tires, and paints contribute to microfiber emissions. Consequently, an increase in microfiber pollution is expected due to the demand for synthetic fibers, as seen in Fig. 1.7 [45].

1.2.2 Microfibers Released from Man-Made or Regenerated Fibers

Although man-made and regenerated fibers are often biodegradable, their environmental effect is considerable. Regenerated fibers emit microfibers after washing, which decompose more swiftly than synthetic fibers in soil or compost; however, they may last in aquatic settings, especially under low-oxygen circumstances. Chemicals utilized in their manufacture, such as caustic soda and carbon disulfide, may seep into ecosystems, resulting in localized toxicity and potentially jeopardizing aquatic species. Functionalization processes such as dyeing and finishing can make these fibers resistant to biodegradation, leading to surface-modified textiles with microfibers that last in the ocean. The possible health impacts are intensified by chemicals in synthetic fabrics, including dyes, flame retardants, and plasticizers, which interfere with endocrine function or encourage carcinogenic activity. Regenerated fibers, although less likely to remain in the body due to their partial biodegradability, may nevertheless provide dangers through the release of leftover processing chemicals.

1.2.3 Microfibers or Fibrous Fragments from Natural Fibers

Natural fibers such as cotton, wool, silk, and hemp emit fibrous particles into the atmosphere during activities like laundering and usage. Despite being biodegradable, their effects on ecosystems and human health are more intricate than previously assumed. Natural fibers may deteriorate more slowly in aquatic environments due to restricted oxygen levels or less microbial activity, and they can also emit coloring agents, insecticides, and other additives during cultivation and processing.

Aquatic species may confuse natural fibers with food, resulting in physical obstructions, diminished nutritional absorption, and physiological stress. The deterioration of natural fibers in aquatic environments frequently leads to the liberation of minute particles, such as cellulose, which can induce inflammation and cellular harm. Human exposure to natural fibers is a significant problem, since research indicates that natural fiber microfibers are more prevalent in both indoor and outdoor air, typically released from clothes and textiles due to friction and wear. This inhalation may result in respiratory complications, especially in industrial environments where natural fibers such as cotton are handled. Certain natural fibers undergo treatment with chemical agents during manufacturing, potentially presenting toxicological hazards with extended exposure. The role of natural fibers in microfiber contamination underscores wider concerns associated with textile manufacturing and consumer behavior. The escalating desire for fast fashion has resulted in heightened textile waste and increasing washing frequency, intensifying the discharge of microfibers into the environment. The ecological impact of natural fibers is significantly influenced by the techniques employed in their production, processing, and disposal.

1.3 Microplastics in Air Stream – Case Studies with Some Cities

Microplastics, present in water, soil, and air, provide a health hazard owing to their propensity for respiratory and oral exposure. Contaminated urban fountains emit these particles, along with mist and water aerosols, into the environment, potentially leading to airborne microplastics. These particles, measuring between 2 μm and 1 mm, have enhanced adsorption capabilities for copper and tetracycline. The weathering processes and attributes of airborne microplastics contrast with those of terrestrial or aquatic plastics, with aged microplastics presenting a heightened environmental and health risk. They can absorb and release pollutants such as polycyclic aromatic hydrocarbons (PAHs), polychlorinated biphenyls (PCBs), antibiotics, and leachates such as BPA, phthalates, and other endocrine disruptors. Temperature profoundly influences the adsorption and desorption of chemicals in the environment. Research on the air transport of microplastic particles is scarce, especially concerning the mechanisms and pathways involved. Comprehending airborne deposition, transport, scavenging, and precipitation of microplastics is crucial for identifying probable sources and evaluating their effects on susceptible ecosystems. Identifying vulnerable locations would enhance comprehension of microplastic impacts on air pollution. The existing literature on the transference of airborne microplastics from urban to distant regions is insufficient, underscoring the necessity for future research to explore the atmospheric transfer processes of both primary and secondary microplastics. A simplified visual representation of the airborne transport of microplastics via wet and dry deposition is illustrated in Fig. 1.8.

1.3.1 Indoor Air

The indoor environment is crucial for modern human life, with up to 90% of people spending time indoors. The quality of interior settings significantly

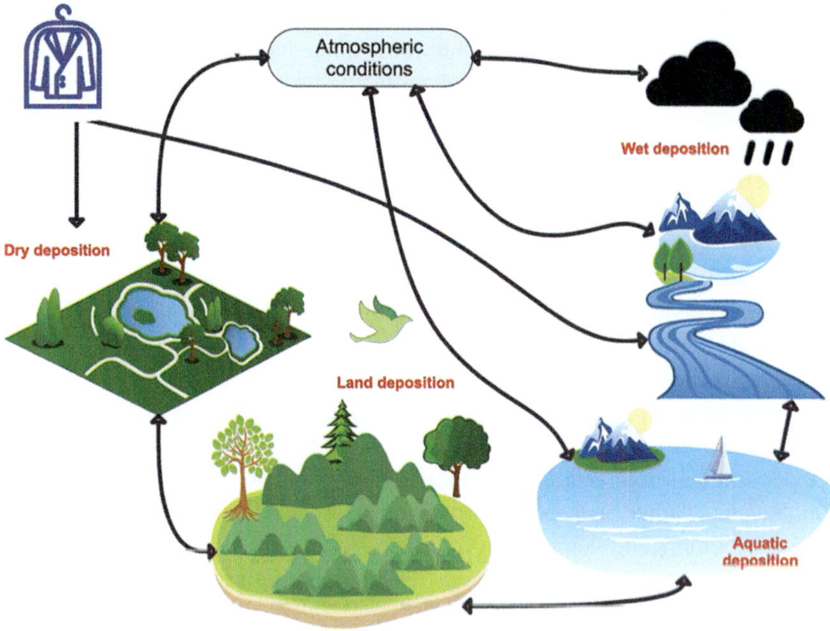

Figure 1.8 Analysis of the airborne transport of microplastics by wet and dry deposition.

influences human health and well-being. The COVID-19 outbreak has led to a rise in remote work and expanded opportunities for various everyday activities within indoor settings. Furthermore, lockdown measures adopted to curb the transmission of COVID-19 may increase exposure to novel contaminants, such as microplastics, in interior settings. The current understanding of microplastic concentrations in environmental media relevant to indoor exposure includes dust, air, food, and drinking water [46].

Indoor dust is widely acknowledged as a major conduit for human exposure to a wide array of contaminants. This includes residues from construction materials, paint particulates, microbes, heavy metals (particularly lead), and other organic chemical contaminants like flame retardants and plasticizers. Recent research has broadened this issue to encompass microplastics and other plastic particles that last in interior settings [46]. Research has demonstrated that indoor dust may harbor microplastics in various amounts, forms, and dimensions, particularly microfibers derived from textiles, furniture, and synthetic carpets. Pivotal research was conducted in three indoor settings in Paris – two private residences and one university office examined dust samples obtained from vacuum cleaner bags. The results indicated microplastic fiber concentrations varying from 1 to 60 fibers m^{-3} of air. The indoor concentration was much greater than the outside dust samples, which varied from 0.3 to 1.5 fibers m^{-3} [46]. The heightened concentrations of microplastics in interior settings indicate that indoor air and dust constitute a more significant exposure pathway for people than outside surroundings. The ongoing release of synthetic fibers from routine activities such as wearing garments, cleaning, and ventilation contributes to this accumulation. Considering that the majority of individuals allocate significant time indoors, particularly in urban environments, the health ramifications of prolonged exposure to airborne and dust-associated microplastics encompassing possible respiratory, inflammatory, and toxicological effects are increasingly concerning.

Research conducted in Shenzhen, China, revealed that anthropogenic activities considerably enhance the emission of atmospheric microplastics (AMPs), averaging 2.22 particles m^{-3} [47]. The primary varieties of microplastics were polyethylene, polypropylene, and PET, predominantly in fibrous forms. Research in Greater Paris, France, emphasized the consumption of microplastics by diverse species and their interaction with other contaminants [48]. A swift evaluation in Philadelphia, United States, concentrated on active air sampling methodologies to delineate outdoor air microplastics employing sophisticated analytical techniques such as FTIR and Raman spectroscopy. The results emphasized the necessity for thorough human risk evaluations [48, 49].

1.4 Conclusion

Microplastics, defined as plastic particles smaller than 5 mm, are a considerable environmental and public health issue in the 21st century. They derive from several sources, including synthetic fabrics, tire abrasion, packaging materials, personal care items, and the decomposition of bigger plastic debris. Textiles, especially those composed of synthetic fibers such as polyester and nylon, significantly contribute to microplastic pollution through manufacture, use, and household laundry. Airborne microplastics represent an escalating issue, identified in both indoor and outdoor environments. Indoors, dust contaminated with microplastics collects from synthetic textiles, upholstery, carpets, and washing machines, rendering inhalation a notable and underreported pathway of human exposure. In outdoor environments, airborne fibers can be carried over considerable distances by wind, subsequently settling in terrestrial and marine ecosystems, hence exacerbating environmental pollution. Immediate measures are required to address microplastic contamination, encompassing the creation of safer textile materials, enhanced filtering systems in washing machines, improvements in fabric design, and more public awareness. A comprehensive, lifecycle-oriented strategy is essential to tackle the origins, routes, and final disposition of microplastic contamination in the environment.

References

1 Barnes, D.K.A., Galgani, F., Thompson, R.C., and Barlaz, M. (2009). Accumulation and fragmentation of plastic debris in global environments. *Philosophical Transactions of the Royal Society, B: Biological Sciences* 364 (1526): 1985–1998.

2 Plastics Europe (2019) Plastics Europe, 2019. Plastics – the facts 2019. An analysis of European latest plastics production, demand and waste data. https://plasticseurope.org/knowledge-hub/plastics-the-facts-2019/#:~:text=An%20analysis%20of%20European%20plastics%20production%2C%20demand%20and,as%20employment%20and%20turnover%20in%20the%20plastics%20industry.

3 Periyasamy, A.P. (2021). Evaluation of microfiber release from jeans: the impact of different washing conditions. *Environmental Science and Pollution Research* 28 (41): 58570–58582.

4 Prata, J.C. (2018). Airborne microplastics: consequences to human health? *Environmental Pollution* 234: 115–126.

5 Prata, J.C., da Costa, J.P., Lopes, I. et al. (2020). Environmental exposure to microplastics: an overview on possible human health effects. *Science of the Total Environment* 702: 134455.

6 Li, Y., Tao, L., Wang, Q. et al. (2023). Potential health impact of microplastics: a review of environmental distribution, human exposure, and toxic effects. *Environment & Health* 1 (4): 249–257.

7 Cole, M., Lindeque, P., Halsband, C., and Galloway, T.S. (2011). Microplastics as contaminants in the marine environment: a review. *Marine Pollution Bulletin* 62 (12): 2588–2597.

8 Robertson, R. (2018) *Are Microplastics in Food a Threat to Your Health?* Healthline (9 May).

9 Sun, A. and Wang, W.-X. (2023). Human exposure to microplastics and its associated health risks. *Environment & Health* 1 (3): 139–149.

10 Belzagui, F., Crespi, M., Álvarez, A. et al. (2019). Microplastics' emissions: microfibers' detachment from textile garments. *Environmental Pollution* 248: 1028–1035.

11 Carney Almroth, B.M., Åström, L., Roslund, S. et al. (2018). Quantifying shedding of synthetic fibers from textiles; a source of microplastics released into the environment. *Environmental Science and Pollution Research* 25 (2): 1191–1199.

12 Dris, R., Gasperi, J., Mirande, C. et al. (2017). A first overview of textile fibers, including microplastics, in indoor and outdoor environments. *Environmental Pollution* 221: 453–458.

13 Geyer, R., Jambeck, J.R., and Law, K.L. (2017). Production, use, and fate of all plastics ever made. *Science Advances* 3 (7): e1700782.

14 Derraik, J.G.B. (2002). The pollution of the marine environment by plastic debris: a review. *Marine Pollution Bulletin* 44 (9): 842–852.

15 Periyasamy, A.P. and Tehrani-Bagha, A. (2022). A review on microplastic emission from textile materials and its reduction techniques. *Polymer Degradation and Stability* 199: 109901.

16 Periyasamy, A.P. (2024). Sustainable approaches in textile finishing to control microfiber releases. In: *Microfibre Pollution from Textiles*. London UK: Taylor & Francis Group.

17 Periyasamy, A.P. (2023). Environmentally friendly approach to the reduction of microplastics during domestic washing: prospects for machine vision in microplastics reduction. *Toxics* 11 (7): 575.

18 Huang, W., Song, B., Liang, J. et al. (2020). Microplastics and associated contaminants in the aquatic environment: a review on their ecotoxicological effects, trophic transfer, and potential impacts to human health. *Journal of Hazardous Materials* 405: 124187.

19 Hernandez, L.M., Xu, E.G., Larsson, H.C.E. et al. (2019). Plastic teabags release billions of microparticles and nanoparticles into tea. *Environmental Science & Technology* 53 (21): 12300–12310.

20 Lambert, S. and Wagner, M. (2016). Characterisation of nanoplastics during the degradation of polystyrene. *Chemosphere* 145: 265–268.

21 Boddy, J. (2017) Are we eating our fleece jackets? Microfibers are migrating into field and food. NPR. https://www.npr.org/sections/thesalt/2017/02/06/511843443/are-we-eating-our-fleece-jackets-microfibers-are-migrating-into-field-and-food.

22 Murphy, F., Ewins, C., Carbonnier, F., and Quinn, B. (2016). Wastewater treatment works (WwTW) as a source of microplastics in the aquatic environment. *Environmental Science & Technology* 50 (11): 5800–5808.

23 Periyasamy, A.P. (2023). Microfiber emissions from functionalized textiles: potential threat for human health and environmental risks. *Toxics* 11 (5): 406.

24 Periyasamy, A.P. (2025). Functionalized textile microplastics: a closer look at the issues, strategy, and legislation on the microplastic reduction. *Kuwait Journal of Science* 52 (3): 100395.

25 Thompson, R.C., Olsen, Y., Mitchell, R.P. et al. (2004). Lost at sea: where is all the plastic? *Science (1979)* 304 (5672): 838–838.

26 Koelmans, A.A., Gouin, T., Thompson, R. et al. (2013). Plastics in the marine environment. *Environmental Toxicology and Chemistry* 33 (1): 5–10.

27 Thompson, R.C., Moore, C.J., vom Saal, F.S., and Swan, S.H. (2009). Plastics, the environment and human health: current consensus and future trends. *Philosophical Transactions of the Royal Society, B: Biological Sciences* 364 (1526): 2153–2166.

28 Wu, P., Huang, J., Zheng, Y. et al. (2019). Environmental occurrences, fate, and impacts of microplastics. *Ecotoxicology and Environmental Safety* 184: 109612.

29 Cole, M., Lindeque, P.K., and Galloway, T.S. (2011). Microplastics as contaminants in the marine environment: a review microplastics as contaminants in the marine environment: a review. *Marine Pollution Bulletin* 62 (12): 2588–2597.

30 Frias, J.P.G.L. and Nash, R. (2019). Microplastics: finding a consensus on the definition. *Marine Pollution Bulletin* 138: 145–147.

31 Common Objective (2018) Synthetics & sustainable synthetics: global production *Common Objective.*

32 Mondal, I.H. (2014). *Textiles: History, Properties and Performance and Applications.* Nova Publishers.

33 Beverley, H., Kirsi, L., and Grimstad, K.I. (2018) Microplastic pollution from textiles: a literature review. https://oda.oslomet.no/oda-xmlui/handle/20.500.12199/5360

34 Peller, J.R., Nelson, C.R., Babu, B.G. et al. (2020). A review of microplastics in freshwater environments: locations, methods, and pollution loads. In: *Contaminants in Our Water: Identification and Remediation Methods, ACS Symposium Series*, vol. 1352, 65–90.

35 Venkatesan, H. and Periyasamy, A.P. (2019). Eco-fibers in the textile industry. In: *Handbook of Ecomaterials*, vol. 3, 1413–1433. Cham: Springer International Publishing.

36 Bunsell, A.R. (2018). Introduction to the science of fibers. In: *Handbook of Properties of Textile and Technical Fibres*, 1–20. Elsevier.

37 Overview of greenhouse gases (2018). https://www.climate.gov/news-features/understanding-climate/reporting-state-climate-2018

38 Thiemens, M.H. and Trogler, W.C. (1991). Nylon production: an unknown source of atmospheric nitrous oxide. *Science (1979)* 251 (4996): 932–934.

39 Mishra, S., Singh, R.P., Rath, C.C., and Das, A.P. (2020). Synthetic microfibers: source, transport and their remediation. *Journal of Water Process Engineering* 38: 101612.

40 Boucher, J., and Friot, D. (2017) Primary microplastics in the oceans: a global evaluation of sources, IUCN International Union for Conservation of Nature.

41 Kristen, B. (2017) What are microfibers and why are our clothes polluting the oceans? Greenpeace International.

42 Wright, S.L., Thompson, R.C., and Galloway, T.S. (2013). The physical impacts of microplastics on marine organisms: a review. *Environmental Pollution* 178: 483–492.

43 Van Cauwenberghe, L., Devriese, L., Galgani, F. et al. (2015). Microplastics in sediments: a review of techniques, occurrence and effects. *Marine Environmental Research* 111: 5–17.

44 Ye, S. and Andrady, A.L. (1991). Fouling of floating plastic debris under Biscayne Bay exposure conditions. *Marine Pollution Bulletin* 22 (12): 608–613.

45 Lamichhane, G. (2018) Analysis of microfibers in waste water from washing machines. https://www.theseus.fi/bitstream/handle/10024/141278/Lamichhane_Ganesh.pdf?sequence=1.

46 Ageel, H.K., Harrad, S., and Abdallah, M.A.-E. (2022). Occurrence, human exposure, and risk of microplastics in the indoor environment. *Environmental Science. Processes & Impacts* 24 (1): 17–31.

47 Chang, K., Yuan, Y., Ma, Y. et al. (2025). Characterization of atmospheric microplastics: a case study in Shenzhen City, a southern coastal area of China. *Environmental Monitoring and Assessment* 197 (3): 294.

48 Dris, R., Gasperi, J., Rocher, V. et al. (2015). Microplastic contamination in an urban area: a case study in Greater Paris. *Environmental Chemistry* 12 (5): 592.

49 Han, I., Lee, C., Belchez, C. et al. (2024). Microplastics in urban ambient air: a rapid review of active sampling and analytical methods for human risk assessment. *Environments* 11 (11): 256.

2

Sources of Microplastics: Textile and Technical Textiles

2.1 Introduction

Textile fibers, crucial to producing yarns and textiles, are classified into natural (e.g. cotton and wool) and synthetic (e.g. polyester and nylon) categories. Synthetic fibers, originating from petrochemicals, present environmental issues owing to their propensity to release microplastics over their lifespan [1]. Although they are preferred in the textile sector for their durability, lightweight characteristics, and resistance to shrinkage and wrinkles, their non-biodegradability results in prolonged environmental persistence [2, 3]. The characteristics of fibers, such as length, diameter, and surface texture, influence their propensity to fragment, with longer, smoother fibers exhibiting reduced susceptibility, as has been well explained in *Chapter* 1. This chapter examines the origins of fibrous fragments from textiles and technical textiles, their effects on aging, and the correlation between aging and the production of fibrous fragments. Typically, all textile fibers emit microplastic fibers or fibrous fragments, and however, these fibrous fragments derived from natural cellulose-based and protein-based fibers tend to degrade rapidly under environmental conditions, posing minimal issues unless present in high concentrations, such as in large-capacity warehouses [4–7]. Their biodegradability may be influenced by surface modifications or functionalizations [8, 9]. The disintegration of fibrous microplastics from synthetic fibers is gradual, occurring over tens to hundreds of years under appropriate climatic circumstances, leading to their buildup and the development of significant issues [4–7].

Surface abrasion, fatigue-induced degradation, and the deterioration of the fibrillar structure are all factors that are related to the formation of microplastics at the fiber level (Fig. 2.1a). Additionally, there is a greater chance as a result of chemical modifications of fibers, such as oxidation (Fig. 2.1b), as well as physical activities, such as surface abrasion (Fig. 2.1c). Microplastics at the fibrous structural level are associated with the extraction of projecting (broken) fibers,

Textile Microplastics in the Environment: Sources, Impacts, and Solutions,
First Edition. Aravin Prince Periyasamy.

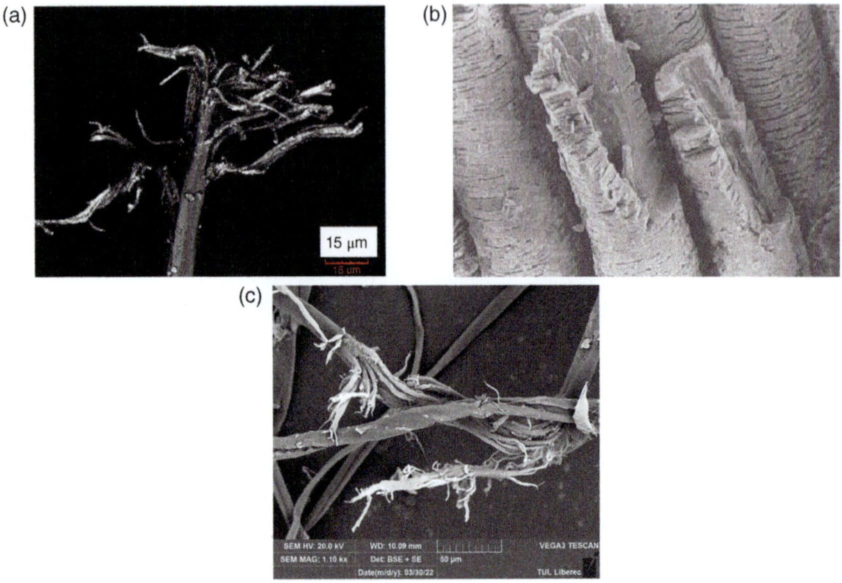

Figure 2.1 Confocal microscope image of cotton fibers having the fibrillation at the end of the fibers (a), frequent degradation of polypropylene fibers due to oxidation (b), and the surface rupture of cotton fibers due to abrasion activities (c). (*Source:* Militký et al. [10] / MDPI / CC BY 4.0.)

which pertain to hairiness. The emission of fibrous microplastics can therefore be regulated by careful selection of the composition, design, and fabrication techniques of the fibrous structures.

Fiber-forming synthetic polymers are typically linear thermoplastic backbones of substantial side groups. The breakdown temperature of the polymer must exceed the melting point by at least 30 °C (melting temperature > 30 °C) for melt spinning. In the manufacturing of synthetic fibers, the polymer melt is initially extruded via apertures in a spinning nozzle. This is succeeded by the solidification of the liquid jet and distortion by drawing. The transition to solid fiber occurs upon cooling below the melting point, whereas the cool air- and water-cooling bath is used for solidification. This is succeeded by tensile deformation – the elongation of the fiber – during which the alignment of the chains and polymer segments shifts along the fiber axis, and crystallization occurs with the development of a fibrillar structure, characteristic of both natural and synthetic fibers [10]. The last procedure in which the development of a fibrillar fibrous structure is finalized is the stability of the fibers, achieved by heating or swelling, which facilitates the relaxation of internal tensions and recrystallization. During isometric fixing (at constant lengths), the orientation of the polymer chains remains unchanged, and their mutual slippage transpires. Shrinkage transpires during isotonic fixation. In both instances, internal pressures dissipate [10].

Degradation mechanisms induce depolymerization and random chain scission of the polymers. Radical depolymerization predominantly occurs during thermal degradation (i.e. during production, domestic washing under higher temperature), whereas the intensity of photodegradation is contingent upon the quantity of light energy (i.e. washing and wearing phases). The energy of photons having a wavelength of 400–300 nm is 300–390 kJ mol^{-1}. The bond energy of a C—C bond is approximately 420 kJ mol^{-1} [10]. The initial step involves severing the main connections within the chains and facilitating depolymerization. Hydrolytic deterioration results from the influence of water in a hostile liquid environment. The fundamental procedure involves chain disruption. Mechanical degradation occurs as shear force fractures polymeric chains and produces radicals. Certain forms of mechanical deterioration, such as abrasion, do not influence the length of the polymer chains. Diverse degradations occur concurrently. Microplastics are frequently generated by hydrolysis, particularly from synthetic fibers, during the consumer phase of washing and wearing. The conventional fragmentation of **polyethylene terephthalate** (PET) fibers during hydrolysis is seen schematically in Fig. 2.2. Nonetheless, the influence of laundering and its correlation with microplastic production is thoroughly examined in *Chapter* 3.

The hydrolysis of PET in boiling water yields terephthalic acid (TPA) and ethylene glycol (EG), which possess unique characteristics. As temperature rises, the proteolytic constant escalates, resulting in a greater availability of protons, independent of acid or base catalysts (see Fig. 2.2). Moreover, the dielectric constant of water diminishes with temperature, enhancing the solubility of organic molecules, such as TPA, which is almost insoluble at ambient temperature. Upon cooling, TPA crystallizes from the solution (see Fig. 2.2) [11].

In textile manufacturing, degradation processes may result in damage from high heat and mechanical stress, causing structural alterations in fibers and the creation of polyester fibers and oligo-ester. Challenges encompass heightened dust levels and complications in spinnability resulting from the deterioration

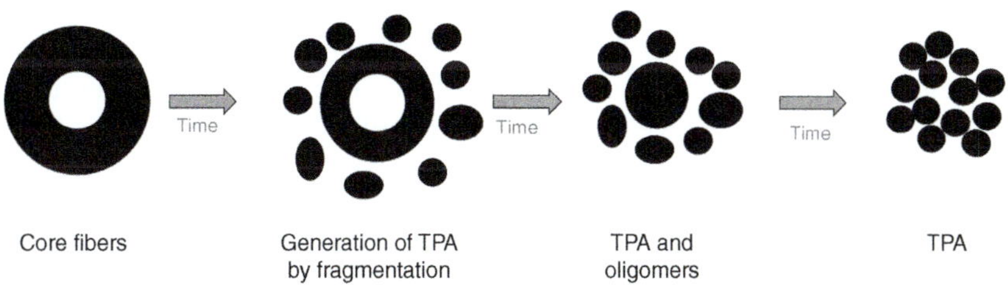

Figure 2.2 Hydrolysis of PET during domestic washing with harsh conditions.

of guides and needles. To reduce microplastic development in synthetic fiber manufacture, the use of protein or fat-based lubricants is recommended. These can mitigate wear and thermal accumulation during processing. Methods like surface coating to provide protective film layers on fibers can also reduce the emission of fibrous microplastics.

2.1.1 Yarns

The manufacturing of yarn and fabric, which produces deteriorated fibers, is defined by rapid procedures like rotor spinning, air jet texturing, and high-speed weaving. Fabric brushing, surface trimming, and fleece preparation are essential stages in this procedure. A study by Cai [12] revealed that polyester rotor yarns had a heightened quantity of extracted fibrous microplastics, indicating that rotor spinning might be a pivotal process in the creation of microplastics. Fibrous microplastics are emitted from textiles by many mechanisms, including the emission of dust from the surfaces of abraded fibrous materials and the shedding of short fibers due to the surface's texture. Mechanical damage to the loose ends of fibers results in their fracture, separation, and disintegration into fibrous microplastics. Air or water flow is characteristic during standard usage or drying, whereas water flow is usual in cleaning procedures. The generation of fibrous microplastics may be governed by the judicious selection of fiber-forming polymers, regulation of supramolecular and surface structures of fibers, optimal design of fibrous architectures, and effective maintenance techniques. Comprehending the impact of textile material properties is essential for elucidating the generation and discharge of microplastics during laundering. Elements like fiber material composition, yarn structure, material structure, construction parameters, and treatment type can significantly influence the generation and emission of microplastics (see Fig. 2.3).

2.2 Textile Characteristics as a Source

The length of staple fiber (i.e. usually 32 mm is synthetic or regenerated fibers) is a critical factor in microfiber production, as longer yarns exhibit reduced slippage within the yarn structure [13]. Short fibers may readily displace during mechanical action during the washing and wearing situation which results in microfiber releases. The structure and type of yarn used in textiles significantly influence the amount of microplastics shed during washing, wear, and environmental degradation. Factors such as fiber composition, yarn spinning technique, twist level, and surface treatment influence the shedding behavior of microplastics. Understanding these factors can help design textiles that reduce microplastic pollution. Generally, filament or staple yarns influence the release of microfibers during the

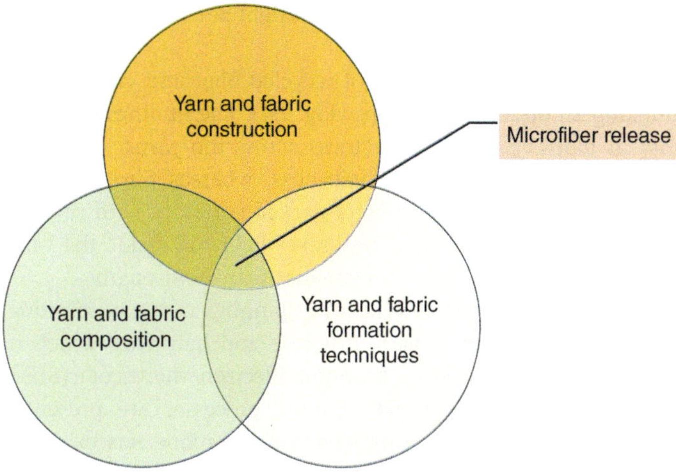

Figure 2.3 Textile characteristics affecting microplastic emission.

washing process [14–16]. Synthetic or synthetic blends release higher microfibers than natural fibers, and the release depends on the blend proportion [15, 17]. 100% synthetic fabrics release higher microfibers than blends, and the higher contribution of synthetic fibers in the blend proportion may cause higher releases of microfibers. However, the results from previous works are more generalized [15], and the releases of microfibers cannot be generalized to blended proportions or 100% synthetic yarn.

2.2.1 Yarn Structure and Microplastic Emission

2.2.1.1 Spinning Methods and Fiber Orientation

The inclination of fibers to release microplastics is influenced by the manner in which they are spun into yarn. Most natural fibers and synthetic fibers are accessible in the form of short-cut fibers, each of which has a unique length. The yarn that is formed from these short-cut synthetic fibers requires spinning and twisting in order to keep the fibers together. Because individual fibers may easily detach from one another, especially when subjected to mechanical stress, this material is more prone to shedding. When yarn is formed from such filaments, it does not require an excessive amount of twisting, and in general, all of the fibers stick together, and there is less of a chance that they would slip. In a similar manner, synthetic and man-made cellulosic fibers have the potential to develop the continuous filament form. Overall, this results in a reduction in the likelihood of microplastics being shed as compared to staple fiber yarns. This is due to the fact that the fibers stay continuous and do not have any exposed loose ends.

There is still the possibility that microplastic release might be caused by excessive mechanical wear and deterioration.

The research investigated the morphology of recycled fiber and virgin cotton mélange yarns utilizing an optical microscope (Fig. 2.4). The findings indicated a notable disparity in hairiness and compactness across the yarns. Ring yarn had the lowest compactness and elevated hairiness, whereas Siro, compact, and compact-Siro yarns demonstrated a closely packed structure with reduced hairiness. This indicates that the fibers are securely integrated inside the yarn, leading to less unevenness, fewer defects, and enhanced tensile strength.

Additionally, various spinning processes (i.e. ring spinning, compact spinning, and Siro spinning) significantly affect yarn structure and hairiness, which in turn greatly influence microfiber shedding. Scanning Electron Microscope (SEM) pictures of yarns produced using various spinning processes are presented in Fig. 2.5. The yarn produced by the ring spinning technique has excessive protruding fibers, which are possibly released during the wash and wear situation and considered as a microfiber or fibrous fragments. This is attributable to the formation of the spinning triangle that occurs soon after the drafted fiber ribbon exits the front roller nip. In this rotating triangle, fibers experience differing degrees of strain based on their locations. The outermost fibers in the spinning triangle are markedly deflected and may be partially incorporated into the yarn structure, resulting in considerable hairiness on the yarn surface (Fig. 2.5a).

2.2.1.2 Yarn Twist and Strength

A yarn's level of twist influences the amount of fiber that sheds. For example, low-twist yarns are made up of looser fiber binding, which makes it simpler to remove the fibers from the yarn. There is a greater likelihood of microfibers shedding when they are washed and worn. For high-twist yarns, perhaps there is a tighter fiber binding that gives superior resistance to fiber loss due to the increased fiber density. High twist, in general, reduces the amount of microplastics that are released during the process of friction and laundering. Furthermore, synthetic fibers such as polyester, polyamides, and acrylic do not possess the natural convolution that cotton does. This convolution allows for greater friction between the fibers, which helps to keep the yarn together after it has been formed. As a result, yarn that is made from synthetic fibers releases a greater quantity of microplastics than yarn that is made from natural fibers. Increased twisting of the yarn diminishes fiber slippage from the yarn structure, hence reducing microfiber release, as noted by Carney Almroth et al. [19]. Specifically, clothing constructed from highly twisted continuous filaments with minimal hairiness tends to emit less microfibers during laundering.

The processes of spinning, twisting, and compacting in yarn manufacture are essential for influencing the behavior of fibers inside the yarn structure. These strategies affect the protrusion of fibers and, thus, the formation and release

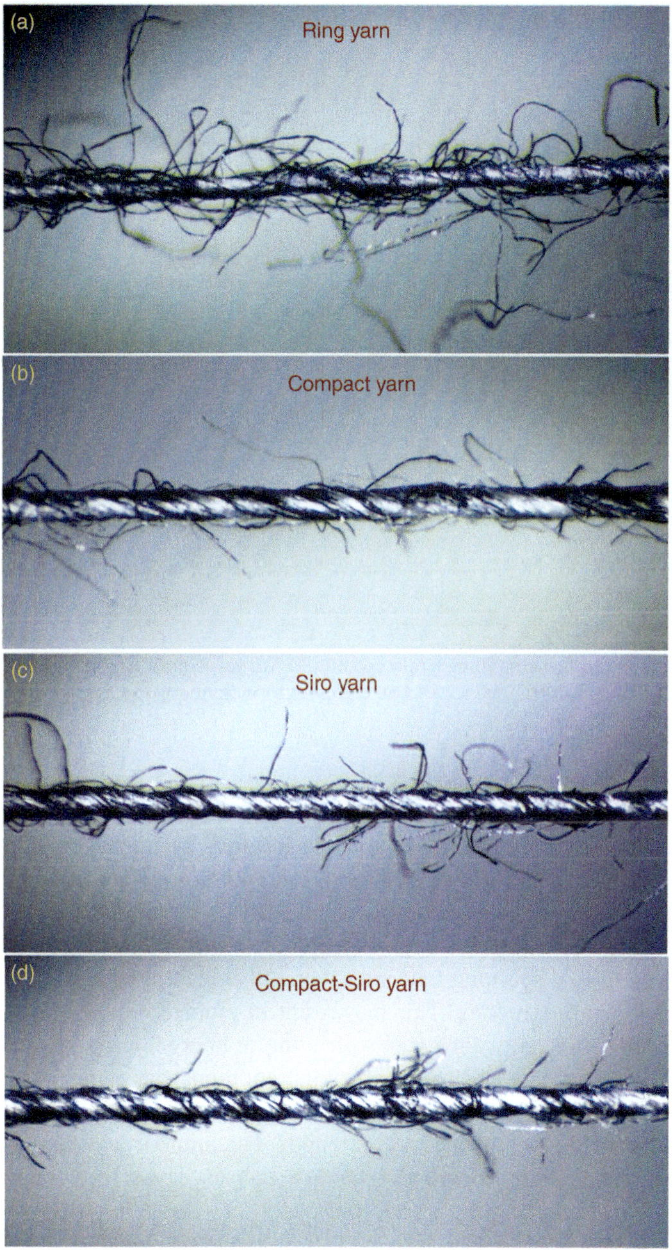

Figure 2.4 Optical microscopic images of recycled fiber/virgin cotton (50/50) yarns produced in (a) conventional ring, (b) compact, (c) Siro, and (d) compact-Siro spinning. (*Source:* Uddin and Roy [18] / Elsevier / CC BY 4.0.)

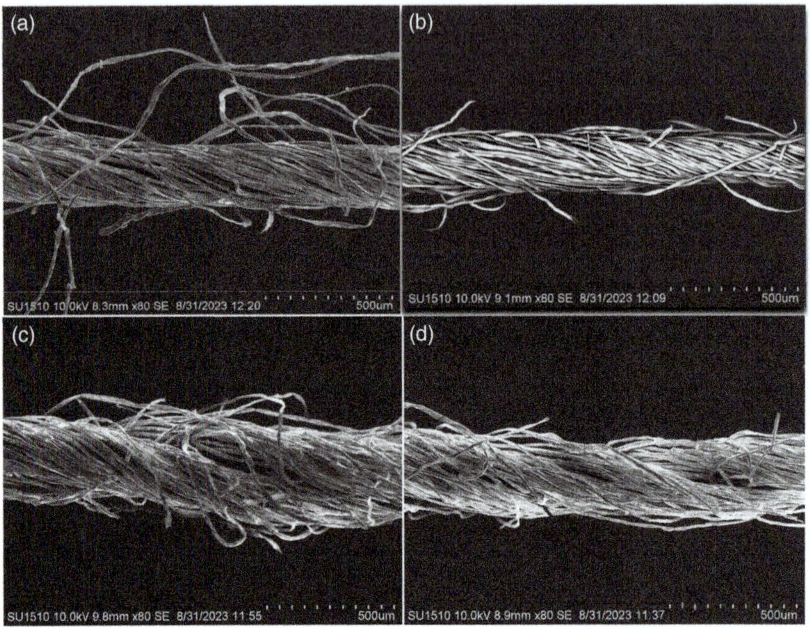

Figure 2.5 SEM images of recycled fiber/virgin cotton (50/50) yarns produced in (a) conventional ring, (b) compact, (c) Siro, and (d) compact-Siro spinning. (*Source:* Uddin and Roy [18] / Elsevier / CC BY 4.0.)

of microfibers. The spinning triangle is a pivotal region in the ring spinning process when the drafted fiber strand converts into a twisted yarn. This triangle area is delineated by the front roller nip and the location where the twist is introduced into the yarn. The dimensions and configuration of the rotating triangle markedly affect the yarn's characteristics, such as its tensile strength and fuzziness. Figure 2.6 shows the spinning triangle, twist, and compact of yarn. Fiber slippage in ring spinning transpires when fibers fail to be uniformly incorporated into the yarn structure. This may result in problems such as yarn rupture and inconsistent yarn quality. Factors leading to fiber slippage include insufficient twist, inappropriate tension, and the inherent properties of the fibers. Proper drafting and twist insertion can mitigate slippage. The twist in yarn combines strands, providing coherence and strength. The amount and kind of twist can significantly affect the properties of the yarn. Increased twist levels often reduce hairiness by more efficiently integrating fibers into the yarn structure; however, excessive twisting can lead to increased stiffness and potential breakage. Compact spinning is an evolution of the traditional ring spinning method. It reduces the size of the spinning triangle, leading to improved fiber integration and decreased hairiness. This process compresses the fiber strand before twisting,

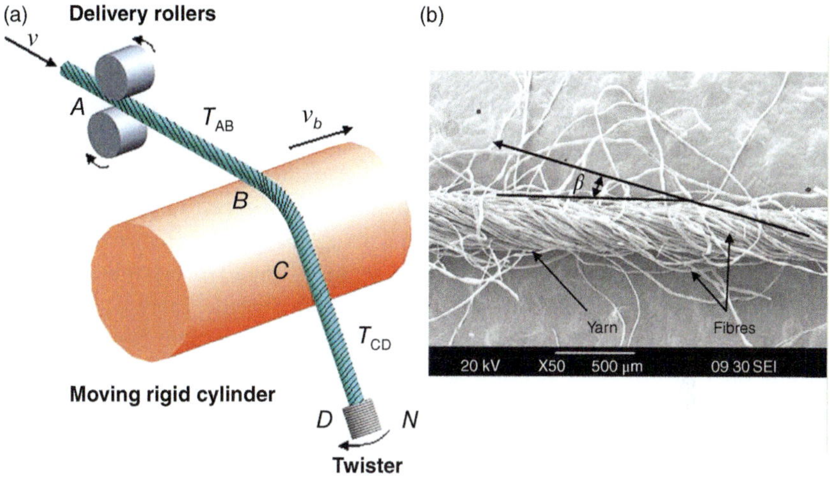

Figure 2.6 Generalized twisting system (a) and SEM micrograph of a 20Ne cotton yarn (b). (*Source:* Yin et al. [20] / Springer Nature / CC BY 4.0.)

so minimizing the spinning triangle and producing a smoother, more durable yarn with fewer protruding fibers. It is anticipated that these compact spun yarns will generate fewer microfibers than ring and rotor spun yarns.

2.2.1.3 Yarn Count and Fineness

When it comes to calculating the amount of microplastics that are emitted from textiles, the yarn count and fineness are two of the most important elements. The structural integrity, durability, and shedding behavior of synthetic fibers are all affected by these elements at various stages of their deterioration, including washing, wear, and environmental degradation. It is crucial to understand this connection in order to create low-shedding fabrics in order to reduce the amount of microplastic contamination.

The hairiness, which is crucial, prevents the generalization of these phenomena. The hairiness of the yarn structure is shown in Fig. 2.4a. Generally, hairiness is predominantly evident in yarn produced from staple fibers, while longer cut lengths tend to result in reduced hairiness. Conversely, yarn manufactured from filament fibers exhibits the least propensity for hairiness production. Yarn hairiness is a parameter that contributes to the formation of projecting fibers or, in certain instances, creates fuzz on the fabric surface [16]. Consequently, all these factors directly affect microfiber emissions. The use of longer fibers (i.e. cut length) in the yarn diminishes the projecting fibers and hairiness of the fabrics. Ultimately, the emission of microfibers is diminished by the cumulative effect of these factors. Typically, fabrics exhibiting elevated rupture work (i.e. resistance to abrasion), high tensile strength, and low hairiness values demonstrate a reduced propensity for fuzz or fibrillation, which subsequently results in the detachment and release of microfibers from the fabric.

The thickness or fineness of the yarn is referred to as the yarn count, and it may be described using a variety of different numerical systems. When compared to coarser fibers, fine yarns, which have a smaller denier per filament (DPF), are considerably more brittle and prone to shedding. As a result of their ultra-fine nature, microfiber textiles, such as those with a DPF of less than one, have a greater propensity for shedding. It is made up of more finely woven strands, which results in a greater surface area. In general, it is more fragile, more flexible, and more likely to shatter. These materials are utilized in lightweight fabrics, which include high-performance textiles, sportswear, and activewear. Fine yarns, also known as high yarn count or low denier or tex, are utilized in the production of lightweight fabrics such as sportswear, activewear, and high-performance textiles. However, as compared to coarser yarns, fine yarns release a substantially greater quantity of microplastics after each wash. After being washed, textiles created from finer yarns release a substantially greater quantity of microplastics than those made from coarser yarns, according to studies. It is possible to optimize yarn count by using medium-count yarns, to minimize synthetic fiber shedding by blending coarser natural fibers with synthetic ones, to increase weave density in order to reduce fiber exposure and shedding, and to apply anti-shed coatings to fine yarn fabrics in order to reduce microfiber loss. These are some of the implications for textile design and sustainability.

Evenness is a measurement of yarn's uniformity in terms of mass and cross-section variation per unit length [21–24]. Unevenness is indicated by the higher coefficient of variation percentages, leading to poor appearance and lower yarn strength. The appearance of textiles is affected by the high hairiness and maximizing the surface friction of yarn and fabrics and their pilling tendency; henceforth, the microfiber emission is affected by it during laundering [25–30]. In this case, similar evenness is observed in the yarn samples (i.e. 65/35, 60/40), excluding 50/50 polyester/cotton (PC) yarn, which has a higher coefficient of variation [31]. The twisting levels and the diameter of the yarn, and the rigidity, elongation, tenacity, and length of the fibers are several features depending on the hairiness [21, 24, 26–28, 32]. The mitigation of microfibers follows the order as 100% cotton spun yarn > 100% polyester spun yarns > 100% viscose rayon spun yarns > 50/50 PC spun yarns. This is due to the lower uniformity in the fiber length in the cotton yarns. For polyester and viscose rayon, static charge and cohesion in the interfibers in the yarn structure are very high. In addition, low wet strength of viscose rayon causes the higher microfiber generation, similar to cotton. Polyester yarns have comparatively similar hairiness to the viscose rayon, since the fibers are much stronger, which does not emit a huge microfiber count during laundering.

2.2.1.4 Yarn Types and Microplastic Emission

The yarn type utilized in fabric manufacturing markedly affects the emission of microplastics after laundering, usage, and environmental deterioration. This article analyzes several yarn varieties, including staple fiber yarns, filament yarns, textured yarns, and mixed yarns, focusing on their shedding behavior and impact on microplastic pollution. Comprehending these linkages is essential for developing sustainable fabrics that reduce microplastic emissions. The preceding portion of this chapter thoroughly addressed staple fiber yarns and filament yarns. Consequently, we will examine the relationship between textured yarns, blended yarns, and their shedding behavior. Textured yarns are often produced by altering filament yarns to enhance bulk, elasticity, and softness. The predominant varieties comprise false-twist textured, air-jet textured, and bulk continuous filament (BCF) yarns. It is often utilized in the production of carpets, athletic apparel, and high-performance fabrics. Typically, textured yarns provide potential. Moderate to significant shedding resulting from an expanded fiber surface area. Textured fibers exhibit increased protrusions, rendering them prone to fiber separation. Used fleece textiles are a recognized source of significant microplastic discharge during laundering. Furthermore, blended yarns (e.g. polyester-cotton blends) have a significant reliance on shedding, since they can reduce shedding in comparison to pure synthetic yarns (see Fig. 2.7). Nonetheless, the breakdown of synthetic materials continues to result in the production of microplastics. Shedding behavior varies based on the blend ratio and fiber bonding. Nonetheless, the deterioration of the natural element might reveal synthetic fibers, resulting in heightened shedding over time.

The researcher utilized five distinct textile samples derived from three principal yarn structures (flat filament, textured filament, and staple yarn) made from the two most prevalent fibers (polyethylene terephthalate and cotton) [33]. These samples underwent repeated simulated washing processes to measure the shed fragmented fiber and the length distribution of the collected fragments. This differs from the inclination observed in the other samples. Fabrics made solely of PET had a noticeable downward trend. The subsequent washing cycles for cotton and PC combination demonstrated a consistent amount of FF shedding. Results from PC blend samples indicated the influence of the close integration of ingredients on fiber shedding. Although the PC blend sample included 52:48 (PC), the FF release profile results were comparable to those of the 100% cotton sample.

2.2.1.5 Yarn Count and Fineness

Finer yarns possess a greater yarn count and a lower denier per filament, leading to thinner and more abundant fibers. They produce softer, more delicate fabrics; yet they possess disadvantages like weaker fibers, heightened shedding, and environmental contamination. Higher-quality yarns can emit microplastics during laundering and use, impacting marine organisms and ecosystems. Coarser yarns

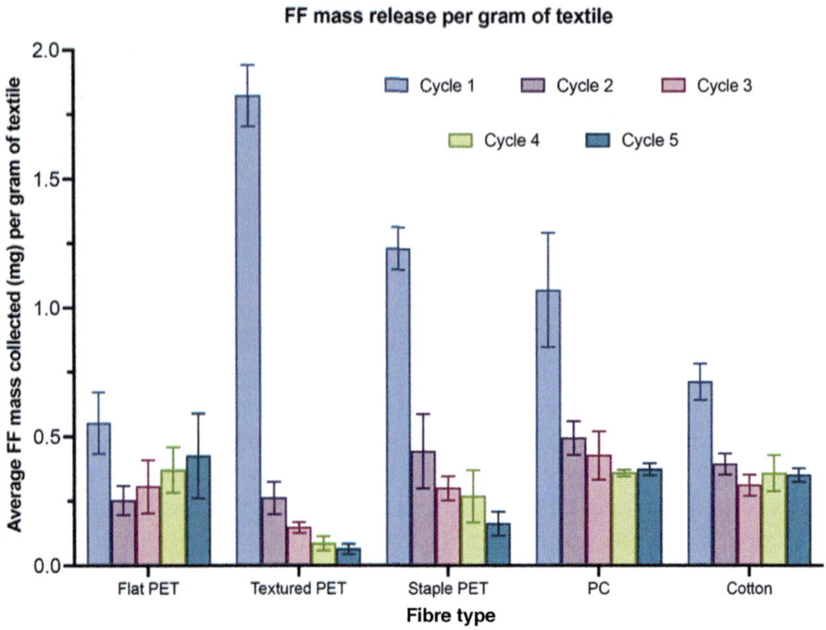

Figure 2.7 Average fragmented fiber (FF) mass recovered from washing effluent throughout washing cycles per gram of textile. (*Source:* Palacios-Marín [33] / with permission of SAGE Publications / CC BY-NC.)

possess a reduced yarn count and an elevated denier per filament, characterized by thicker and fewer fibers. They provide enhanced durability, enabling textiles to endure greater wear and tear. Furthermore, coarser yarns emit less microplastics, making them a more environmentally friendly option. However, this is not true in many instances, since most coarser yarns have thicker fibers with reduced fiber length, which results in increased microfiber release owing to their shorter length and inability to bond well with the yarn structure. In conclusion, it is essential to evaluate the advantages and disadvantages of finer and coarser yarns when selecting textile applications.

2.2.2 Fabric Structure Versus Fibrous Fragments

The fabric's weaves, knit pattern, and physical properties significantly influence microfiber release during washing. Microfiber release also depends on other factors such as fabric weight. In general, heavier fabric releases more microfibers than lighter fabrics. Nevertheless, fabrics made from synthetic or synthetic blends release higher microfibers as compared to the natural fibers and a similar result was observed from earlier studies [14–16]. Floating weaves are threads that

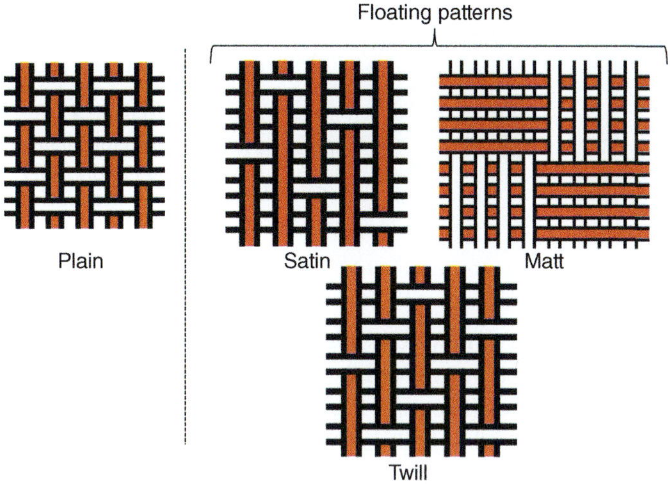

Figure 2.8 Different woven fabric structures.

traverse over others prior to interlacing (see Fig. 2.8), resulting in a smooth, shiny surface, particularly the weave structures, namely satin, sateen, and matt weaves. The configuration of these weaves can affect microfiber production during laundering. Extended floats are more susceptible to abrasion, resulting in increased microfiber release. Floating weaves typically produce softer fabrics, rendering them more prone to fiber loss. Moreover, washing parameters like temperature, cycle time, and load size also influence microfiber release, with lower temperatures and shorter cycles mitigating microfiber production [34].

De Falco [35] conducted an analysis of the various materials, which included a knitted T-shirt made of 100% PET, a knitted blouse made of 100% PET, a knitted fabric made of 100% PET, and a knitted sweatshirt made of 50% PET and cotton. According to the types of fabrics, the microfiber releases in the following order: 100% PET woven blouse, 100% PET knitted T-shirt, 100% PET knitted fabric, and 50% PET/50% cotton knitted sweatshirt. Among them, staple yarns are used to make knitted sweatshirts composed of 50% cotton and 50% PET, and it has been noted that there is no substantial difference in the amount of microfibers that are released [35]. In comparison, it has been discovered that continuous filament fibers emit a lower number of microfibers into the atmosphere compared to short-staple fibers [35]. The knitted sweatshirt made of 50% polyester and 50% cotton had the maximum microfiber emission, which was measured at 80% (in water) and 97% (in air). A portion of the knitted fabric is seen in Fig. 2.9, which is oriented in the parallel direction of the rows and columns. The framework of the fabric was made up of three distinct kinds of filament yarn: basic yarn, which included face

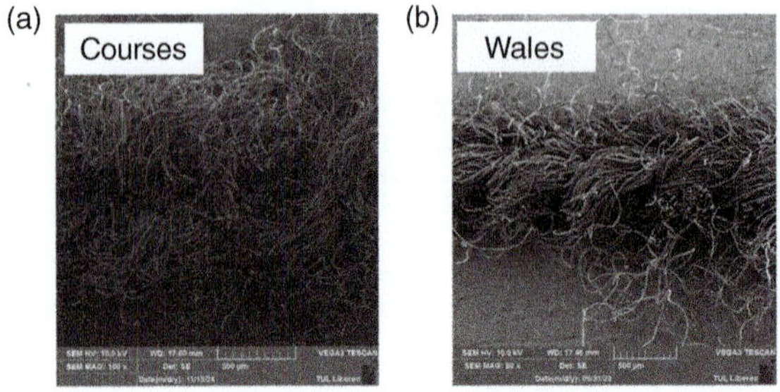

Figure 2.9 Microscopic images of the cross-sectional view of the single jersey fleece fabric's row (courses) (a) and column (wales) (b). (*Source:* Novotna and Militky [36] / American Chemical Society / CC BY-NC-ND 4.0.)

yarn and binding yarn, and loop yarn, which served as the filler layer. Using these SEM images, it can be understood that the number of microfibers that are outside and slide away from the structure of the yarn. Even within the scope of this study, the authors have utilized commercial single jersey fleece textiles, and images are taken before it washed.

The loosely structured fleece of knitted fabrics results in increased microfiber shedding compared to jerseys. This is due to the raising process, where loose yarns are extracted using an emery roller to enhance softness and provide insulation, mimicking the fur of animals in cold environments [16]. Moreover, fleece textiles, frequently composed of recycled polyester, significantly contribute to microplastic contamination. Knitted fabrics exhibit greater susceptibility to microplastic release compared to woven fabrics, attributable to their loose structure, diminished yarn twist, and increased surface area exposure. The open and flexible configuration of knitted fabrics renders fibers more susceptible to friction and rupture, resulting in increased microplastic release. The reduced twist in knitted yarns facilitates the detachment of individual fibers. The looped configuration of the fabric enhances its surface area, facilitating the breakage and shedding of fibers. Knitted fabrics experience increased mechanical stress during laundering and wear, resulting in greater fiber degradation and microplastic emission. This contrasts with woven materials, which are more rigid and densely packed. Additionally, knitted fabrics often use synthetic fibers like polyester and nylon, which do not biodegrade and release microplastics more readily. Consequently, knitted fabrics exhibit greater susceptibility to microplastic releases than woven fabrics. In addition, continuous filaments were used in the yarn, which is further used to produce the knitted fabric. Microfleece is a kind of knitted terry fabric that is napped and/or cut by mechanical process. In contrast, single yarn is used in the weft and doubled yarn is used for warp. These microfleeces released significantly higher microfibers than single jersey knitted fabrics.

Furthermore, the emission of microfibers is mostly contingent upon the blend ratio, as fabrics composed entirely of synthetic materials emit a greater quantity of microfibers compared to blended fabrics. The increased number of synthetic fibers in the blend may result in greater releases of microfibers compared to a lesser proportion of the same fibers in the mix ratio [15, 17]. The findings from the prior studies [15] are broader for the microfiber generations and their blending ratios. Nonetheless, we cannot extrapolate the release of microfibers based on blended proportions or 100% synthetic fabrics. However, parameters such as weave structure, fabric porosity, fabric density (including both warp and weft density), the existence of a singeing process, the number of wet processes, and weaving speed are crucial in the formation of microfibers. The release of microfiber is elevated due to the projecting threads on the fabric surface and the hairiness of the yarn structure. The propensity for pilling originates from the production of fuzz, which detaches during the washing process in the machine owing to mechanical action. The pill formation exhibits a comparable mechanism for both synthetic fibers, such as polyester, and cellulosic fibers, including cotton and lyocell [37]. The projected mechanism for microfiber generation is illustrated in Fig. 2.10. The creation and release of microfibers primarily result from the surface rupture of staple fibers under mechanical stress.

2.2.3 Coloration and Garmenting as a Source

A significant contributor to the release of microplastics is the dyeing and finishing procedures, which entail the application of chemical stress and the destruction of fibers. In addition, the huge volumes of water that are utilized in these processes are responsible for transporting loose fibers and microplastics into wastewater systems. Additionally, dye powders and additives contain binders that are based on microplastics, which have the potential to degrade and enter water systems [38]. Plastisol inks, which are based on polyvinyl chloride (PVC) and contribute to the shedding of microplastics for an extended period, are frequently used in printing operations. When washed and subjected to friction, printed surfaces deteriorate, resulting in the release of minute plastic pieces. In addition, waste from screen and digital printing contributes to the pollution caused by microplastics. In addition, the process of cutting and sewing garments results in the release of loose fiber particles, which can then become microplastics in the air or infiltrate water systems through washing. Fraying of the edges and industrial dust from firms that cut and handle huge quantities of synthetic textiles are further factors that contribute to the pollution caused by microplastic fragments [9]. Figure 2.11 illustrates the desirable properties of textiles for the reduction of microfibers.

Yang et al. [39] studied the surface modifications of knit textiles, especially rib fabric and plain fabrics, which are investigated before and after the specific abrasion cycles. In terms of microplastic release, rib fabric has the lowest release (Fig. 2.12), while plain fabric has the highest release. There is a distinction between the two fabrics in terms of the development of microplastic fibers and

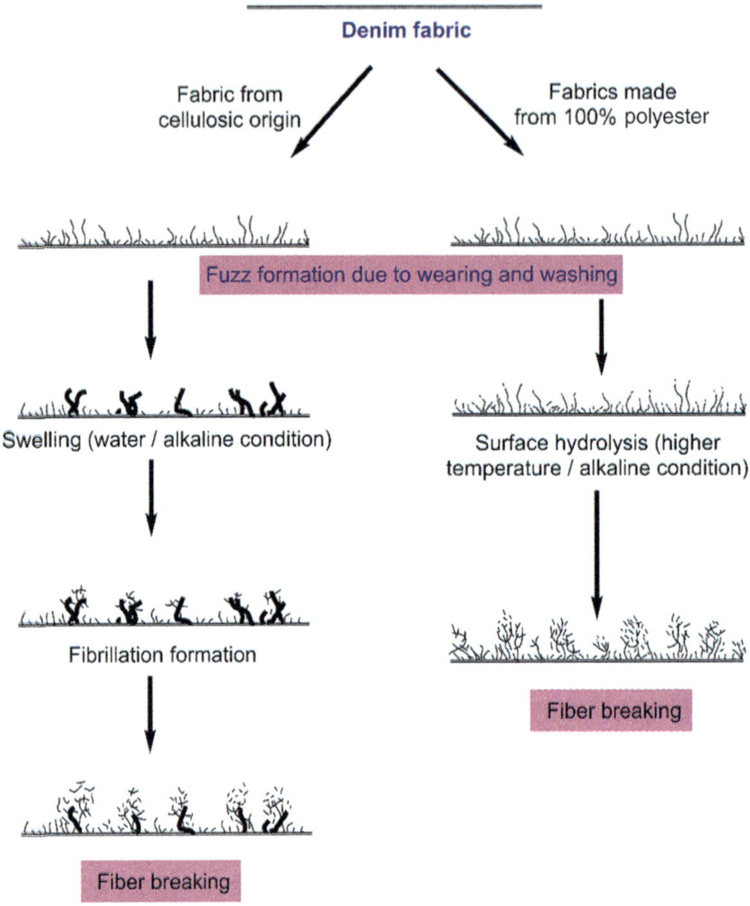

Figure 2.10 A proposed mechanism of microfiber generation on denim fabric made from polyester and cellulosic fibers. (*Source:* Adapted from Periyasamy [37].)

fibrils (Fig. 2.12). In the plain fabric, the warp yarns are not affected, but loose fiber ends are visible at the point where the warp and weft connect (i.e. interlacement). In addition, fibrillation is found near the ends of the fibers, which is most likely the origin of the fibril release. On the other hand, surface modifications fluctuate depending on the type of woven cloth.

2.3 Technical Textiles as a Source of Microplastics

Growing within the textile industry is the field of technical textiles, which are textiles that are created for utilitarian uses. Geotextiles, medical textiles, protective

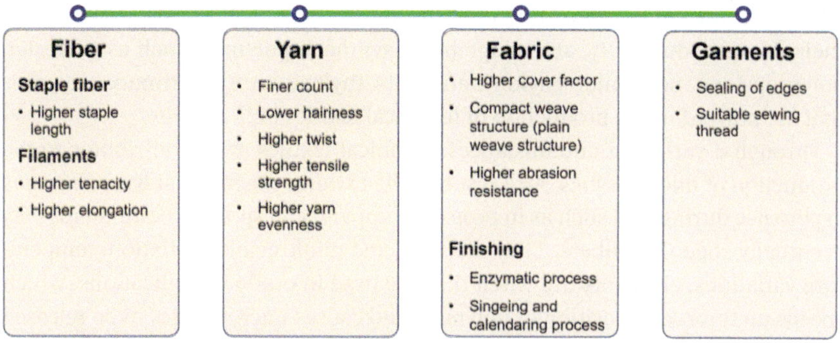

Figure 2.11 Desirable textile properties for microfiber reduction.

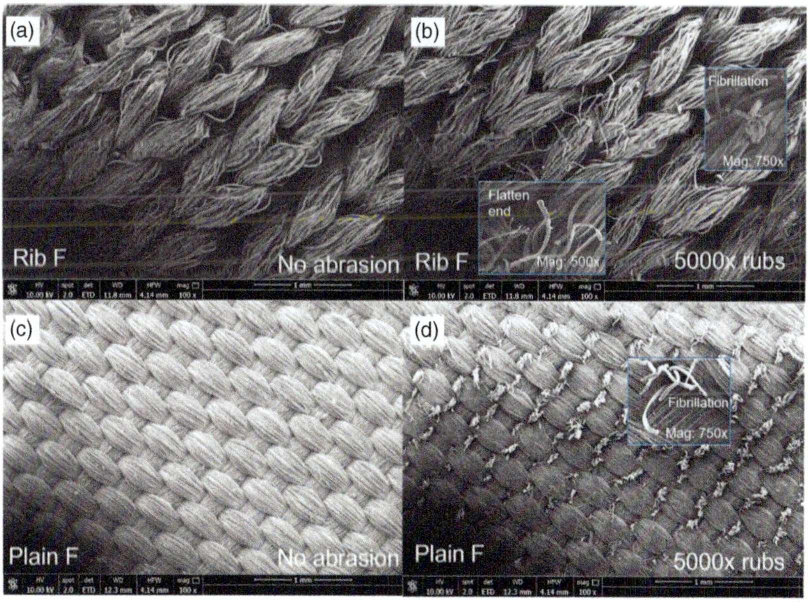

Figure 2.12 SEM images of rib fabric before abrasion (a) and after 5000 abrasion cycles (b), plain fabric before abrasion (c) and after 5000 abrasion cycles (d). (*Source:* Yang et al. [39] / Elsevier / CC BY 4.0.)

textiles, automotive textiles, and filtration fabrics are all examples of different types of technical textiles. These materials are designed to fulfill purposes across a wide range of sectors, with performance, durability, and efficiency being the primary concerns from the beginning [40, 41]. For instance, geotextiles are utilized in the construction industry, medical textiles are used for surgical masks, industrial textiles are utilized for conveyor belts, hoses, and safety gear, and high-performance fabrics are utilized for utilization in sports and outdoor gear. To take advantage of their strength, durability, and adaptability, synthetic polymers such as polyester, polypropylene, polyamide (nylon), and polyethylene are the primary materials that are utilized in the production of technical textiles [42].

Through a variety of circumstances, technical textiles are a contributor to the production of microplastics. As a result of the extreme mechanical wear that they experience during use, such as in protective apparel and sporting equipment, they eventually shed their fibers. They are subjected to ultraviolet radiation, temperature variations, and moisture when they are used in outdoor applications, which speeds up the decomposition of polymers and causes microplastics to be released into the environment. The use of geotextiles in soil reinforcement results in the release of fibers into the soil, which can then leach into water systems in the surrounding area. By washing and cleaning these fabrics, microfibers, which are largely synthetic, are released into wastewater. Microfibers are a significant contributor to the pollution caused by microplastics. The degradation of technical textiles occurs in landfills as well, where they are broken down into microplastics that are then absorbed by the soil and water systems. Especially in large-scale manufacturing plants, industrial activities like production and cutting can leak offcuts and tiny synthetic particles into the air or water. This is especially true in the case of manufacturing plants. The employment of filtration fabrics in the purification of air or water may result in the production of microplastics as a byproduct of wear or inappropriate disposal, so contributing to the problem of pollution rather than helping to alleviate it [43].

The microplastics that are produced by technical textiles have substantial effects on the environment, including the contamination of soil, the degradation of marine environments, the release of microplastics into the air, and toxicity. The release of microplastics into the environment by geotextiles has the potential to have an impact on the fertility of soil, the life of microorganisms, and the retention of water. Runoff, wastewater, and inappropriate disposal are all potential entry points for these particles into aquatic systems, which can result in bioaccumulation and cause harm to marine life. Microplastics that are airborne can become airborne during the production and usage of the material, which poses a threat to the health of animals and humans who breathe in [43]. Microplastic particles can be inhaled during the manufacturing process, cutting, or usage, which can result in respiratory problems. This poses a risk to human health. The contamination of food chains by microplastics that originate from technical textiles might potentially lead to the absorption of these microplastics by people, which can result in inflammation, disturbance of the endocrine system, and other health problems.

To reduce the amount of microplastic pollution caused by technological textiles, many initiatives have been implemented. These strategies include, but are not limited to, the following: innovation in material design; enhanced manufacturing processes; advanced wastewater treatment; sustainable disposal practices; policy and regulation; and awareness among consumers and industry. There are several ways in which the environmental effect of technological textiles can be reduced, including the development of innovative material designs, the improvement of manufacturing processes, and the implementation of sophisticated wastewater treatment and disposal procedures. Since the demand for technical textiles is expected to continue to increase, particularly in the construction, healthcare, and sports industries, it is becoming increasingly necessary to address the role that these fabrics play in the pollution caused by microplastics. Their environmental footprint can be reduced through the implementation of more stringent environmental legislation, an increase in awareness, and advancements in environmentally responsible textile production. Industries can contribute to the reduction of microplastic pollution and the protection of ecosystems for future generations by rethinking the life cycle of technical textiles, which includes the creation of the textiles as well as their disposal.

2.3.1 Plastic Mulch Can Be a Major Source of Microplastics in Agricultural Soils and the Environment

Agrotextiles encompass a variety of products, including nonwoven materials used for mulching. Most agrotextile products are manufactured using synthetic materials derived from petroleum, which can contribute to environmental pollution like artificial waste. Inadequate and improper disposal of synthetic mulch waste, influenced by ultraviolet radiation, physical abrasion, thermal oxidation, and microbial activity, results in the breakdown into fragmented plastic particles known as microplastics. Recently, concerns have been raised regarding the contamination of agroecosystems by microplastics. Synthetic fiber–based nonwoven mulching significantly contributes to microfiber pollution in agricultural lands. Natural fibers possess numerous beneficial attributes, including biodegradability, renewability, widespread availability, and neutral CO_2 emissions. These fibers exhibit excellent mechanical strength, modulus, and moisture absorption, complemented by low density, elongation, and elasticity. Their production process is relatively straightforward and cost-effective, making them economically significant, particularly for developing nations [44, 45]. Figure 2.13 shows the different nonwoven mulches produced by using natural fibers.

2.3.2 Mattress

Phenol-formaldehyde resin has been used by the floristry industry since the 1950s to make blocks of floral foam. This technique is used to provide support for the

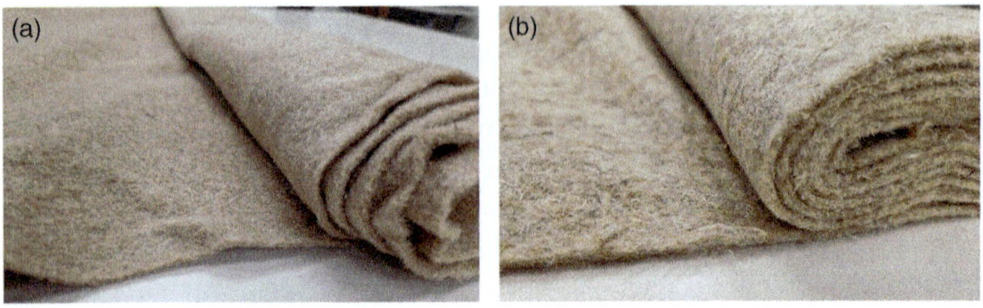

Figure 2.13 The nonwoven mulches produced by (a) jute and (b) hemp fibers. (*Source:* Kopitar et al. [44] / MDPI / CC BY 4.0.)

stems of flowers that are used in floral arrangements [46]. Floral foam users are responsible for the release of phenol-formaldehyde microplastics into the environment through a variety of mechanisms (Fig. 2.14a). Observations have shown that foam microplastics are discharged into water streams, rivers, and oceans through the wastewater systems that are used in residential settings [47–50]. However, the biofoam cell walls are joined to the adjoining struts using a thin glue, and there are fewer attachment points than in the previous picture. As a result of this attribute, it is recommended that bio-foam become more friable in comparison to ordinary form (Fig. 2.14b,c). During the process of crushing the foam, secondary microplastics such as the thick resinous struts and the membranous walls were present.

2.4 Aging of Textiles Versus Microplastics

Natural fibers, such as cotton and wool, biodegrade naturally over time, producing organic waste that is less harmful to the environment. However, they are more susceptible to biological degradation and significantly weaken under UV exposure. Their recyclability is limited and depends on the type and contamination level of the fibers. On the other hand, synthetic fibers like polyester and nylon break down into persistent microplastics, generating microplastic pollution. Although they are more resistant to wear, they shed microfibers and gradually break down under UV light. Synthetic fibers are recyclable, but the processes are often inefficient. While natural fibers biodegrade more easily, their degradation can still contribute to environmental issues, particularly when dyed or chemically treated. Synthetic fibers, however, pose a more persistent threat due to their long-lasting microplastic residues.

Apart from the fabric composition, old clothing emits 25% more microfibers than fresh garments, as observed by Hartline et al. [51]. The mechanical integrity of fibers degrades over time owing to different factors. The reduction in strength of fibers when exposed to sunshine. Moreover, the absorption of sunlight may

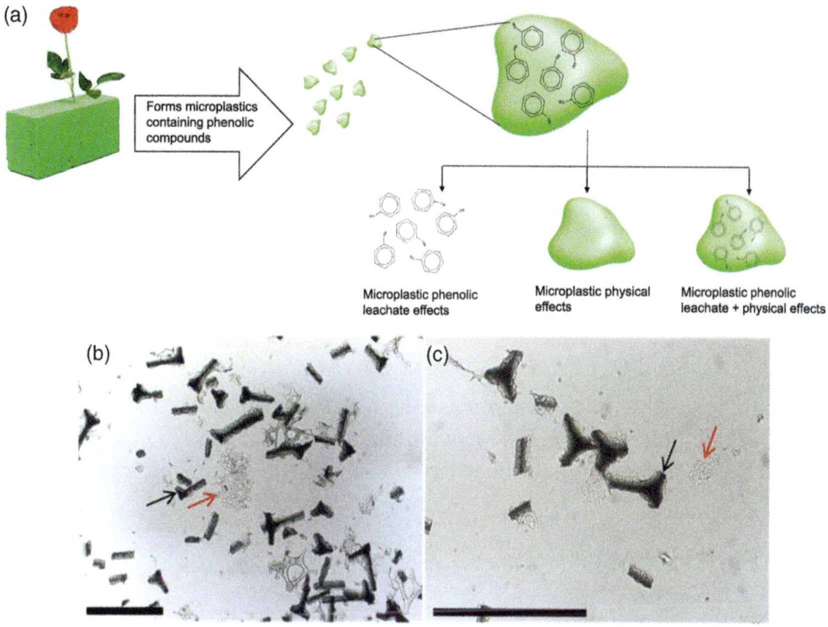

Figure 2.14 Sources of secondary microplastics from foam (a), regular foam, and (b) biofoam (c). (*Source:* Trestrail et al. [46] / with permission of Elsevier.)

elevate the temperature of fabric structures, resulting in the oxidation of cellulosic fibers. Ultraviolet radiation induces photochemical reactions in synthetic fibers. Additionally, fibers degrade owing to mechanical stress, repetitive laundering and wear cycles, abrasion, friction during use, perspiration, and exposure to gases such as CO_2, NO_X, and O_3. Textile aging is the gradual degradation of materials due to wear, light, heat, moisture, and chemicals. It affects fiber structure and properties in several ways:

- Physical aging: Loss of strength and elasticity.
- Photochemical aging: Discoloration and weakening from UV radiation.
- Thermal aging: Brittleness from high temperatures.
- Chemical aging: Breakdown from pollutants or cleaning agents.
- Mechanical aging: Abrasion and wear from use and washing.

The aging of synthetic textiles, such as polyester and nylon, can contribute to the generation of microplastics through several mechanisms. As textiles age and degrade, fibers may detach and become microplastics. This phenomenon is particularly prevalent during washing, where mechanical action and detergents facilitate fiber shedding. The exposure of textiles to chemicals during washing and environmental pollutants can weaken fibers, increasing their susceptibility to

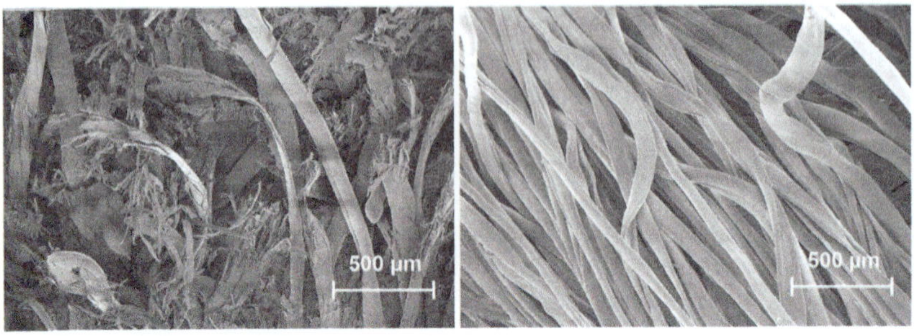

Figure 2.15 Microscopical images of aged and virgin textiles. (*Source:* Raiskio et al. [52] / Elsevier / CC BY 4.0.)

fragmentation. Ultraviolet radiation can degrade synthetic fibers, resulting in the release of microplastic particles. Here are microscopic images (Fig. 2.15) of aged and new textiles, showing how two-year-old garments generate microplastics.

2.5 Conclusion

The textile and technical textile sectors are becoming recognized as major contributors to microplastic contamination. This chapter has provided several origins of microplastics from textiles, highlighting that the problem is complex and inherent in the design and processing of textile materials, including the type of yarn utilized and the fabric structure, whether woven, knitted, or nonwoven, which significantly influence the likelihood of microplastic emission. Fabrics made from staple fibers are more susceptible to shedding because of their short fiber lengths and diminished inter-fiber cohesiveness. Similarly, loosely woven textiles or those with exposed structures are more vulnerable to abrasion and deterioration, leading to heightened microfiber release. The methods of coloration, such as dyeing and printing, also affect the release of microplastics. The use of certain dyes, finishing agents, and binding chemicals can modify fiber surface characteristics, potentially rendering them more brittle or susceptible to mechanical breakdown. Moreover, the severe chemical and heat processes often utilized in dyeing might compromise synthetic fibers, heightening the risk of microplastic release during washing and usage.

Technical textiles, designed for utility in industries such as construction, transportation, healthcare, and agriculture, provide unique challenges. These materials frequently incorporate high-performance synthetic fibers and owing to their use in challenging situations, endure significant mechanical stress, UV radiation, and variable weather conditions. The aforementioned aging mechanisms intensify polymer degradation, rendering technical textiles a persistent and frequently neglected contributor to microplastic emissions.

Furthermore, when fabrics age, particularly when subjected to frequent washing, UV degradation, and mechanical wear, their structural properties have declined, hastening the release of microfibers. The aging process influences not just consumer items such as clothing and household textiles but also affects durable technical textiles utilized in outdoor and industrial environments, where deterioration occurs continuously and is mostly unobserved. In conclusion, addressing microplastic contamination from textiles necessitates a comprehensive strategy that begins with material and structural design. Advancements in yarn engineering, fabric structure, surface treatment, and dye technologies must emphasize durability and minimal shedding properties. Furthermore, regulations and industry standards must advance to govern not just emissions from textile use but also the whole life cycle of textiles, encompassing manufacturing, usage, and end-of-life phases. Comprehending and tackling the intricate impacts of textile and technological textile systems on microplastic contamination is crucial for developing sustainable materials and safeguarding environmental and human health.

References

1 Periyasamy, A.P. (2021). Evaluation of microfiber release from jeans: the impact of different washing conditions. *Environmental Science and Pollution Research* 28 (41): 58570–58582.

2 Periyasamy, A.P. and Tehrani-Bagha, A. (2022). A review on microplastic emission from textile materials and its reduction techniques. *Polymer Degradation and Stability* 199: 109901.

3 Periyasamy, A.P. (2024). Sustainable approaches in textile finishing to control microfiber releases. In: *Microfibre Pollution from Textiles*. London UK: Taylor & Francis Group.

4 Periyasamy, A.P. (2022). Natural dyeing of cellulose fibers using syzygium cumini fruit extracts and a bio-mordant: a step toward sustainable dyeing. *Sustainable Materials and Technologies* 33: e00472.

5 Periyasamy, A.P. and Militky, J. (2020). Sustainability in textile dyeing: recent developments. *Sustainability in the Textile and Apparel Industries*, 37–79.

6 Sadannavar, M.K., Dong, X., Periyasamy, A. et al. (2024). Eco-friendly sustainable dyeing of tencel textiles with plant-based bio-mordant using dewberry fruit extract as a source of natural dye. *Fibers and Polymers* 25 (10): 3829–3839.

7 Sadannavar, M.K., Periyasamy, A., Islam, S.R. et al. (2024). Natural dyeing and antimicrobial functionalization of wool fabrics dyed with chinese dragon fruit extract to enhance sustainable textiles. *Sustainability* 16 (16): 6832.

8 Periyasamy, A.P. (2023). Microfiber emissions from functionalized textiles: potential threat for human health and environmental risks. *Toxics* 11 (5): 406.

9 Periyasamy, A.P. (2025). Functionalized textile microplastics: a closer look at the issues, strategy, and legislation on the microplastic reduction. *Kuwait Journal of Science* 100395.

10 Militký, J., Novotná, J., Wiener, J. et al. (2024). Microplastics and fibrous fragments generated during the production and maintenance of textiles. *Fibers* 12 (7): 51.

11 Militký, J. (2018). Tensile failure of polyester fibers. In: *Handbook of Properties of Textile and Technical Fibres*, 421–514. Elsevier.

12 Cai, Y., Mitrano, D.M., Hufenus, R., and Nowack, B. (2021). Formation of fiber fragments during abrasion of polyester textiles. *Environmental Science & Technology* 55 (12): 8001–8009.

13 Mcloughlin, J. and Hayes, S. (2013). Types of fabric and their joining requirements. In: *Joining Textiles*, 1–44. Elsevier.

14 Sillanpää, M. and Sainio, P. (2017). Release of polyester and cotton fibers from textiles in machine washings. *Environmental Science and Pollution Research* 24 (23): 19313–19321.

15 Napper, I.E. and Thompson, R.C. (2016). Release of synthetic microplastic plastic fibres from domestic washing machines: effects of fabric type and washing conditions. *Marine Pollution Bulletin* 112 (1–2): 39–45.

16 Pirc, U., Vidmar, M., Mozer, A., and Kržan, A. (2016). Emissions of microplastic fibers from microfiber fleece during domestic washing. *Environmental Science and Pollution Research* 23 (21): 22206–22211.

17 Zambrano, M.C., Pawlak, J.J., Daystar, J. et al. (2019). Microfibers generated from the laundering of cotton, rayon and polyester based fabrics and their aquatic biodegradation. *Marine Pollution Bulletin* 142: 394–407.

18 Uddin, A.J. and Roy, P. (2024). Transforming melange fabric waste into mélange yarn employing compact, Siro, and compact-Siro spinning: a cleaner and sustainable strategy. *Cleaner Waste Systems* 8: 100142.

19 Carney Almroth, B.M., Åström, L., Roslund, S. et al. (2018). Quantifying shedding of synthetic fibers from textiles; a source of microplastics released into the environment. *Environmental Science and Pollution Research* 25 (2): 1191–1199.

20 Yin, R., Tao, X.M., and Xu, B.G. (2016). Mathematical modeling of yarn dynamics in a generalized twisting system. *Scientific Reports* 6 (1): 24432.

21 Li, J.J., Wang, C., Tu, W.X., and Zuo, B.Q. (2014). A new measurement system of yarn appearance based on machine vision. *Advanced Materials Research* 945–949: 1810–1814.

22 Soydan, A.S., Karakan Günaydın, G., and Palamutcu, S. (2019). Comparison of fabric properties knitted by pin spacer compact and conventional compact yarns. *The Journal of The Textile Institute* 110 (11): 1519–1528.

23 Gulsevincler, E., Usal, M.R., and Yilmaz, D. (2020). Yarn evenness parameters optimization in jetring spinning process. *The Journal of The Textile Institute* 111 (4): 540–549.

24 Slater, K. (1986). Yarn evenness. *Textile Progress* 14 (3–4): 1–90.

25 Barella, A. (1983). Yarn hairiness. *Textile Progress* 13 (1): 1–57.

26 Majumdar, A. (2010). Yarn hairiness and its reduction. In: *Technical Textile Yarns*, 112–139. Elsevier.

27 Xia, Z., Liu, X., Wang, K. et al. (2019). A novel analysis of spun yarn hairiness inside limited two-dimensional space. *Textile Research Journal* 89 (21–22): 4710–4716.

28 Ghosh, S.N., Das, D.K., Battacharya, G.K. et al. (1988). An electronic instrument for measuring the hairiness of jute yarn. *Journal of the Textile Institute* 79 (4): 634–640.

29 Xia, Z. and Xu, W. (2013). A review of ring staple yarn spinning method development and its trend prediction. *Journal of Natural Fibers* 10 (1): 62–81.

30 Wang, X. (1998). Testing the hairiness of a rotor-spun yarn on the Zweigle G565 hairiness meter at different speeds. *Journal of the Textile Institute* 89 (1): 167–169.

31 Raian, S., Khan, M.H.K., Shanzid, M.L.R., and Tumpa, T.R. (2020). A pairwise comparison of cotton-polyester blend yarn's characteristics produced with vaying percentage of polyester by analytical hierarchy process. *International Journal of Scientific and Technology Research* 9 (2): 4790–4793.

32 Barella, A. and Manich, A.M. (1997). Yarn hairiness update. *Textile Progress* 26 (4): 1–29.

33 Palacios-Marín, A.V., Jabbar, A., and Tausif, M. (2022). Fragmented fiber pollution from common textile materials and structures during laundry. *Textile Research Journal* 92 (13–14): 2265–2275.

34 Lim, J., Choi, J., Won, A. et al. (2022). Cause of microfibers found in the domestic washing process of clothing; focusing on the manufacturing, wearing, and washing processes. *Fashion and Textiles* 9 (1): 24.

35 de Falco, F., Cocca, M., Avella, M., and Thompson, R.C. (2020). Microfiber release to water, via laundering, and to air, via everyday use: a comparison between polyester clothing with differing textile parameters. *ACS Applied Materials & Interfaces* 54 (6): 3288–3296.

36 Novotna, J. and Militky, J. (2025). Release of microplastic fibers from polyester knit fleece during abrasion, washing, and drying. *ACS Omega* 10 (14): 14241–14249.

37 Periyasamy, A.P. (2020). Effects of alkali pretreatment on lyocell woven fabric and its influence on pilling properties. *The Journal of The Textile Institute* 111 (6): 846–854.

38 Wennman, M., Hellberg, M., Svagan, A.J., and Hedenqvist, M.S. (2022). Plastic-free chitosan and cellulose binder providing dry and wet strength to paper and nonwoven. *Carbohydrate Polymer Technologies and Applications* 4: 100240.

39 Yang, T., Gao, M., and Nowack, B. (2023). Formation of microplastic fibers and fibrils during abrasion of a representative set of 12 polyester textiles. *Science of the Total Environment* 862: 160758.

40 Elmogahzy, Y.E. (2020). Performance characteristics of technical textiles: part III: healthcare and protective textiles. In: *Engineering Textiles*, 399–432. Elsevier.

41 Gulrajani, M.L. (2013). *Advances in the Dyeing and Finishing of Technical Textiles*. Woodhead Publishing.

42 Miraftab, M. (2016). Technical fibres. In: *Handbook of Technical Textiles*, Second E.e (ed. A.R. Horrocks and S.C. Anand), 21–41. Elsevier.

43 Hernandez, E., Nowack, B., and Mitrano, D.M. (2017). Polyester textiles as a source of microplastics from households: a mechanistic study to understand microfiber release during washing. *Environmental Science & Technology* 51 (12): 7036–7046.

44 Kopitar, D., Marasovic, P., Jugov, N., and Schwarz, I. (2022). Biodegradable nonwoven agrotextile and films – a review. *Polymers (Basel)* 14 (11): 2272.

45 Marasović, P., Kopitar, D., Brunšek, R., and Schwarz, I. (2023). Performance and degradation of nonwoven mulches made of natural fibres and PLA polymer – open field study. *Polymers (Basel)* 15 (22): 4447.

46 Trestrail, C., Walpitagama, M., Hedges, C. et al. (2020). Foaming at the mouth: ingestion of floral foam microplastics by aquatic animals. *Science of the Total Environment* 705: 135826.

47 Murphy, F., Ewins, C., Carbonnier, F., and Quinn, B. (2016). Wastewater treatment works (WwTW) as a source of microplastics in the aquatic environment. *Environmental Science & Technology* 50 (11): 5800–5808.

48 Mason, S.A., Garneau, D., Sutton, R. et al. (2016). Microplastic pollution is widely detected in US municipal wastewater treatment plant effluent. *Environmental Pollution* 218: 1045–1054.

49 Lebreton, L.C.M., van der Zwet, J., Damsteeg, J.-W. et al. (2017). River plastic emissions to the world's oceans. *Nature Communications* 8 (1): 15611.

50 Mougel, C., Garnier, T., Cassagnau, P., and Sintes-Zydowicz, N. (2019). Phenolic foams: a review of mechanical properties, fire resistance and new trends in phenol substitution. *Polymer (Guildf)* 164: 86–117.

51 Hartline, N.L., Bruce, N.J., Karba, S.N. et al. (2016). Microfiber masses recovered from conventional machine washing of new or aged garments. *Environmental Science & Technology* 50 (21): 11532–11538.

52 Raiskio, S., Periyasamy, A., Hummel, M., and Heikkilä, P. (2025). Transforming mechanically recycled cotton and linen from post-consumer textiles into quality ring yarns and knitted fabrics. *Waste Management Bulletin* 3 (1): 76–86.

3

Domestic Washing and Drying as the Source of Microplastics from the Textile and Garments

3.1 Introduction

The formation of microfibers and microplastics is substantially influenced by the activity of washing or laundering clothes in the home. The temperature of the water, the length of time spent washing, the spin speed, the type of detergent used, the type of washing machine, and the load size are all factors that influence the release of microplastics [1, 2]. The purpose of this chapter is to investigate the elements that influence the formation of microfibers in textiles during the washing process. Figure 3.1 illustrates the many elements that contribute to the production of microplastics during domestic washing.

3.2 Factors Influencing the Generation of Microplastics During Domestic Washing

The production of microplastics during domestic washing is influenced by several different factors. This chapter explores the numerous parameters that have an influence on the creation of microplastics [3]. For example, when compared to natural materials, synthetic fabrics such as polyester, nylon, and acrylic are associated with a higher likelihood of shedding microfibers. The structure of fabrics, such as knitted or loosely woven textiles, is another factor that influences the release of microfibers [4]. Yarn type, which includes characteristics such as twist, evenness, and hairiness, is also a factor. Microfiber release is also affected by washing circumstances, which include the temperature of the water, the length of time spent washing, and the kind of detergent used. Wear and tear over time causes older clothing or garments that have been aged to emit a greater quantity of microfibers. Additionally, mechanical and chemical finishes, such as coatings or softeners, have the potential to affect the amount of microfibers that shed.

Textile Microplastics in the Environment: Sources, Impacts, and Solutions,
First Edition. Aravin Prince Periyasamy.

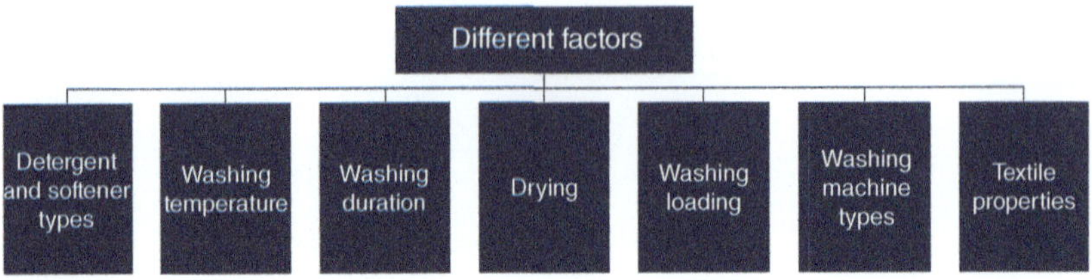

Figure 3.1 Factors influencing the generation of microplastics during domestic washing.

The development of measures to minimize the amount of microfiber pollution caused by home washing can be aided by an understanding of these aspects.

3.2.1 Influence of Detergent and Softener

A variety of detergents are utilized for the purpose of home washing [5–9]. These detergents have a significant impact on the creation of microplastics during the process of domestic washing due to the fact that their chemistry, alkalinity, and other qualities associated with the process are influenced. There is a comparison made between powder and liquid detergents, detergents that include enzymes and those that do not contain enzymes, and ecologically friendly options. Several studies [5–9] observed that the use of detergent significantly increased the microfibers emission. When compared to liquid detergents, powder detergents can weaken synthetic fibers due to their higher alkalinity and abrasive additives (due to the insoluble particles). Hernandez et al. [5] and De Falco et al. [6] observed that powder detergent causes higher emissions as compared to the liquid detergent. Nevertheless, powder detergent contains an inorganic substance like zeolite, which is insoluble in water and generates higher friction between the garments and the machines, resulting from the higher possibilities of microfiber generation. In general, the alkali-based detergents usually hydrolyze the surface of fibers (e.g. Polyethylene terephthalate (PET)) and facilitate the release of microfibers. So, it can be concluded that the powder detergent plays a vital role in the shedding of microfibers during domestic washing, however, the chemical formulation in the detergent is important rather than the state of detergent [3].

This might result in an increase in the number of microfibers that are shed due to the detergent. In addition, they could leave behind residues that contribute to the loss of fiber. Liquid detergents are ideal for limiting the release of microplastics since they are less harsh on textiles, they dissolve fully, and they typically cause less harm to fibers. The difference between enzyme-based and non-enzyme detergents is that the detergents based on enzymes can successfully remove organic stains; nevertheless, they have the potential to weaken synthetic fibers over time, which can result in more shedding of microplastics.

Detergents that do not include enzymes are gentler on textiles, but they may be less successful at removing stains. Additionally, they may need longer wash cycles, which may result in an increase in the release of fibers. Detergents that are both environmentally friendly and free of microplastics. Choosing solutions that are verified to be free of microplastics is essential to reduce the impact on the environment. To reduce the amount of microplastics that are released into the environment, some of the best practices include using liquid and non-enzyme detergents, utilizing products that are biodegradable, and washing in cold water with shorter cycles [10]. In general, the selection of detergent has a considerable impact on the amount of microplastics that are shed throughout the washing process. Individuals have the capacity to contribute to the reduction of microplastic pollution and the promotion of environmental sustainability by selecting detergents that are appropriate and switching to more effective washing techniques [3].

In another study, detergent reduces the mass of particles released from synthetic garments [11]; thus the detergent behavior influences, as the detergent contains the surfactant that reduces the surface tension and lubricates the fibers to protect them from releases. Furthermore, considering current low understanding regarding releases and fiber affinity to water, more research is needed to understand the influence of detergent chemical action and methodology type. Furthermore, while counting the microfiber emission when the garment was washed with powder-based detergent caused a substantial error (underestimation) due to the entanglement of several microfibers [6]; this was due to the insoluble nature of inorganic substances. Fabric softeners reduce the release of microfibers during washing [9], since they reduce more than 35% of emissions [6]; however, some studies state that there is no impact of fabric softeners [5, 12]. Moreover, the ratio of the garment to washing liquor in the washing machine plays a vital role in the emission of microfibers [13]. The authors observed 65 mg of microfibers per kilogram of garment, and it increases to 125 mg of microfibers when the wash water is used twice the amount. In addition, the higher water volume increases the mechanical stress to the garments during washing process.

New additives, such as builders and softeners, for washing liquor were studied by Piñol et al. [14] to reduce microfiber emissions. The primary function of builders is to chelate positive ions (Ca^{2+}, Mg^{2+}) from hard water during the washing process. Although anionic surfactants efficiently remove soil and stains, they are sensitive to multivalent ions in hard water. Positive ions bind to anionic surfactants, causing precipitation that deposits on fabric surfaces. Additionally, calcium ions act as glue between stains and fabric surfaces, making them harder to remove and thereby reducing cleaning efficiency. This necessitates a greater quantity of detergents or more friction to achieve satisfactory cleaning, which can increase microfiber release [3].

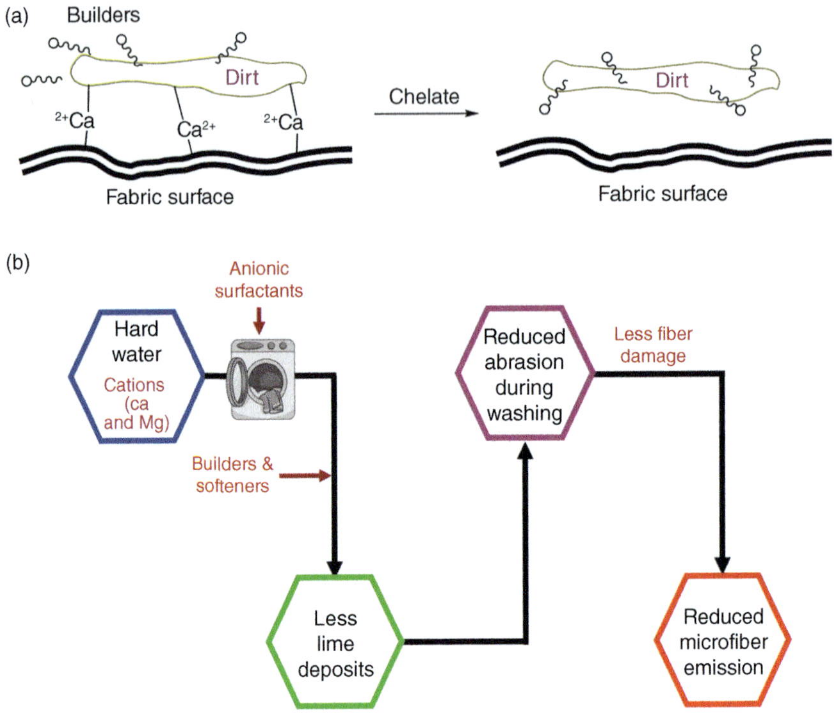

Figure 3.2 Builders and their impacts on stain removal under hard water (a), action on microfiber reduction (b).

Generally, detergents containing nonionic surfactants are easily soluble in water and perform well in hard water; however, they are less effective at removing soil compared to anionic surfactants. Nonionic surfactants possess hydrophilic groups that provide lubrication to garments, potentially reducing microfiber release [15]. This hypothesis, however, lacks support from research studies. Figure 3.2a illustrates the chelation reaction of calcium ions by builders on fabric surfaces and their effect on stain removal. Lime deposits on fabrics caused by hard water lead to abrasion and significant fiber damage through continuous washings, generating more microfibers. Builders mitigate these adverse effects by preventing hard water from causing problems (Fig. 3.2b). While builders may deposit during the washing cycle, the quantity is minimal compared to cycles without builders.

3.2.2 Fabric Softeners

Fabric softeners are a material that is frequently used in contemporary washing regimens. They provide softer garments, a reduction in static, and additional fresh fragrances. Nevertheless, they are also responsible for the contamination caused by microplastics, which is a covert environmental problem. The process by which fabric softeners function involves coating fabrics with a very thin layer of chemicals. These chemicals often include silicones, compounds derived from petroleum, and synthetic polymers. Over time, this coating degrades the structure of the fabric, which increases the amount of fiber shedding, particularly in synthetic fabrics such as polyester, nylon, and acrylic. However, it does momentarily lower the amount of friction that occurs between the fibers. Fabric softeners that are sold commercially typically contain quaternary ammonium compounds (also known as quats) and synthetic polymers, both of which do not biodegrade completely. After being removed by rinsing, these additives are sent into wastewater systems, where they directly contribute to the pollution caused by microplastics. It is inevitable that microplastics will ultimately find their way into rivers, seas, and even the food chain. This is because wastewater treatment plants are not completely successful in screening out microplastics [3].

The use of fabric softeners results in an increase in the amount of fibers that are lost, which in turn makes the fabric more fragile and prone to breaking. Therefore, clothing sheds more microfibers over time. This is especially upsetting for things that are already prone to significant microfiber shedding, such as fast fashion and sportswear, which are already sensitive to this. Apple cider vinegar, wool dryer balls, baking soda, and environmentally friendly softeners are all examples of sustainable alternatives to fabric softeners. Fabric softeners improve the comfort of clothes, yet it is impossible to overlook the fact that they contribute to the pollution caused by microplastics. By reducing the strength of synthetic fibers, introducing substances that are not biodegradable into wastewater, and increasing the amount of fiber shedding, they contribute to the deterioration of the environment in a way that is both subtle and important. By choosing natural alternatives, one may contribute to the reduction of microplastic pollution while yet maintaining the softness and freshness of one's clothing.

3.2.3 Influence of Temperature

The washing temperature [16] has a strong influence on the release of microfibers. Certainly, a higher level of microfibers is released during the washing process at higher temperatures even without adding any detergent or conditioner [16]. This is due to the thermal degradation and mechanical agitation of the fibers under the higher washing temperatures. Thermal degradation causes synthetic fibers to become more fragile, which therefore makes them more prone to breaking. At higher temperatures, the aging process of polymers is accelerated, which results in decreased fabric strength and increased fiber loss. Also, warmer water decreases the stiffness of the fabric, which in turn increases the movement and friction between the fiber bundles. As a result, the mechanical stress that is placed on textiles is increased, which results in a higher detachment of microplastics. Muthusamy et al. [17] studied the release of microfibers increased with increasing washing temperature; at 30 °C 2,877,278 microfibers are released per 1 kg of jeans, and it increased to 3,121,692 and 3,283,341 microfibers per kilogram of PET jeans when the washing temperature is at 45 °C and 60 °C, respectively; these results are observed in constant washing treatment [3, 17]. Hernandez et al. [5] observed there was no significant difference in the microfiber releases during washing process without addition of detergent and conditioner at different washing temperatures (25 °C, 40 °C, 60 °C, and 80 °C); perhaps, addition of detergent predominantly varies the result of microfiber releases. Similar results were obtained, where the PET garments released higher microfibers at higher washing temperature [18]. This is due to the surface hydrolysis of PET [19]. Henceforth, it is essential to wash the synthetic fabric/garments at low temperatures (30 °C), whereas higher temperatures may tend to release more microfibers.

The washing process is often conducted at temperatures between 30 and 90 °C. That depends on the type of garments, the level of contamination, and the nature of constituent fibers (e.g. cotton or PET). It is possible to lessen the release of microplastics by washing materials at varying temperatures. Due to low temperatures (≤30 °C), the fiber damage and energy consumption are reduced, resulting in an environmentally beneficial process. Most materials may be cleaned effectively while minimizing the loss of fiber when temperatures are moderate (40 °C). At temperatures exceeding 60 °C, the breakdown of fibers is accelerated, synthetic polymers such as polyester and nylon are weakened, and more mechanical stress is caused. Additionally, these temperatures result in a greater use of power, which contributes to a greater carbon footprint. As a result, washing at temperatures lower than 30 °C also helps save energy and drastically reduces the amount of microplastics that is produced. In order to preserve the structure of the fabric, it is recommended to wash the garments at temperatures below 30 °C, employ cold water detergents for the removal of stains, reduce the frequency of washing by airing out the garments and spot cleaning, utilize washing bags or filters to trap microfibers, and avoid excessive heat drying after washing to prevent the fabric from becoming weak [3].

The formulation of suitable detergents, washing temperature, pH, and duration can be slightly different for various natural or synthetic fibers. The washing temperature and addition of detergent increase the swelling action of cellulosic fibers. Water acts as a plasticizer and can penetrate the amorphous regions of fibers and reduce the intermolecular interactions between the fibers. As an example, viscose fibers get weaker when they are wet due to the interruption of hydrogen bonds between the cellulose chains. Water can also reduce the glass transition temperature, stiffness, and resiliency of fibers that can indirectly affect the release of microfibers during the washing process [20–23]. The swelling of cellulosic fibers promotes free space for the mobilization of the shattered fibers from the fabric structure. Due to the shear force, the fuzz formation is enhanced as the swollen fabric structure. The swollen state of fibers leads to the fuzz formation which is responsible for the shedding capacity of the fabrics, and the mechanical action of the washing machine and tumble dryers were broken this fuzz on the surface of the fabric before it develops like pills, it can be well described in Fig. 2.10. However, swelling is not a serious issue for hydrophobic synthetic fibers as compared to the cellulosic fiber under higher temperatures. The strong influence on the microfiber formation and release is the breakage of the short staple fibers. In general, the short-staple fibers (length between 2.5 and 6.0 cm) and long-staple fibers (length > 6.0 cm) [24–26] are the types of natural staple fibers, generally it is very difficult to obtain the uniformity in the natural fibers. Nevertheless, combed yarn (i.e. combing process) certainly provides uniformity. For staple yarn productions, synthetic fibers like PET, PA, and PAN and regenerated fibers like lyocell and Viscose rayon were cut uniformly ≤32 mm [26–31]. The friction, shape, thickness, stiffness, yarn hairiness, and abrasion resistance of the fibers are the sources of the fuzz formation [32–37]. During washing, short-staple fibers (also known as protruding fibers) and higher hairiness in the yarn are accountable for a huge amount of microfiber emission [38], which further possible to form fuzz (see Fig 2.10).

The tendency of pilling is initiated from the fuzz formation [37], which detaches from the surface due to the mechanical action during the washing and wearing cycles. The pill formation delivers a similar mechanism for both synthetic fibers like polyester and cellulosic fibers like cotton/lyocell. The predicted mechanism for the microfiber generations is shown in Fig. 2.10. The formation of microfiber and the release are mainly due to the surface rupture of staple fibers under mechanical actions. As per the reference [10], fabric conditioners increase the possibilities of pilling generations on synthetic fabrics. In summary, the washing temperature plays a crucial role in microplastic generation. Lower temperatures (30 °C or below) are the most effective way to reduce fiber shedding, protect fabrics, and conserve energy. By making simple adjustments to laundry habits, they could significantly minimize their contribution to microplastic pollution while maintaining effective cleaning results.

3.2.4 Influence of Washing Duration

The release of microplastics is a critical concern, since prolonged wash cycles result in enhanced fiber degradation, increased water exposure, and heightened friction among textiles. Brief wash cycles (≤30 min) reduce mechanical stress and conserve water, whereas conventional cycles (40–60 min) efficiently cleanse lightly dirty garments but may fail to eliminate stubborn stains. Extended cycles

(90+ min) optimize fabric deterioration and increase microplastic emissions, rendering them less sustainable. To mitigate microplastic release, it is suggested to utilize rapid or environmentally friendly wash cycles, launder just when essential, employ mild detergents, enclose synthetic fabrics in microfiber-catching bags, and air dry if feasible. These measures can mitigate mechanical stress, preserve fabric integrity, capture loose fibers, and diminish the effects of microplastic contamination in wastewater. In summary, abbreviated wash cycles, safe laundering methods, and air drying can markedly decrease microplastic contamination while ensuring garments remain clean and fresh. By selecting abbreviated wash cycles, employing preventive laundering techniques, and cleaning just when essential, individuals may markedly diminish their contribution to microplastic contamination while preserving the cleanliness and freshness of their garments [3].

3.2.5 Influence of Drying on Microfiber Releases

In developed nations, the laundering process frequently entails the use of a tumble clothes dryer, which may be indispensable or different from the washing machine. Moreover, the release of fibers during tumble drying was approximately 3.5 times higher than washing process [12]. The authors observed for PET blanket, the average relative fiber emissions in weight % for only washing was 0.0014 and it increased to 0.00359% during drying process. Generally, the microfibers were broken during the drying process and attached to the fabric surface and will certainly be freed and moved to the atmosphere (i.e. air), and some portions may be released during subsequent washing process, thence pass to wastewater systems. [39]. In contrast, the atmospheric drying or air drying is not significantly influencing the quantity of microfiber release [5]. Normally, drying temperature, rate of agitation, and time of drying are major criteria that strongly affect the microfiber releases in tumbling dryer. With lower temperature settings, the rate of drying slows down; therefore, the load is exposed to higher amount of agitation, which releases the higher microfibers.

Tumble drying and line drying are two ways that substantially contribute to microplastic emission in laundry effluent. Tumble drying is a mechanical procedure that subjects textiles to heat and friction, hastening fiber degradation and facilitating microplastic production. This process leads to heightened fiber shedding, thermally induced fiber breakdown, and the production of airborne microplastics. Mitigation techniques for tumble drying encompass routine cleaning of lint filters, employing a dryer vent filter, selecting low heat settings, and reducing drying duration. Line drying, or air drying, is a more gentle and environmentally sustainable method that markedly decreases microplastic production [4]. It imposes no mechanical stress, incurs no thermal damage, and

Table 3.1 Impacts of different drying methods and their influence on the microplastic generation.

Factors	Tumble drying	Line drying
Microplastic emission	High	Low
Heat damage on the fibers	Yes	No
Energy use (electricity)	High	None
Airborne microplastics	Possible	None

operates with energy efficiency, utilizing no electricity [3]. Nevertheless, line drying presents disadvantages, including ultraviolet exposure and wind-induced deterioration of materials over time. To efficiently dry garments, utilize shaded regions, shake clothing prior to hanging to eliminate loose fibers, and employ indoor drying racks in high-pollution environments to avert dust contamination. In summary, both tumble drying and line drying possess distinct advantages and drawbacks for the mitigation of microplastic emission in laundry wastewater. A comparison of the effects of various drying techniques and how they affect the production of microplastics is shown in Table 3.1.

3.2.6 Overloading or Underloading

Overloading or underloading induces mechanical stress, which exerts significant strain on the surface fibers of textiles and clothing, leading to the formation of microfibers. Overloading and underloading significantly contribute to microplastic pollution in washing machines. Overloading increases fabric friction and mechanical stress, whereas underloading causes excessive agitation and fiber degradation. To reduce microplastic emissions, it is essential to control load size, utilize moderate wash cycles, and implement microfiber filters. Overloading reduces space for movement, leading to increased fiber separation and decreased cleaning efficiency. To reduce microplastic discharge, it is recommended to restrict the drum's capacity to 75%, employ liquid detergent for improved dispersion, and select a gentle wash cycle with little agitation. Conversely, underloading exacerbates mechanical stress and water flow on individual garments, leading to inefficient energy and water use. To reduce microplastic release in small loads, it is recommended to wash synthetic garments with similar materials, employ a delicate cycle, and consider handwashing small loads when possible. In summary, both overloading and underloading intensify microplastic contamination via separate processes. Maintaining an optimal load size, utilizing gentle wash cycles, and including microfiber filters may significantly reduce microplastic pollution

Table 3.2 Influence of washing, loading, and microplastics generation.

Factor	Overloading	Underloading	Optimal load
Fiber friction	High	Low	Balanced
Mechanical stress	High	High	Low
Water & detergent efficiency	Poor	Poor	Efficient
Microplastic shedding	High	High	Low

while prolonging the lifespan of clothing. A comparison of the effects of washing loading, including over, under, and optimal loading, and how they affect the production of microplastics is shown in Table 3.2.

3.2.7 Type of Washing Machines

In general, two types of washing machines are used for domestic purposes: top-loading and front-loading machines. Top-loading washers, especially those equipped with agitators, are recognized for their vigorous mechanical activity, resulting in higher fragmentation of garments that then enter wastewater. This leads to elevated microplastic discharge and augmented fabric deterioration. Mitigation options encompass employing a delicate cycle, laundering synthetic materials in mesh bags, and installing microfiber filters to capture microplastics prior to their entry into the aquatic system. Conversely, front-loading washers employ a gravity-driven tumbling mechanism, leading to less microplastic release, enhanced water efficiency, and diminished detergent waste. Front-loading machines are superior for mitigating microplastic pollution, as they result in diminished fabric abrasion and shedding [40]. To mitigate microplastic pollution, it is advisable to utilize a front-loading washing machine, opt for a delicate wash cycle, launder garments at reduced temperatures to avert fiber degradation, install a microfiber filter to capture released fibers prior to their entry into wastewater, and decrease washing frequency by wearing garments multiple times before cleaning. The type of washing machine substantially affects microplastic formation. Top-loading washers emit a greater quantity of microfibers because of their vigorous mechanical action, but front-loading washers are more fabric-friendly and produce reduced fiber loss. Table 3.3 shows that the primary distinctions between top-loading and front-loading machines are their washing methods, water consumption, and mechanical action, all of which affect fiber shedding [9, 41].

To reduce microfiber emissions during domestic laundering, utilize cold water (15–30 °C), opt for shorter wash cycles, and wash entire loads rather than partial loads. Utilize gentle detergents and refrain from using excessive quantities.

Table 3.3 Difference between washing machines with respect to their microplastic emissions.

Factor	Top-loading washing machine	Front-loading washing machine
Mechanism	Uses an agitator or impeller for movement	Uses a drum rotation with gravity-based tumbling
Mechanical actions	**High** (stronger agitation leads to more fiber breakage)	**Low** (gentler tumbling reduces fiber shedding)
Water consumption	**High** (submerges clothes in more water)	**Low** (efficient water usage with drum rotation)
Detergent efficiency	**Lower** (more dilution in excess water)	**Higher** (better detergent dispersion)
Microplastic shedding	Higher due to the higher mechanical agitation	Lower

Select front-loading washing machines for more delicate outcomes and reduced microfiber release. Refrain from using fabric softeners, since they may deteriorate fibers and exacerbate shedding. Install filters on your washing machine to intercept microfibers prior to their release into wastewater. Reduce washing frequency to minimize total fiber discharge. Air dry garments whenever feasible to prevent further fiber release. The optimal washing parameters for minimizing microfiber emissions are illustrated in Fig. 3.3.

3.2.8 Effect of Textile Properties

The generation and release of microfibers during domestic washing are affected by various factors, including fabric type, geometry (woven, knit, or nonwoven), yarn characteristics (twist, evenness, hairiness, and fiber count), processing history (spinning, knitting or weaving, scouring, bleaching, dyeing, finishing, and drying), and the physicochemical properties of the fibers [5, 42]. The proportion of different components in yarn or fabric construction (e.g. in blended fibers, yarns, fabrics) significantly influences the formation and release of microfibers during laundry. Conversely, textiles composed of 100% synthetic materials or synthetic mixes emit a greater quantity of microfibers than natural fibers, attributable to the friction among the threads during yarn production. Synthetic fibers typically lack natural convolution or crimp. The severe mechanical action of washing and tumble drying, coupled with reduced interfiber friction between synthetic fibers or their blending with natural fibers, may result in the release of embedded short fibers (fragments) often measuring from millimeters to micrometers.

The fabric structure greatly affects microfiber generation; specifically, fleece and pile structures release markedly more microfibers than plain-woven textiles

Figure 3.3 Preferrable washing conditions for the reduction of microfiber emissions.

due to their open configuration of loops or fibers [6, 12, 43]. Additionally, fleece materials are typically composed of PET [44]. De Falco et al. [6] noted that woven polyester fabric releases more microfibers than knitted polyester fabric, while the yarn utilized in knitted polyester fabric exhibits less hairiness relative to the woven fabric. This is further elucidated in *Chapter 2*.

3.3 Conclusion

The extensive utilization of synthetic fibers such as polyester, nylon, and acrylic in contemporary apparel results in the release of microfibers, which ultimately infiltrate water bodies, soil, and the atmosphere, becoming a ubiquitous environmental pollutant with possible health hazards for aquatic organisms and humans. Microfibers and microplastics are generated during domestic laundry as a result of several variables, including water temperature, wash length, spin speed, detergent type, washing machine model, and load size. Elevated water temperatures, prolonged wash cycles, high spin velocities, and certain detergents can undermine fiber integrity and augment mechanical stress, resulting in increased microfiber release. To reduce microplastic contamination, it is crucial to use washing techniques that alleviate these stresses, including utilizing lower water temperatures, shorter wash cycles, softer spin rates, and suitable detergents. Enhancing load capacities and using front-loading washing machines might furthermore diminish microfiber emissions. Additionally, the technological solutions, like in-drum filters, external microfiber traps, and specialized washing bags, provide quick mitigation options for consumers. Advancements in textile engineering, particularly the creation of low-shedding fibers and textiles, are crucial for the sustained mitigation of microfiber emissions. Enhancing consumer understanding about clothing maintenance, laundry frequency, and the ecological consequences of rapid fashion can promote sustainable habits.

References

1 Periyasamy, A.P. and Tehrani-Bagha, A. (2022). A review on microplastic emission from textile materials and its reduction techniques. *Polymer Degradation and Stability* 199: 109901.

2 Periyasamy, A.P. (2024). Sustainable approaches in textile finishing to control microfiber releases. In: *Microfibre Pollution from Textiles*. London, UK: Taylor & Francis Group.

3 Periyasamy, A.P. (2021). Evaluation of microfiber release from jeans: the impact of different washing conditions. *Environmental Science and Pollution Research* 28 (41): 58570–58582.

4 Periyasamy, A.P. (2023). Environmentally friendly approach to the reduction of microplastics during domestic washing: prospects for machine vision in microplastics reduction. *Toxics* 11 (7): 575.

5 Hernandez, E., Nowack, B., and Mitrano, D.M. (2017). Polyester textiles as a source of microplastics from households: a mechanistic study to understand microfiber release during washing. *Environmental Science & Technology* 51 (12): 7036–7046.

6 De Falco, F., Gullo, M.P., Gentile, G. et al. (2018). Evaluation of microplastic release caused by textile washing processes of synthetic fabrics. *Environmental Pollution* 236: 916–925.

7 Sillanpää, M. and Sainio, P. (2017). Release of polyester and cotton fibers from textiles in machine washings. *Environmental Science and Pollution Research* 24 (23): 19313–19321.

8 De Falco, F., Gentile, G., Di Pace, E. et al. (2018). Quantification of microfibres released during washing of synthetic clothes in real conditions and at lab scale★. *The European Physical Journal Plus* 133 (7): 257.

9 Napper, I.E. and Thompson, R.C. (2016). Release of synthetic microplastic plastic fibres from domestic washing machines: effects of fabric type and washing conditions. *Marine Pollution Bulletin* 112 (1–2): 39–45.

10 Chiweshe, A. and Crews, P.C. (2000). Influence of household fabric softeners and laundry enzymes on pilling and breaking strength. *Textile Chemist and Colorist & American Dyestuff Reporter* 32 (9): 41–47.

11 Cesa, F.S., Turra, A., Checon, H.H. et al. (2020). Laundering and textile parameters influence fibers release in household washings. *Environmental Pollution* 257: 113553.

12 Pirc, U., Vidmar, M., Mozer, A., and Kržan, A. (2016). Emissions of microplastic fibers from microfiber fleece during domestic washing. *Environmental Science and Pollution Research* 23 (21): 22206–22211.

13 Kelly, M.R., Lant, N.J., Kurr, M., and Burgess, J.G. (2019). Importance of eater-volume on the release of microplastic fibers from laundry. *Environmental Science & Technology* 53 (20): 11735–11744.

14 Piñol, L., Rodriguez, L., Brouta, M., Escamilla, M., and Escudero, R. (2014) Mapping and estimation of laundry effluent microplastics release. https://static1.squarespace.com/static/5afae80b7c93276139def3ec/t/5b06e71c70a6ad631bc540ee/1527179088560/OCEAN+CLEAN+WASH+GUIDE.pdf (accessed on 28 November 2024).

15 Cheng, K.C., Khoo, Z.S., Lo, N.W. et al. (2020). Design and performance optimisation of detergent product containing binary mixture of anionic-nonionic surfactants. *Heliyon* 6 (5): e03861.

16 Yang, L., Qiao, F., Lei, K. et al. (2019). Microfiber release from different fabrics during washing. *Environmental Pollution* 249: 136–143.

17 Muthusamy, L.P., Periyasamy, A.P., Militký, J., and Palani, R. (2021). Adaptive neuro-fuzzy inference system to predict the release of microplastic fibers during domestic washing. *Journal of Testing and Evaluation* 50 (1): 91–104.

18 de Falco, F., Cocca, M., Avella, M., and Thompson, R.C. (2020). Microfiber release to water, via laundering, and to air, via everyday use: a comparison between polyester clothing with differing textile parameters. *ACS Applied Materials & Interfaces* 54 (6): 3288–3296.

19 Dave, J., Kumar, R., and Srivastava, H.C. (1987). Studies on modification of polyester fabrics I: alkaline hydrolysis. *Journal of Applied Polymer Science* 33 (2): 455–477.

20 Klemm, D., Philipp, B., Heinze, T. et al. (1998). *Comprehensive Cellulose Chemistry*. Hoboken, New Jersey: Wiley Blackwell.

21 Schimper, C.B., Ibanescu, C., and Bechtold, T. (2009). Effect of alkali pre-treatment on hydrolysis of regenerated cellulose fibers (part 1: viscose) by cellulases. *Cellulose* 16 (6): 1057–1068.

22 Öztürk, H.B. (2008) Ph.D Thesis on Regenerated cellulosic fibers – effect of alkali treatment on structure, chemical reactivity and fiber properties. University of Innsbruck: Dornbirn, AT

23 Jaturapiree, A. (2007) Ph.D Thesis on Porosity and transport properties in the system cellulose fiber/aqueous medium-determination and modification of porosity. University of Innsbruck: Dornbirn, AT

24 Negm, M. and Sanad, S. (2020). Cotton fibres, picking, ginning, spinning and weaving. In: *Handbook of Natural Fibres*, 3–48. Elsevier.

25 Babu, K.M. (2015). Natural textile fibres. In: *Textiles and Fashion*, 57–78. Elsevier.

26 Elhawary, I.A. (2015). Fibre to yarn. In: *Textiles and Fashion* (ed. R. Sinclair), 191–212. Elsevier.

27 Erdumlu, N. and Ozipek, B. (2008). Investigation of regenerated bamboo fibre and yarn characteristics. *Fibres & Textiles in Eastern Europe* 16 (4): 43–47.

28 El-Sayed, M.A.M. and Sanad, S.H. (2010). Compact spinning technology. In: *Advances in Yarn Spinning Technology* (ed. C.A. Lawrence), 237–260. Woodhead Publishing.

29 Bhat, G. and Kandagor, V. (2014). 1 – Synthetic polymer fibers and their processing requirements. In: *Advances in Filament Yarn Spinning of Textiles and Polymers* (ed. D. Zhang), 3–30. Woodhead Publishing.

30 Hagewood, J. (2014). Technologies for the manufacture of synthetic polymer fibers. In: *Advances in Filament Yarn Spinning of Textiles and Polymers* (ed. D. Zhang), 48–71. Woodhead Publishing.

31 Alagirusamy, R. and Das, A. (2015). *Textiles and Fashion*. Amsterdam, NL: Woodhead Publishing.

32 Barella, A. (1983). Yarn hairiness. *Textile Progress* 13 (1): 1–57.

33 Barella, A. and Manich, A.M. (1997). Yarn hairiness update. *Textile Progress* 26 (4): 1–29.

34 Majumdar, A. (2010). Yarn hairiness and its reduction. In: *Technical Textile Yarns*, 112–139. Elsevier.

35 Okubayashi, S. and Bechtold, T. (2005). A pilling mechanism of man-made cellulosic fabrics – effects of fibrillation. *Textile Research Journal* 75 (4): 288–292.

36 Okubayashi, S., Campos, R., Rohrer, C., and Bechtold, T. (2005). A pilling mechanism for cellulosic knit fabrics – effects of wet processing. *The Journal of The Textile Institute* 96 (1): 37–41.

37 Periyasamy, A.P. (2020). Effects of alkali pretreatment on lyocell woven fabric and its influence on pilling properties. *The Journal of The Textile Institute* 111 (6): 846–854.

38 Ratnam, T.V. (2010). *SITRA Norms for Spinning Mills.* Coimbatore, India: South India Textile Research Association.

39 Kärkkäinen, N. and Sillanpää, M. (2021). Quantification of different microplastic fibres discharged from textiles in machine wash and tumble drying. *Environmental Science and Pollution Research* 28 (13): 16253–16263.

40 Laitala, K., Klepp, I.G., and Henry, B. (2018). Global laundering practices – alternatives to machine washing. *Household and Personal Care Today* 12 (5): 10–16.

41 Lamichhane, G. (2018) Analysis of microfibers in waste water from washing machines. https://urn.fi/URN:NBN:fi:amk-201802132413

42 Bishop, D.P. (1995). Physical and chemical effects of domestic laundering processes. In: *Chemistry of the Textiles Industry* (ed. C.M. Carr), 125–172. Netherlands, Dordrecht: Springer.

43 Carney Almroth, B.M., Åström, L., Roslund, S. et al. (2018). Quantifying shedding of synthetic fibers from textiles; a source of microplastics released into the environment. *Environmental Science and Pollution Research* 25 (2): 1191–1199.

44 How Sustainable Is Fleece Fabric? (2024). https://kingcason.com/how-sustainable-is-fleece-fabric/. (accessed on 23 February 2025).

4

Microplastics Pollution: Health Hazards

4.1 Introduction

Microplastics, characterized as plastic particles measuring less than 5 mm, are a significant environmental and public health concern. Microplastics, originating from significant plastic waste, synthetic textiles, and microbeads in cosmetics, can remain unchanged as they flow through waterways until they reach the ocean. They manifest in many shapes and forms, including fibers, fragments, spheres, beads, films, flakes, pellets, and foam, as elucidated in *Chapters* 1, 2, *and* 3. Primary microplastics consist of microbeads and abrasive cleaning agents, whereas secondary microplastics arise from the degradation, fragmentation, or incorrect disposal of plastic materials [1, 2]. Paints can emit both primary and secondary microplastics [3]. The possible effects of microplastics differ based on the organism's size and morphology [4, 5]. Larger microplastics are generally not absorbed by most plants and soil creatures; however, smaller microplastics may be taken up by cells, presenting a greater environmental threat [6]. Microplastics serve as a source of greenhouse gas emissions, hinder the ocean's ability to absorb and sequester carbon dioxide, and are ingested by people through air and food intake [6]. Plastic pellets are utilized in manufacturing and are lost across the supply chain, including shipping, packaging, production, and waste management processes. Capturing plastic is challenging, because its discharge into the environment via air and water impacts four dimensions: the environment, climate, human health, and the economy. Policy involvement is necessary to resolve these difficulties. Significant quantities of pellets are generated and managed annually, both worldwide and inside the EU, totaling approximately 57 million tons in the EU in 2021. Estimates indicate that between 52,140 tons and 184,290 tons of pellets were released into the environment in the EU in 2019. This equates to around 2100 to 7300 trucks laden with pellets annually.

Textile Microplastics in the Environment: Sources, Impacts, and Solutions,
First Edition. Aravin Prince Periyasamy.

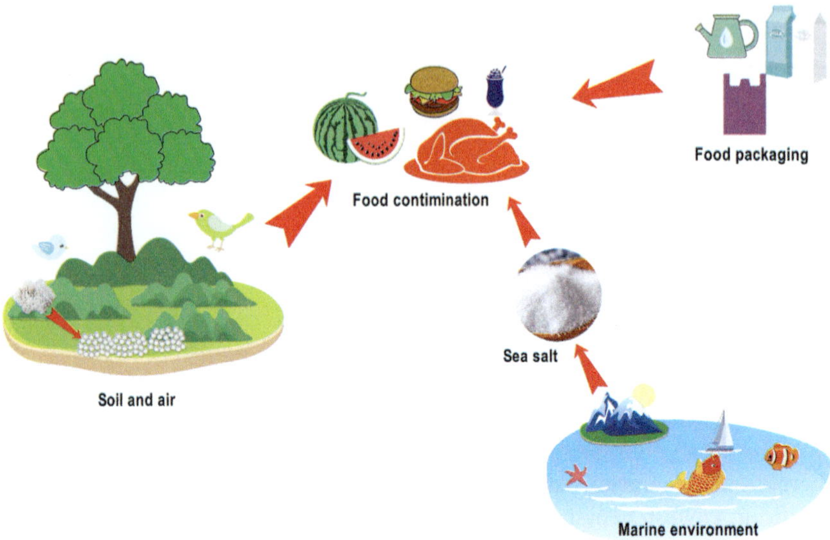

Figure 4.1 Possible routes for food contamination.

Microplastic pollution in food webs is an escalating environmental issue with considerable consequences for marine and terrestrial ecosystems. These little plastic particles, measuring less than 5 mm, can infiltrate food webs by ingestion by plankton, fish, and other aquatic creatures, resulting in toxicological consequences such as DNA damage, oxidative stress, and reproductive toxicity. Contaminated organisms are ingested by predators, facilitating the passage of microplastics along the food chain, ultimately culminating in human exposure. Microplastics are found in several food items, such as seafood, drinking water, honey, and milk. Immediate research and legislative measures are required to alleviate its effects [7]. Figure 4.1 illustrates the pathways of food web contamination, and this chapter thoroughly examines the potential routes of human exposure, textile functionalization, and their implications for human health related to functionalized microplastics, including the associated health issues and toxicity mechanisms [4, 5].

4.1.1 Possible Routes for Human Exposure

Microplastics can be ingested by aquatic animals and substantially threaten their metabolic activities [8, 9]. Microplastics are exposed by three different ways to the human body. Figure 4.2a provides an explanation of the three distinct ways in which microplastics are exposed to the human body [1, 2].

- Inhalation (both indoors and outdoors),
- Ingestion (i.e. contaminated food and water), and

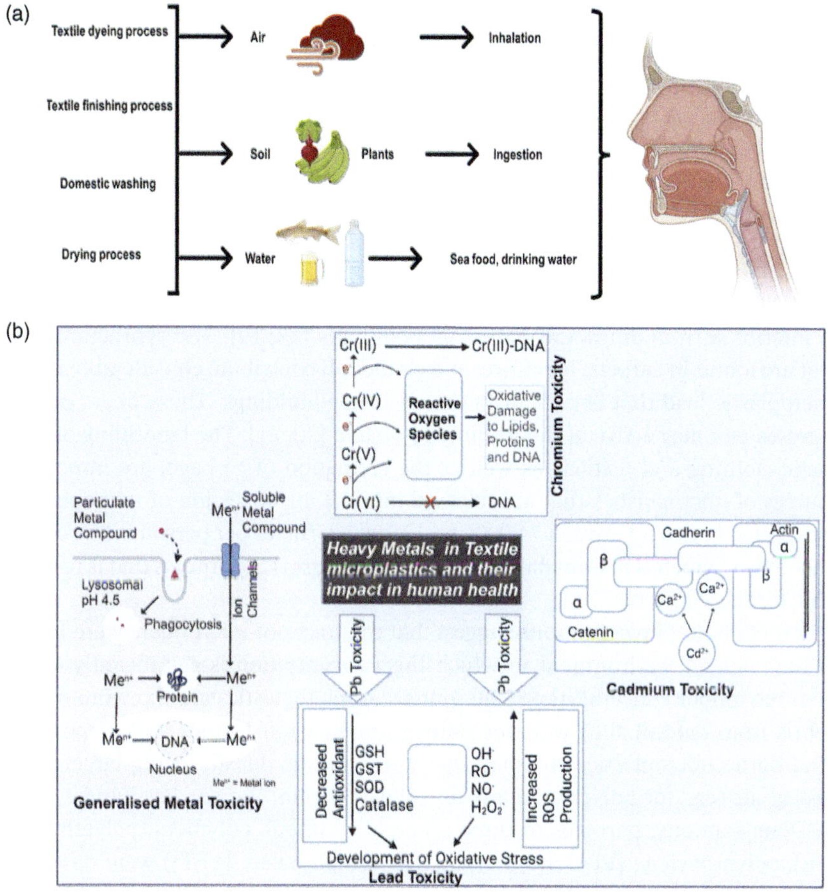

Figure 4.2 Heavy metals originating from textiles (a) and the ways in which they interact with the environment and various human beings (b). (*Source:* Periyasamy [5] / with permission of Elsevier.)

- Direct dermal contact of the particles through personal care products, textiles, or indoor dust triggering health effects.

4.1.1.1 Ingestion

The majority of microplastics are consumed through the consumption of contaminated food and water [10]. Microplastics can be discovered in seafood, table salt [11], honey [12], bottled water [13], beer [14], and fruits/vegetables [15], and canned fish [16]. The prevalence of these substances in the human diet is also influenced by the packing and processing methods used for food that is contaminated. According to studies, the typical individual may take in thousands of microplastic particles on a yearly basis within their body. The most significant sources of microplastics are food and food containers, with the average individual consuming between 39,000 and 52,000 particles of microplastics each day, depending on the diet and lifestyle patterns of the American population [17]. Consumption, on the other hand, differs depending on factors such as age, gender, regional circumstances, dietary habits, and lifestyle choices. From Fig. 4.2a, it is confirmed that food and food containers are the major source of microplastics for human beings.

4.1.1.2 Inhalation

In addition to being found in airborne dust, microplastics may also be detected in interior surroundings and industrial pollutants [18, 19]. The synthetic fibers that are found in carpets, furniture, and clothes all contribute considerably to the microplastic load that is present in the air inside buildings. These fibers can be ingested and may arrive at the respiratory system [20, 21]. The landfilling of synthetic clothing and textiles, as well as the cremation of garbage, are important sources of microplastics that are released into the air. Ingestion of microplastics is expected to reach between 74,000 and 121,000 particles per person in the city of Vancouver, which is in Canada [17]. Overloading dust is the process that is responsible for these effects.

Dris et al. [22] investigations suggest that the traces of microfibers were found in the open-air environment in which this concentration is significantly lower than the indoor concentration. The sources of microplastics are the wind-blown debris from landfill sites with microfibers from castoff textiles with deposition of airborne microplastics common due to the plastic density being larger than that of air (e.g. for polyester, approximately $1.39\,g\,cm^{-3}$ at sea level and 15 °C). In Dongguan city, particles of three different synthetic polymers (polyethylene [PE], polypropylene [PP], and polyethylene terephthalate [PET]) were observed in the outdoor environment; the concentration of these particles is from 175 to 313 particles per square meter per day [23]. In general, airborne microplastic particles can be observed in both synthetic and natural fibers. Shanghai people daily inhale approximately 21 particles of microplastics from open air [24]. Pauly et al. [25] studied the lung cancer patients who are out of 114 human lungs and fibers were observed 99, which is equivalent to 87%. The observed fiber dimension is 50 μm length, and it is likely to spread longer than 250 μm. The inhaled fibers stay in the lungs for prolonging time, even if it creates life threat. Further, the inhaled fibers contain toxic substances like dyes, mordants, and chemicals. It was surprising to know that the private apartment located 10 km away from Paris city center found 190–670 microfibers [22]. The air samples were collected inside the private apartment at 1.2 m height which is standard breathing height of an adult. Quartz fiber GF/A Whatman filters were used to collect the deposited microfibers, and it normalized then expressed number of microfibers per square meter per day. Overall, 1–60 fibers m^{-3} microfibers were observed in the indoor environment, in contrast, the outer environment is much better, and it ranges from 0.3 to 1.5 fibers m^{-3}. The observed fibers in the indoor environment, 67% of fibers have belonged to cellulosic region and 33% are synthetic fibers such as PP, which is predominantly used in the carpets, sofa, and other furnishings [22].

4.1.1.3 Dermal Contact

When skin comes into touch with microplastics, it can be using cosmetics, personal care items, or even through water sources that are polluted. Even though the skin serves as a natural barrier, nanoparticles can enter hair follicles and skin that are damaged. It is less significant to encounter nanoplastics through the skin; yet there is a need for evidence to evaluate whether or not nanoplastics (<100 nm) may pass through the dermal barrier and induce toxicity [1].

4.2 Textile Functionalization as Source of Microfibers Toxicity

When it comes to boosting the value-added element of the final product, the manufacturing process of textiles, which includes the dyeing and finishing processes, is essential since it provides specific functionality and brilliant colors [26]. On the other hand, due to restricted production capacity, natural dyes and finishing compounds derived from natural sources are insufficient [27]. This results in the utilization of synthetic dyes and functional chemicals. To get a vibrant hue, the woolen and silk industries make use of chromium or mordant dyes as well as metal complex dyes. Heavy metals, including lead (Pb), arsenic (As), chromium (Cr), nickel (Ni), copper (Cu), cadmium (Cd), mercury (Hg), and zinc (Zn), are found in the majority of the textile dyes [28, 29]. These heavy metals are potentially harmful to both the environment and human health due to their inability to break down by biological processes [30]. As a result of their intricate chemical makeup, heavy metals that have a density of more than 5 mg cm^{-3} are rather prevalent and provide difficulties in the process of cleanup. Heavy metals are found in the biosphere and the larger ecosystem, which includes the air, water, and soil. As shown in Fig. 4.2b, the interaction between metal ions and biological macromolecules that results in toxicity is illustrated. It demonstrates how cadmium disrupts the cadherin cell–cell adhesion system, the toxic effects of lead, which are illustrated in a graphic, the process of mercury toxicity in cells, and the metabolism of intracellular chromium, which includes the creation of reactive oxygen species (ROS) and its effects on DNA.

4.2.1 Fiber Production

The textile industry has been responsible for a significant amount of environmental degradation, including the release of pollutants that have not been treated well, the emission of air, and other forms of pollution, which has resulted in the normal lives of people being made trouble [28, 29]. This is especially true in coastal areas, where polymeric chemicals are widely recognized as being known to disturb the endocrine system [31]. In the meantime, the textile industry is comprised of several process value chains, and each step requires a significant amount of chemicals to be consumed to enhance the processability and to achieve certain functionality [32]. For example, during the production process of fibers, several chemicals are utilized to improve the fiber's processability and usefulness. These additives include ultraviolet (UV) stabilizers, titanium dioxide, diisobutyl phthalate (DIBP), and phthalates [5]. It is standard practice to include phthalates into coatings that are based on polyvinyl chloride (PVC). Nevertheless, there is a possibility that these compounds might have adverse effects on humans, particularly on the functions of reproduction and development. A complete inventory of these additives may be seen in Table 4.1.

Table 4.1 Typical additives are used in man-made fiber productions.

Type	Function	Examples
Processing aids	Antioxidant	Hindered phenols and amines, phosphites
	Hydrolysis stabilizer	Carbodiimide
	Nucleating agent	Talcum powder, boron nitride, organic phosphate salts
	Lubricant	Stearates, low molecular wax
	Polymer processing aid	Fluoropolymers
	Surfactant	Stearates, polyethylene glycols (PEG)
Enhancing additives	Plasticizer	Tributylcitrate, acetyltributylcitrate
	Chain extender	Difunctional acid derivatives, anhydrides, and epoxides
	UV stabilizer	Hindered amine light stabilizers (HALS), TiO_2, ZnO, carbon black
	Flame retardant	Phosphorus and halogen derivatives, HALS
	Thermal protection	Zirconia
Functional additives	Colorant	Pigments, dyes, carbon black
	Delustering	TiO_2, ZnO, mica, optical brightening agents
	Antistatic	Carbon black, carbon nanotubes, graphene, ZnO
	Antimicrobial	TiO_2, ZnO, Ag^+, Cu^{2+}, Zn^{2+}, plants extracts, phenol
	Water/oil repellent	Silicone and fluorine compounds

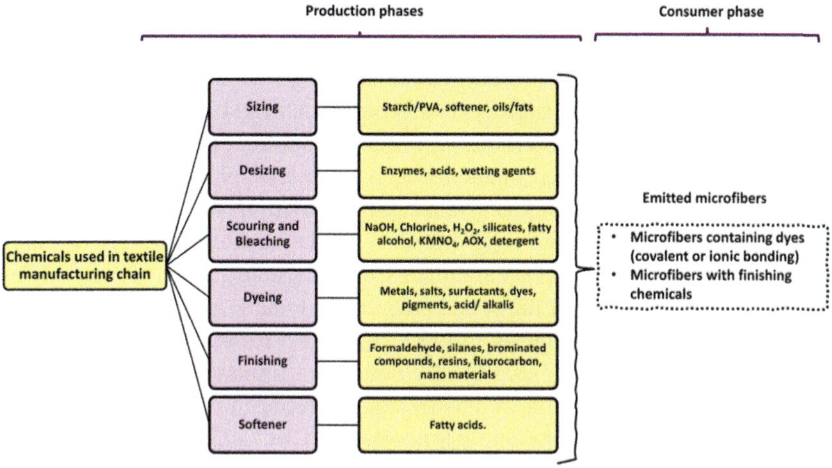

Figure 4.3 Various chemicals used in the textile manufacturing chain.

Furthermore, microplastics that include these chemicals are another series of problems that need to be addressed. Without a doubt, the emission of these untreated chemicals is a significant cause for worry since it poses a threat to both the environment and human health [5]. Therefore, the discharge of functionalized microplastics from a variety of sources is a global problem, notably in the liberation of microfibers from textile materials that contain heavy metals that prevent them from biodegrading of emitted fibrous fragments in both natural and synthetic origins. Consequently, this presents considerable difficulties for the ecosystem [33]. The use of conductive chemicals on fiber surfaces for the purpose of electromagnetic shielding and electrical conduction has resulted in the toxicity of heavy metals and the adverse effects that they have on the health of humans, aquatic organisms, and animals. When compared to the microfibers that are discharged into the environment, the presence of organic and inorganic additives, as well as residues of monomers, metals, or other chemicals, offers a bigger threat to human health [34]. It is possible to understand the quantity of dangerous and hazardous compounds that are employed in this industry by referring to Fig. 4.3, which depicts the wide variety of chemicals that are utilized in the textile production process.

Heavy metals such as antimony, aluminum, zinc, bromine, arsenic, lead, titanium, and chrome are commonly used as additives in various types of polymers. Antimony is found in polyester cotton or polyester wool fabric as flame retardants and biocides and is associated with metal-estrogenic effects and breast cancer. Aluminum, used in polyester cotton or polyester wool fabric as stabilizers, inorganic pigments, and flame retardants, also has metal-estrogenic effects leading to breast cancer [35]. Zinc is utilized in PVC and PE as heat stabilizers, flame retardants, anti-slip agents, and inorganic pigments. Bromine, present in PE and PP as flame retardants, causes apoptosis and genotoxicity. Arsenic, used as biocides in PET, is a known carcinogen affecting the lungs, skin, liver, bladder, and kidneys, as well as causing gastrointestinal damage and death. Lead, found in PVC and various plastics with red pigments such as heat stabilizers, UV stabilizers, and inorganic pigments, is linked to anemia, hypertension, miscarriages, nervous system disruptions, brain damage, infertility, oxidative stress, and cell damage [5]. Titanium, used in PVC as UV stabilizers and inorganic pigments, causes cytotoxicity in human epithelial lungs and colon cells. Finally, chrome, employed in dyes for silk and metal complexes in PVC, PE, and PP, leads to allergic reactions, nasal septum ulcers, and severe cardiovascular, respiratory, hematological, gastrointestinal, renal, hepatic, and neurological effects, and possibly death [35].

In order to create a vibrant cloth, it is necessary to dye or print it using a variety of dyes, such as azo, nitro, nitroso, indigo, phthalein, triphenylmethane, and anthraquinone, among others. Colorants are utilized in a wide range of sectors, including but not limited to the following: food, cosmetics, paper, polymers, leather, paint, medicines, and textiles. More than ten thousand distinct hues are utilized in the production of these goods [36, 37]. For a wide range of applications, the textile industry makes use of a wide variety of dyes, including vat dyes, reactive dyes, direct dyes, and cationic dyes, among others. The textile dyeing industry, on the other hand, has been responsible for environmental problems due to the discharge of colored contaminants. Natural-based dyes have been embraced by industry as a reaction, following the advice of nongovernmental organizations [38]. Natural dyes are frequently employed in combination with conventional mordants such as copper, hydrated potassium aluminum sulfate, ferrous sulfate, tin, chrome, and copper. This is the case even though natural colors have low binding affinities [39]. The health consequences of many chemicals that are used in the textile industry are depicted in Fig. 4.4, with a particular emphasis not only on the immediate effects (represented by the green tick) but also on the long-term effects (marked by the pink tick). The breakdown of these dyes can result in the production of hazardous aromatic compounds such as azoamine, which puts individuals at risk for a variety of potential health consequences. These complications include cutaneous hypersensitivity, respiratory impairment, endocrine disturbances, and certain cancers. There are several health consequences that may be brought on by exposure to hazardous colors. Some of these complications include skin hypersensitivity, pulmonary impairments, endocrine abnormalities, and some cancers. Figure 4.5 presents a complete account of the influence that different dyes have on textiles, as well as the consequences that these colors have for the health of humans and the environment.

The finishing process of textile materials entails chemical and physical treatments to attain desirable properties such as softness, aesthetic appeal, and

Chemistry	Flame Retardants	PFAS	Lead	Chromimum	Phthalates	Chlorine	Azo dyes	VOCs
Probable Carcinogenes	✓ (prolonged) ✓ (short-term)	✓ (prolonged) ✓ (short-term)	✓ (prolonged) ✓ (short-term)	✓ (prolonged) ✓ (short-term)	✓ (prolonged) ✓ (short-term)	✓ (prolonged) ✓ (short-term)	✓ (prolonged)	✓ (prolonged)
Skin irritants	✓ (prolonged) ✓ (short-term)	✓ (prolonged) ✓ (short-term)	✓ (prolonged) ✓ (short-term)	✓ (prolonged) ✓ (short-term)	✓ (short-term)	✓ (prolonged) ✓ (short-term)	✓ (prolonged)	✓ (prolonged)
Harmone disruptors	✓ (prolonged) ✓ (short-term)	✓ (prolonged) ✓ (short-term)	✓ (prolonged) ✓ (short-term)	✓ (prolonged) ✓ (short-term)	✓ (prolonged)	✓ (prolonged)	✓ (prolonged)	✓ (prolonged)
Environment water pollution	✓ (prolonged) ✓ (short-term)	✓ (prolonged) ✓ (short-term)	✓ (prolonged) ✓ (short-term)	✓ (prolonged) ✓ (short-term)	✓ (prolonged)	✓ (prolonged) ✓ (short-term)	✓ (prolonged)	✓ (prolonged)
Environment soil pollution	✓ (prolonged) ✓ (short-term)	✓ (prolonged) ✓ (short-term)	✓ (prolonged) ✓ (short-term)	✓ (prolonged) ✓ (short-term)	✓ (prolonged)	✓ (prolonged) ✓ (short-term)	✓ (prolonged)	✓ (prolonged)

✓ (prolonged) Prolonged exposure lead severe health impacts

✓ (short-term) Shorterm exposure lead to health impacts

Figure 4.4 Health impacts of different chemicals used in the textile industry.

functional features like flame retardancy, water repellence, and crease resistance. To improve the flame retardancy of textiles, many ingredients are often utilized in textile production, such as hexafluorotitanate salts, TiO_2 nanoparticles, and brominated and phthalate compounds [40]. These chemicals have been identified as exhibiting reproductive and developmental toxicity. Cellulose-based textiles often utilize formaldehyde-based resins to improve their crease-healing properties. Likewise, polyfluoroalkyl compounds are frequently utilized to improve the water repellency of fabrics. The unintentional inclusion of various chemical substances in the final textile product. Moreover, the textile industry includes pesticide residues from natural fibers like cotton, fungicide triclosan, and colors derived from several chemical groups, all of which have been recognized as having detrimental effects. The release of textile microfibers into the water leads to their photo- and biodegradation, consequently producing plastic debris at micro- and nanoscale levels. The release of dissolved organic carbon (DOC) is substantially affected by the degradation of polymers, which are composed of oligomers with varying degrees of oxidation [41]. The imperceptible contamination from microfibers derived from textiles, identified in marine sediments and organisms, can be ascribed to the textile industry.

Acrylic polymers and quaternary ammonium compounds are commonly used as wrinkle-resistant finishes due to their ability to reduce the need for ironing and help fabrics retain their shape [42]. Polydimethylsiloxane and polyvinyl acetate (PVA) are utilized for moisture management finishes, making them suitable for activewear and outdoor clothing. Polyethylene glycol (PEG) is another polymer used for moisture management finishes. Polytetrafluoroethylene (PTFE) is applied to create hydrophobic barriers that repel water in water-repellent finishes. Phosphorus-containing polymers are integral for flame-retardant finishes, which aim to reduce the flammability of textiles [43]. Conductive polymers are employed to reduce the buildup of static electricity in antistatic finishes. Antimicrobial finishes, which inhibit the growth of bacteria and fungi and control odor, make

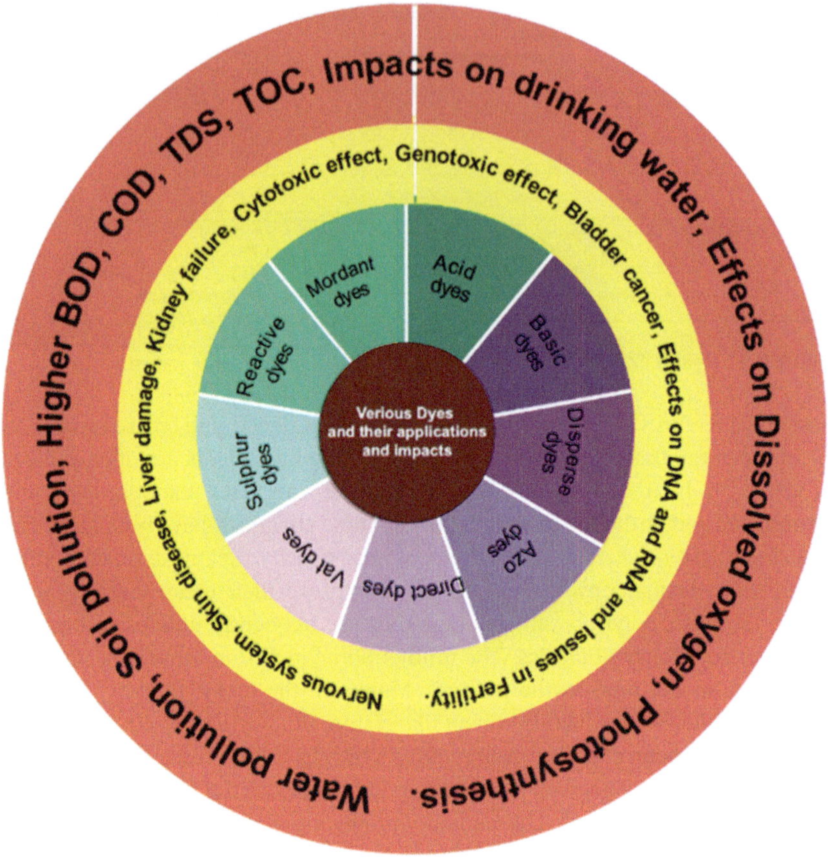

Figure 4.5 Using of various dyes in textile industry and their health and environmental impacts. *Note: BOD: biological oxygen demand; COD: chemical oxygen demand; TOC: total organic carbon.* (*Source:* Periyasamy [5] / with permission of Elsevier.)

use of polypyrrole, polythiophene, and polyaniline. Polyurethane and epoxy are polymers that improve the abrasion resistance of textiles, thereby extending their lifespan. Silicone polymers, PE emulsions, and cationic polymers are used in softening finishes to improve tear strength, abrasion resistance, wrinkle resistance, stretch recovery, and to avoid shrinkage. Acrylic polymers and quaternary ammonium compounds also serve as antipilling finishes by reducing friction between fibers and minimizing pilling. Finally, acrylics, polyurethanes, and epoxies are utilized in UV-protective finishes to absorb or reflect harmful UV radiation, helping to prevent sunburn and skin damage [44].

Cellulose-based textiles can undergo easy-care or crease-resistant finishing processes, enhancing their capacity to recover from creases. Urea derivatives

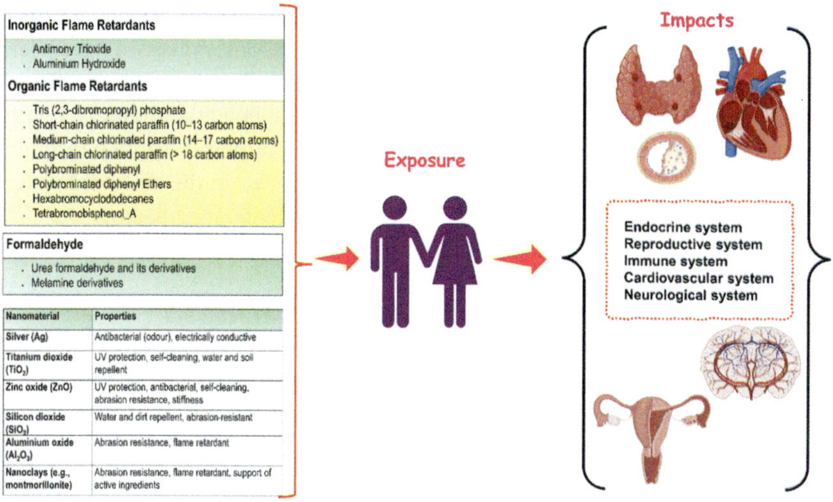

Figure 4.6 Various categories of flame-retardant chemicals, including formaldehyde and nanomaterials, are utilized in the textile industry. (*Source:* Periyasamy [5] / with permission of Elsevier.)

such as urea-formaldehyde and melamine-formaldehyde are often utilized in industrial applications; nevertheless, their environmental discharge and human contact may result in allergic responses and dermal conditions, thereby heightening the risk of skin cancer [45] (Fig. 4.6). Durable flame-retardant treatment, utilizing phosphorus and halogen chemicals, preserves efficacy over several laundering cycles. Nonetheless, specific flame retardants, like bromine, chlorine, and halogens, exhibit endocrine-disrupting characteristics, presenting potential hazards to human health. Microplastics infused with flame-retardant chemicals adversely impact the health of marine organisms, contaminating aquatic food webs (Fig. 4.6).

The utilization of nano-textile materials has demonstrated potential in a variety of applications, including the manufacture of fibers and the processes of finishing. To improve the uniform dispersion of the fibers, they are frequently added by mixing with polymers prior to the spinning process. Additionally, they are utilized in the finishing operations of fabric, such as dipping, cushioning, printing, or coating. The abrasion of fiber, on the other hand, might result in the release of nanoparticles that are already present in the molecular composition of the fiber [46, 47]. This can result in both health concerns and environmental hazards. It depends on the characteristics of the organic polymer and its capacity to stick to the fabric as to whether or not nanoparticles on fabric surfaces will remain there for an extended period (Fig. 4.6). In maritime environments, the durability of nanomaterials is contingent upon factors such as the temperature of the water, the salt of the water, and the possibility of the release of microplastic fibers. Several issues have been expressed regarding the possible adverse consequences on both the health of humans and the environment. Nanomaterials have the potential to interact with marine creatures through a variety of mechanisms, including absorption through the food chain, absorption through exposure to chemicals in the dissolved water phase, and liberation from microfiber detritus that individuals consume.

4.3 Microplastic Contamination in Human Health

Microplastics, which may be found in the environment, are known to provide considerable health hazards to people. These risks include cancer, immunotoxicity, intestinal disorders, lung diseases, cardiovascular disease, inflammatory diseases, pregnancy, and mother exposure to offspring. Microplastics enter the human body through the consumption of contaminated food and drink, which is the principal route of exposure to these particles. It is estimated that between 0.1 and 5 g of microplastics are consumed per week by inhalation [48], via food and beverages [17, 49], and through other means [48]. Microplastics may be found in a variety of foods and beverages, including milk, tea bags, shellfish, bottled water, salt, and sugar. The most common way that people are exposed to microplastics is by oral consumption. Consuming shellfish puts Europeans at risk of being exposed to around 11,000 microplastic particles per person on a yearly basis. The human body is responsible for the absorption of between 39,000 and 52,000 particles of plastic on a yearly basis. Microplastic contamination is caused by food containers composed of polymers such as PP, polystyrene (PS), PE, and PET [49]. There is a possibility that those who consume takeaway meals 4–7 times per week might potentially consume anything from 12 to 203 microplastic bits. The disintegration of the surface of silicone rubber baby teats that are sterilized with steam leads to the emission of microplastic particles into the environment that is around the teats. It is estimated that around 0.66 million microplastic particles [50] are introduced into the body of a newborn over the course of a year when the infant is fed from a bottle on a constant basis [51].

Microplastics, which contain harmful chemicals such as nanoparticles, perfluoroalkyl derivatives, and formaldehyde, can also be experienced by inhalation, which is another route of exposure to these substances. The effects of polystyrene nanoparticles (PS-NPs) on human lung epithelial cells have been investigated in several laboratory studies [11]. The possibility of exposure to PE microplastics may be increased by a number of factors, including the usage of consumer items that contain microplastics, such as cleansers and face creams. Additionally, protective coatings for mobile phones have the potential to generate microplastic particles, which can subsequently be passed to the hands of humans [16]. Animals, particularly aquatic organisms such as fish and marine mussels, have been shown to contain microplastics, which represent a substantial threat to the animals' ability to survive. Because of their toxic properties, there is cause for worry over the possible influence that they might have on animal populations, which are an essential component in preserving the environmental integrity of the ecosystem. To better understand the underlying processes of toxicity, it will be helpful to conduct a comprehensive investigation of the possible toxicity of microplastics at a variety of higher levels [16]. The physical characteristics, chemical makeup, and capacity to function as carriers for dangerous compounds are the components that are responsible for the health hazards that are linked with exposure to microplastics. A comprehensive discussion of the many health problems that are brought on by microplastics is presented here in this section.

4.3.1 Inflammatory and Immune Responses

Inhalation or ingestion of microplastics may provoke immune reactions, potentially leading to inflammation and chronic diseases [52]. Upon inhalation or ingestion, microplastics can induce oxidative stress, metabolic disruptions, inflammation, and immunological responses, resulting in the development of tumor microenvironment and immunosuppression, which may facilitate carcinogenesis. The effects are facilitated by cytokine overexpression and the activation of critical signaling pathways (Fig. 4.7), leading to inflammation and immunological responses that promote cancer [53].

Microplastics may build up in the body, elevating the risk of tissue injury, fibrosis, and carcinogenesis. This risk is linked to the activation of immunological responses, including oxidative stress, apoptosis, and necrosis in the liver, intestines, brain, and other organs [54]. The pervasive presence of microplastics in ecosystems and their accumulation in the human body have elicited concerns over their possible health risks. Inhaled microplastics may induce respiratory inflammation and contribute to pulmonary conditions such as asthma and fibrosis. Inflammation is a potent defense response that can be triggered by exposure to xenobiotics or harmful substances. It is very efficacious in safeguarding the body against pathogens and other harmful toxins. Engineered nanomaterials, frequently identified in living organisms, have the capacity to engage with the immune system and can be considered a particular example of "particulate xenobiotics." It is essential to meticulously regulate and balance inflammatory processes across varying degrees of complexity, as they possess the capacity to inflict damage. Research has shown that micro- and nanoplastic particles can induce inflammation at the molecular, cellular, and tissue levels, likely due to oxidative stress and impaired lysosomal function. Both in vivo and in vitro studies

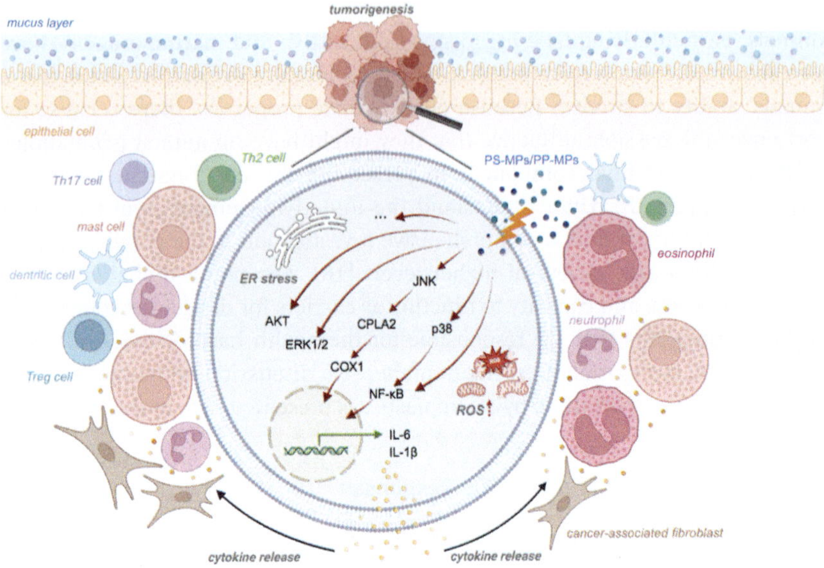

Figure 4.7 Microplastic induces chronic inflammation and potential tumor micro-environment formation. *Note: IL-1β:Interleukin-1; IL-6: Interleukin-6; JNK: C-Jun N-terminal kinase; NF-κB: Nuclear factor kappa-light-chain-enhancer of activated B cells.* (*Source:* Reproduced with permission of Cheng et al. [53] / Springer Nature / CC-BY-NC-ND 4.0.)

have demonstrated indications of inflammatory tissue injury in zebrafish and murine models. The data indicate that exposure to micro- and nanoplastics may induce inflammation at all three levels. Additional data is required for a more comprehensive understanding of the processes involved [55].

Moreover, exposure to microplastics has been extensively shown to induce the overexpression of ROS and inflammatory mediators, leading to DNA damage, oxidative stress, immunological responses, and chronic inflammation. Furthermore, these impacts should not be examined just at the individual organism level; increased focus must be directed toward the alterations in the population structure of certain species and the potential for these changes to eventually disturb the dynamics of entire ecosystems [56]. The release of diverse proinflammatory cytokines and chemokines by inflammatory cells fosters an environment favorable to cancer cell proliferation. Inflammation can expedite cancer progression and facilitate all phases of carcinogenesis [57]. Moreover, the exposure to microplastics and nanoplastics can directly enhance tumor cell proliferation and affect tumor start and progression by modulating inflammatory responses [57], and the precise mechanisms by which microplastics induce tumor growth are limited; nonetheless, the shift from chronic inflammation to cancer, linked to immune regulation from microplastics exposure, provides a very interesting avenue for research [57].

4.3.2 Gastrointestinal and Metabolic Disruptions

Ingested microplastics can disrupt gut flora, which is essential for digestion and immunity. Certain studies have associated microplastic exposure with heightened gut permeability, potentially permitting hazardous chemicals and germs to infiltrate the circulation, resulting in metabolic problems. Numerous studies indicate that microplastics can influence the composition and variety of microbial communities in the gastrointestinal tract (GIT) of humans, mice, chickens, and aquatic organisms, as elucidated. Microplastic exposure has been studied across various organisms, revealing diverse impacts on microbiota and host health. In humans, PET microplastics (particle size $160 \pm 110\,\mu m$) at doses of 4–5 intakes of 166 mg have increased Firmicutes (*Phascolarctobacterium*, *Lachnoclostridium*, *Megasphaera*), Proteobacteria, and Desulfobacterota (*Escherichia*/*Shigella*, *Bilophila*), while decreasing Bacteroidetes (*Bacteroides*, *Parabacteroides*, *Alistipes*) and Actinobacteria (*Bifidobacterium* spp.) [58]. Similar studies in mice have shown that PS microplastics (0.5–50 μm) altered gut microbiota composition, decreasing Firmicutes, α-Proteobacteria, and Bacteroidetes and increasing Proteobacteria and Actinobacteria [59]. Notably, PE microplastics (particle size 45–53 μm) in mice increased Actinobacteria, *Adlercreutzia*, *Butyricimonas*, and *Parabacteroides*, impacting gut health [60]. In chickens, exposure to PS microplastics (1–100 mg L^{-1}, 5 μm) led to changes in caecum content, with increases in Peptostreptococcaceae and decreases in Butyricicoccaceae, Erysipelatoclostridiaceae, Lachnospiraceae, and Lactobacillaceae, causing inflammation and tissue damage. Studies on zebrafish larvae exposed to PS microplastics (1000 mg L^{-1} of water, 5 and 50 μm) indicated microbiota alterations, including increases in Firmicutes and decreases in Bacteroidetes, Proteobacteria, and Methylobacterium, affecting energy, glucose, and lipid metabolism [61]. Moreover, crabs exposed to PS microplastics (40 mg L^{-1}, 5 μm) experienced increases in Fusobacteria, Proteobacteria (*Pseudomonas*), Cyanobacteria, and Chloroflexi, while Firmicutes and Bacteroidetes (*Dysgonomonas*) decreased, alongside reductions in alkaline phosphatase, phenoloxidase, lysozyme, and acid phosphatase activities [62]. This extensive research highlights the pervasive influence of microplastics on various organisms, emphasizing the urgent need for further investigation and mitigation strategies.

4.3.3 Respiratory System

Although microplastics are eliminated by the immunological clearance process of the body, it is nevertheless possible for them to be maintained in the respiratory tract, which can result in harmful effects on the respiratory system [48, 63]. Researchers have shown that workers who are exposed to plastic fiber and particles have a higher risk of developing airway and interstitial lung illnesses [48]. These diseases include asthma, emphysema, bronchitis, granuloma, pneumonia, interstitial fibrosis, and even an increased risk of developing cancer. The inhalation of microplastics has the potential to adversely affect respiratory function, trigger an inflammatory response in the lungs, exacerbate oxidative stress, and lead to the development of lung fibrosis [64]. For instance, the presence of PS microplastics in the lung tissue of rats was discovered when they were exposed to the substance through their respiratory system. This resulted in the integrity of the respiratory epithelial barrier being disrupted, as well as lung damage and inflammation. Microplastics exposure was linked with transcriptional modifications in functional gene clusters related to cellular stress response, immunological response, and cell death in asthmatic settings. This was accompanied by an increase in mucus output and an exacerbation of airway inflammation. Additionally, clear granulomatous inflammation was observed in the lungs of mice that were given PE-MPs through their mouths orally on many occasions [65].

4.3.4 Cardiovascular System

In zebrafish, rats, and chickens used as models, it was discovered that microplastics had a toxicological reaction on the cardiovascular system [66]. It has been demonstrated via research that the accumulation of microplastics in the pericardium of zebrafish can result in disorders of the cardiovascular system. Microplastics have been shown to produce decreased cardiac output, pericardial edema, and endothelial dysfunction in zebrafish embryos. These effects have the potential to promote ROS development as well as systemic inflammation. Microplastics, on the other hand, do not have a substantial impact on the heart rate of zebrafish embryos and do not cause atrioventricular block. Microplastics are responsible for the dysfunction and structure of the heart in Wistar rats, as well as the impairment of mitochondrial integrity, apoptosis, and pyroptosis in the myocardial tissue inside the heart [67]. It has been shown that myocardial damage, apoptosis, interstitial hyperplasia, and collagen deposition are all present in the cardiac tissue of rats. Microplastics are responsible for inducing myocardial pyroptosis, inflammation, and mitochondrial dysfunction in chickens [68].

4.3.5 Nervous System

There is a high likelihood that microplastics will accumulate in the cerebral cortex, hippocampus, and cerebellum because of their tiny size and distinctive surface activity [69]. This is because microplastics can readily pass across the blood–brain barrier. This can result in deficiencies in learning and memory, as well as lower levels of synaptic proteins and neuroinflammation in mice that have been administered the substance. Microplastics inhibit the production of neurotransmitters in the brains of zebrafish and cause behavioral changes that are similar to those associated with anxiety. According to the findings of recent research, microplastics are responsible for neurodegeneration in mice's brains that is like that of Parkinson's disease, as well as disruptions in protein and energy metabolism and intercellular communication [69, 70]. It is possible for a mother to have neurological impairment and delayed brain development because of exposure to microplastics. The deposition of microplastics in human microglia and heterogeneous neurons has been demonstrated through in vitro research. This deposition has been proven to influence cell activity, morphology, phagocytosis processes, cytokine release, and death [69].

4.3.6 Reproductive System

It has been discovered that microplastics are able to penetrate through the tissues of the gonads, which influences the reproductive fitness of medaka [71, 72]. A decrease in the quantity of sperm, an increase in the presence of sperm malformation, structural damage to testicular tissue, and aberrant testosterone levels are all potential outcomes of exposure to these microplastics. There is also evidence of DNA damage in the tests of rats that have been exposed to microplastics. Furthermore, microplastics have been shown to reduce the generation of follicular cells, impede the maturation of oocytes, and inhibit follicular development and atresia [73]. Microplastics are responsible for the apoptosis and necroptosis of ovarian granulosa cells, which in turn causes oxidative stress, inflammation, and an imbalance in calcium homeostasis. This is the reason why this occurs. In addition, microplastics and amino-modified microplastics have the potential to produce transgenerational reproductive toxicity, gonadal dysplasia, and germline apoptosis in the organism *Caenorhabditis elegans* [74]. The results of this study shed light on the possible adverse effects that microplastics may have on male reproductive systems, as well as the necessity of implementing effective preventative and mitigation strategies.

4.3.7 Immune System

Both vertebrates and invertebrates have been shown to suffer from a weakened immune system because of exposure to microplastics. Ingestion of microplastics through seafood or water can cause damage to the immune systems of the local digestive tract, which can result in the activation of neutrophils, hypersecretion of mucus, and immunological reactions [75]. As a result of reduced mucus production and gut microbiota creation in mice, the immunological barrier against pathogens was impaired, which contributed to the development of gut inflammatory diseases. Microplastics could cause microglial polarization and altered immunological responses, which can result in altered serum levels of cytokines and lower percentages of *Th17* and *Treg cells* in the spleen [60]. Oral ingestion of PE microplastics has the potential to disrupt the immune response and cause tissue damage, with a specific impact on the lymphocyte subsets in the spleen of the offspring. Studies conducted in vitro have demonstrated that nanoplastics are rapidly absorbed by murine splenocytes, which leads to a decrease in cell viability and an increase in apoptosis. In addition, microplastics interfere with the function of neutrophils, cause damage to macrophages, and hinder the complement system as well as the actions of enzymes involved in the immune system [76].

4.3.8 Microplastic-Induced Cancer

Microplastics are associated with several health issues, as owing to their diminutive size, they possess a significant surface area-to-volume ratio, which can inflict DNA damage within cells and potentially result in cancer. The unregulated dispersion of microplastics in aquatic environments sequesters hydrophobic organic contaminants, hence facilitating carcinogenesis. Heavy metals utilized in textile manufacturing (i.e. mostly in dyes) are also hazardous. Epidemiological studies indicate that prolonged exposure to textile microplastics is strongly correlated with cancer growth in people and animals [77]. Marine species can directly ingest microplastics, leading to contamination of the human food chain through bioaccumulation. Research indicates that human ingestion of microplastics may contribute to the onset of several malignancies [78]. The challenges associated with microplastic ingestion encompass prolonged inflammation and irritation, resulting in deoxyribonucleic acid damage and the production of proinflammatory mediators that induce angiogenesis. Polycyclic aromatic hydrocarbons in food and water have emerged as a significant worry, with microplastic leaching posing roughly 1000 times more risk than benzo[a]pyrene [79]. Microplastics are predominantly ingested via the stomach, presenting a potential danger for cancer development. Prolonged exposure to microplastics may elevate the risk of gastric cancer, as microplastics enhance the expression of asialoglycoprotein receptor 2 (ASGR2), which signifies the presence of cancer hallmarks including CD44, N-cadherin, programmed death ligand 1, and cellular proliferation. Prolonged exposure to microplastics can diminish survival rates and enhance tumor proliferation [80]. Microplastics have differing toxicity levels in humans, with elevated toxicity noted in microplastics measuring 0.3 mm, 0.5 mm, and 6 mm [81]. Nonetheless, reduced toxicity was noted in microplastic sizes of 1 mm and

3 mm. The absorption rates of microplastics were elevated for smaller particle sizes and diminished for larger particle sizes, suggesting that a greater surface area results in heightened cellular oxidative stress [79, 81]. Bisphenol A, a plasticizer utilized in microplastic at the nanoscale, has demonstrated synergistic toxicity on Caco-2 cells [82]. Exposure to bisphenol A may induce localized inflammation and influence the permeability of colon cells, mediated by increased levels of interferon-g, interleukin-17, and immunoglobulin A. These proteins are important to human physiology and immune defense, notifying the immune system upon the detection of pathogens or neoplastic cells within the body [82].

Exposure to bisphenol A in rats resulted in similar effects on colon cells and localized inflammation. The procedure pertained to the binding of bisphenol A to estrogen receptor beta, predominantly located in the human intestines. Colon cancer cells overexpressed owing to bisphenol A exposure in estrogen receptor beta were linked to the proliferation of colon cancer. Bisphenol A appears to inhibit the activities of estrogen mediated by its receptor, hence compromising the protective effects of endogenous estrogen hormone against colon cancer cell proliferation [82]. The digestive tract is a possible entrance site for microplastics in humans, with research indicating the presence of microplastics in lung tissues measuring less than 5.5 mm [83]. Studies have demonstrated that microplastics in human lungs influence cell proliferation and induce morphological alterations. Tri-o-cresyl phosphate, an isomer of plasticizers, has neurotoxic effects and induces liver and reproductive damage. Research indicates that growth deficiencies influence reproduction and fertility in aquatic organisms, implying a significant likelihood of leaching from microplastics that disrupt the endocrine system. Excessive exposure to bisphenol A can damage deoxyribonucleic acid by severing the double helix, leading to genomic instability and chromosomal rearrangements. The modulatory effect of bisphenol A on the cellular epigenome and metabolome poses a potential danger for secondary mutagenesis and tumorigenesis. This discovery parallels those observed in individuals with gastric and ovarian carcinoma [83].

4.3.9 Toxicity Mechanisms

Oxidative stress, resulting from an imbalance between the generation of ROS and antioxidant mechanisms, is a critical contributor to the toxicity of microplastics. Experimental studies have demonstrated that oxidative stress and the failure of the antioxidant defense system are critical factors in toxicity. Inhibiting oxidative stress can mitigate learning and memory impairments generated by microplastic mice. in vitro studies have demonstrated that exposure to both microscale and nanoscale PS elevates intracellular ROS levels and facilitates the production of superoxide radical anion. Microplastics can cause mitochondrial damage and

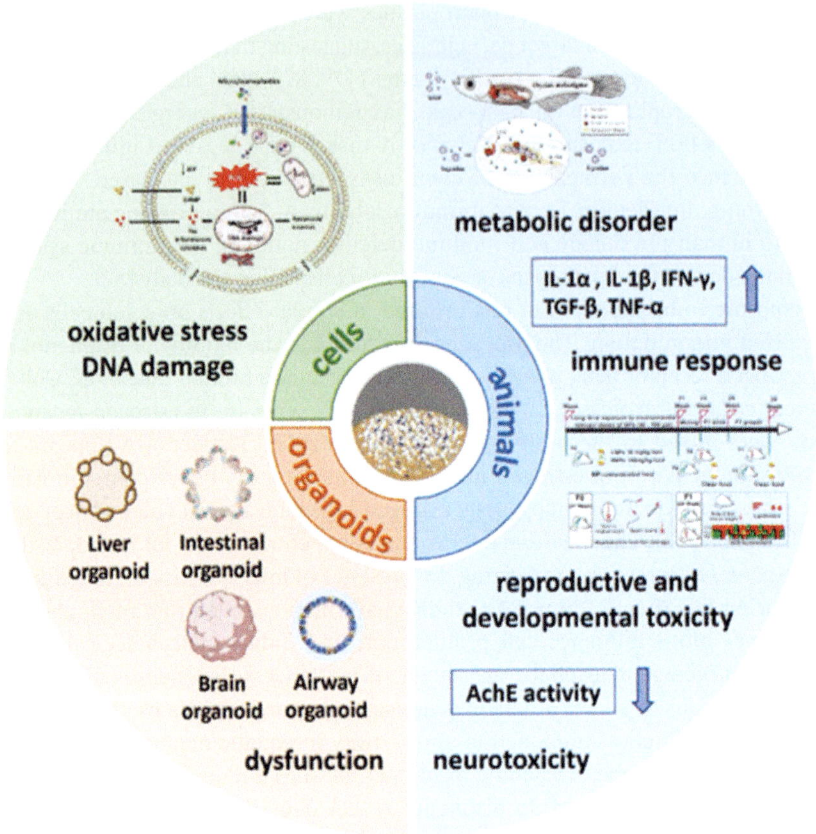

Figure 4.8 Toxicity mechanism of microplastics on humans and animals with oxidative stress and DNA damage. (*Source:* Reproduced with permission of Li et al. [50] / American Chemical Society / CC-BY-NC-ND 4.0.)

malfunction, resulting in membrane potential disruption, respiratory inhibition, and excessive generation of mitochondrion-derived ROS [81, 83, 84]. Figure 4.8 depicts the several toxicity mechanisms of microplastics on both humans and animals, including DNA damage and oxidative stress.

Damage to cell membranes, necrosis, and structural and functional abnormalities in different tissues can result from microplastics activating the immune system and inducing an inflammatory response. Evidence suggests that prolonged exposure to microplastics raises myeloperoxidase, a marker of inflammation activity in Gilthead seabream livers and causes renal tubular epithelial cells to produce more inflammatory proteins [68]. An important factor in the inflammatory reaction caused by microplastics is the activation of signaling pathways such as toll-like

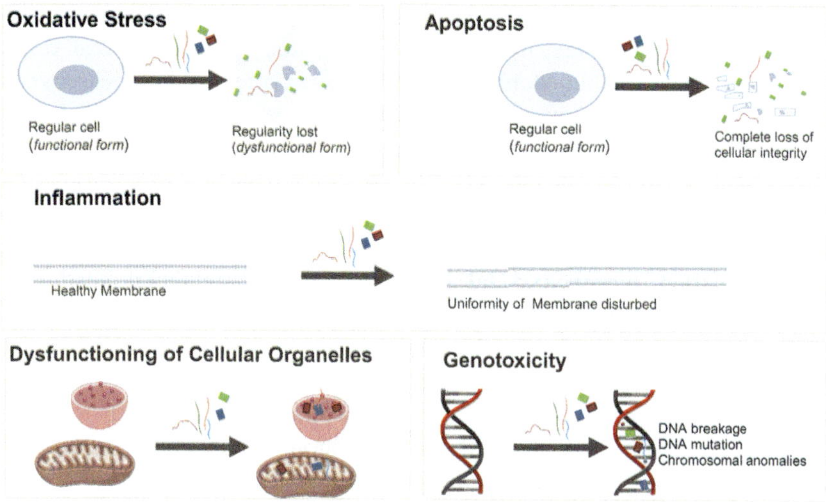

Figure 4.9 Toxicological effects of polystyrene microplastics on cell cultures: oxidative stress, apoptosis, inflammation, mitochondrial and lysosomal dysfunction, and genotoxicity. (*Source:* Periyasamy [5] / with permission of Elsevier.)

receptor (TLR), NF-κB, and nuclear factor erythroid 2-related factor 2. Damage to cell membranes, genetic toxicity, apoptosis, and autophagy are toxicological endpoints that are influenced by the inflammatory response and oxidative stress that are caused by microplastics [68] (Fig. 4.9).

Metabolic abnormalities can result in several health complications, such as obesity, insulin resistance, nonalcoholic fatty liver disease, and coronary artery disease. Disruptions in energy, lipid, glucose, and protein metabolism are linked to exposure to microplastics, which can induce oxidative stress, inflammation, and cellular apoptosis [85] (Fig. 4.9). Research indicates that the oral administration of microplastics can provoke hyperglycemia via the PI3K/Akt signaling pathway, elevate triglycerides, promote cholesterol buildup, and enhance ROS levels in the murine liver [86]. Oral exposure to microplastics may produce hepatic metabolic problems, including apoptosis and ferroptosis of cerebellar cells via the liver-brain axis, eventually resulting in cerebellar damage. Intestinal exposure to microplastics may disrupt metabolic balance by altering microbiota composition. Mice subjected to microplastics for six weeks exhibited a considerable reduction in intestinal actinomycetes and modifications in microbial diversity. Microplastics significantly influenced the primary metabolic pathways of functional genes in gut microbiota, including pyruvate metabolism, tyrosine metabolism, fatty acid production, and bacterial invasion of epithelial cells. Exposure to microplastics in the zebrafish model resulted in a notable reduction of intestinal proteobacteria and an elevation of Fusobacteria. Apoptosis is a significant harmful mechanism in microplastic-induced toxicity. Apoptosis of intestinal epithelial cells can result in intestinal barrier malfunction, whereas apoptosis of ovarian granulosa cells can cause a decrease in ovarian reserve (Fig. 4.9) [87]. Apoptotic markers, including TNF, p53, casp3b, gadd45ba, and ptgs2a, were markedly upregulated in zebrafish treated with microplastics [86, 88]. The genotoxic potential of microplastics has been proven through experimental findings (Fig. 4.9) [89]. Exposure to microplastics significantly elevated the rates of micronucleation, nucleoplasmic bridge formation, and nuclear bud formation in human peripheral blood cells. Microplastics can trigger DNA damage even at low concentrations. Furthermore, other systems may contribute to the possible toxicity triggered by microplastics. Advanced methodologies such as transcriptomics, metabolomics, proteomics, and single-cell sequencing may facilitate the exploration of additional molecular targets and pathways associated with toxicity induced by microplastics [90].

4.4 Conclusion

This chapter examines the intricate pathways by which microplastics, especially those from textile sources, negatively affect human health. Microplastics or fibrous fragments released from clothing and technical textiles during manufacturing, use, or laundering are now prevalent in air, water, and food, resulting in various pathways of human exposure, including ingestion, inhalation, and dermal contact. A primary concern is that textile functionalization processes frequently utilize persistent chemicals and nanoparticles that may adhere to or integrate into textile fibers. Functional additives, hazardous dyes, and auxiliary chemicals can leach from microplastics, resulting in a toxic mixture upon ingestion or inhalation. Microplastic particles and their associated contaminants can induce oxidative stress, inflammation, endocrine disruption, cytotoxic effects, and may interfere with immune and reproductive functions. Evidence indicates that microplastics can traverse biological barriers and accumulate in essential organs. Their interaction with the gut microbiome, respiratory pathways, and placental tissues raises concerns regarding long-term cumulative health impacts, many of which are inadequately understood and under-researched. Therefore, further investigation is necessary to systematically evaluate the occurrence of microplastics in human body fluids, clarify their effects on human physiology, and address potential sample contamination from airborne microplastics. The health risks linked to microplastic pollution necessitate immediate action at both the policy and industrial levels. Safer textile production practices are essential, encompassing the development of biodegradable or low-shedding fibers, the elimination of hazardous functional chemicals, and the improvement of filtration systems during laundering. Enhancing public awareness and conducting further scientific research on the toxicological pathways and long-term effects of microplastics are crucial for developing informed regulations and protective strategies. Without prompt intervention, human exposure to microplastics is likely to increase, potentially leading to irreversible effects on public health.

References

1 Periyasamy, A.P. and Tehrani-Bagha, A. (2022). A review on microplastic emission from textile materials and its reduction techniques. *Polymer Degradation and Stability* 199: 109901.

2 Periyasamy, A.P. (2024). Sustainable approaches in textile finishing to control microfiber releases. In: *Microfibre Pollution from Textiles*. London UK: Taylor & Francis Group.

3 Hidalgo-Ruz, V., Gutow, L., Thompson, R.C., and Thiel, M. (2012). Microplastics in the marine environment: a review of the methods used for identification and quantification. *Environmental Science & Technology* 46 (6): 3060–3075.

4 Periyasamy, A.P. (2023). Microfiber emissions from functionalized textiles: potential threat for human health and environmental risks. *Toxics* 11 (5): 406.

5 Periyasamy, A.P. (2025). Functionalized textile microplastics: a closer look at the issues, strategy, and legislation on the microplastic reduction. *Kuwait Journal of Science* 52 (3): 100395.

6 Hale, R.C., Seeley, M.E., La Guardia, M.J. et al. (2020). A global perspective on microplastics. *Journal of Geophysical Research, Oceans* 125 (1): e2018JC014719.

7 Benson, N.U., Agboola, O.D., Fred-Ahmadu, O.H. et al. (2022). Micro(nano)plastics prevalence, food web interactions, and toxicity assessment in aquatic organisms: a review. *Frontiers in Marine Science* 9: 851281.

8 Ziajahromi, S., Kumar, A., Neale, P.A., and Leusch, F.D.L. (2018). Environmentally relevant concentrations of polyethylene microplastics negatively impact the survival, growth and emergence of sediment-dwelling invertebrates. *Environmental Pollution* 236: 425–431.

9 Jemec, A., Horvat, P., Kunej, U. et al. (2016). Uptake and effects of microplastic textile fibers on freshwater crustacean Daphnia magna. *Environmental Pollution* 219: 201–209.

10 Galloway, T.S. (2015). Micro- and nano-plastics and human health. In: *Marine Anthropogenic Litter* (ed. M. Bergmann, L. Gutow, and M. Klages), 343–366. Cham: Springer International Publishing.

11 Peixoto, D., Pinheiro, C., Amorim, J. et al. (2019). Microplastic pollution in commercial salt for human consumption: a review. *Estuarine, Coastal and Shelf Science* 219: 161–168.

12 Mühlschlegel, P., Hauk, A., Walter, U., and Sieber, R. (2017). Lack of evidence for microplastic contamination in honey. *Food Additives & Contaminants: Part A* 34 (11): 1982–1989.

13 Tong, H., Jiang, Q., Hu, X., and Zhong, X. (2020). Occurrence and identification of microplastics in tap water from China. *Chemosphere* 252: 126493.

14 Kosuth, M., Mason, S.A., and Wattenberg, E.V. (2018). Anthropogenic contamination of tap water, beer, and sea salt. *PLoS One* 13 (4): e0194970.

15 Oliveri Conti, G., Ferrante, M., Banni, M. et al. (2020). Micro- and nano-plastics in edible fruit and vegetables. The first diet risks assessment for the general population. *Environmental Research* 187: 109677.

16 Karami, A., Golieskardi, A., Choo, C.K. et al. (2018). Microplastic and mesoplastic contamination in canned sardines and sprats. *Science of the Total Environment* 612: 1380–1386.

17 Cox, K.D., Covernton, G.A., Davies, H.L. et al. (2019). Human consumption of microplastics. *Environmental Science & Technology* 53 (12): 7068–7074.

18 Amato-Lourenço, L.F., Carvalho-Oliveira, R., Júnior, G.R. et al. (2021). Presence of airborne microplastics in human lung tissue. *Journal of Hazardous Materials* 416: 126124.

19 O'Brien, S., Okoffo, E.D., O'Brien, J.W. et al. (2020). Airborne emissions of microplastic fibres from domestic laundry dryers. *Science of the Total Environment* 747: 141175.

20 Morrow, P. (1988). Possible mechanisms to explain dust overloading of the lungs. *Fundamental and Applied Toxicology* 10 (3): 369–384.

21 Tran, C.L., Buchanan, D., and Cullen, R.T. (2000). Inhalation of poorly soluble particles. II. Influence of particle surface area on inflammation and clearance. *Inhalation Toxicology* 12 (12): 1113–1126.

22 Dris, R., Gasperi, J., Mirande, C. et al. (2017). A first overview of textile fibers, including microplastics, in indoor and outdoor environments. *Environmental Pollution* 221: 453–458.

23 Cai, L., Wang, J., Peng, J. et al. (2017). Characteristic of microplastics in the atmospheric fallout from Dongguan city, China: preliminary research and first evidence. *Environmental Science and Pollution Research* 24 (32): 24928–24935.

24 Liu, K., Wang, X., Fang, T. et al. (2019). Source and potential risk assessment of suspended atmospheric microplastics in Shanghai. *Science of the Total Environment* 675: 462–471.

25 Pauly, J.L., Stegmeier, S.J., Allaart, H.A. et al. (1998). Inhaled cellulosic and plastic fibers found in human lung tissue. *Cancer Epidemiology Biomarkers & Prevention* 7 (5): 419–428.

26 Periyasamy, A.P. and Militky, J. (2020). Sustainability in textile dyeing: recent developments. In: *Sustainability in the Textile and Apparel Industries: Production Process Sustainability* (ed. S.S. Muthu and M.A. Gardetti), 37–79. Springer.

27 Lehtonen, J., Johansson, P., Babaeipour, S. et al. (2025). Application of kraft lignin as a colorant for screen printing on cotton. *Industrial Crops and Products* 230: 121056.

28 Periyasamy, A.P. (2024). Recent advances in the remediation of textile-dye-containing wastewater: prioritizing human health and sustainable wastewater treatment. *Sustainability* 16 (2): 495.

29 Periyasamy, A.P. and Militky, J. (2017). Denim processing and health hazards. In: *Sustainability in Denim*, 161–196. Elsevier.

30 Periyasamy, A.P. (2020). Environmental hazards of denim processing-I. *Asian Dyer* 17 (1): 56–60.

31 Periyasamy, A.P. (2021). Evaluation of microfiber release from jeans: the impact of different washing conditions. *Environmental Science and Pollution Research* 28 (41): 58570–58582.

32 Periyasamy, A.P., Ramamoorthy, S.K., Rwawiire, S., and Zhao, Y. (2018). Sustainable wastewater treatment methods for textile industry. In: *Sustainable Innovations in Apparel Production*, 21–87. Springer.

33 Periyasamy, A.P., Venkataraman, M., and Militky, J. (2022). Effect of sol–gel treatment on physical, chemical and mechanical stability of copper-coated conductive fabrics: focus on EMI shielding effectiveness. *Journal of Materials Science* 57 (44): 20780–20793.

34 Periyasamy, A.P. and Periyasami, S. (2023). Critical review on sustainability in denim: a step toward sustainable production and consumption of denim. *ACS Omega* 8 (5): 4472–4490.

35 Byrne, C., Divekar, S.D., Storchan, G.B. et al. (2013). Metals and breast cancer. *Journal of Mammary Gland Biology and Neoplasia* 18 (1): 63–73.

36 Periyasamy, A.P. and Militky, J. (2020). Sustainability in textile dyeing: recent developments. *Sustainability in the Textile and Apparel Industries* 37–79.

37 Periyasamy, A.P. and Negi, A. (2024). Alkoxide-based solvent dyeing: a feasible strategy for pollution minimization and sustainable approach for the reactive dyeing of cellulosic materials. *Cellulose* 31 (12): 7765–7791.

38 Periyasamy, A.P. (2022). Natural dyeing of cellulose fibers using syzygium cumini fruit extracts and a bio-mordant: a step toward sustainable dyeing. *Sustainable Materials and Technologies* 33: e00472.

39 Patel, B.H. (2011). Natural dyes. In: *Handbook of Textile and Industrial Dyeing*, vol. 1, 395–424. Elsevier.

40 Katoueizadeh, E., Zebarjad, S.M., and Janghorban, K. (2018). Investigation of mechanical characteristics of functionalized cotton textiles by N-doped TiO_2 nanoparticles. *Materials Chemistry and Physics* 218: 239–245.

41 Egea, L.G., Brun, F.G., and Jiménez-Ramos, R. (2024). Dissolved organic carbon leaching from microplastics and bioavailability in coastal ecosystems. *Science of the Total Environment* 909: 168673.

42 Zhuang, J., Rong, N., Wang, X. et al. (2022). Adsorption of small size microplastics based on cellulose nanofiber aerogel modified by quaternary ammonium salt in water. *Separation and Purification Technology* 293: 121133.

43 Deh, S., Gähr, F., and Buchmeiser, M.R. (2016). Synergistic effects in the pyrolysis of phosphorus-based flame-retardants: the role of Si- and N-based compounds. *Polymer Degradation and Stability* 130: 155–164.

44 Luzi, B., Carnevale Miino, M., Rada, E.C. et al. (2025). Critical review of microfiber release from textiles: results, comparative challenges, mitigation strategies, and legislative perspectives. *Chemosphere* 378: 144394.

45 Zhou, Y., Yan, Y., Du, Y. et al. (2013). Preparation and application of melamine-formaldehyde photochromic microcapsules. *Sensors and Actuators B: Chemical* 188: 502–512.

46 Almeida, L. and Ramos, D. (2017). Health and safety concerns of textiles with nanomaterials. *IOP Conference Series: Materials Science and Engineering* 254: 102002.

47 Periyasamy, A.P., Militky, J., Sachinandham, A., and Duraisamy, G. (2021). Nanotechnology in textile finishing: recent developments. In: *Handbook of Nanomaterials and Nanocomposites for Energy and Environmental Applications*, 1–31. Cham: Springer International Publishing.

48 Prata, J.C. (2018). Airborne microplastics: consequences to human health? *Environmental Pollution* 234: 115–126.

49 Van Cauwenberghe, L. and Janssen, C.R. (2014). Microplastics in bivalves cultured for human consumption. *Environmental Pollution* 193: 65–70.

50 Li, Y., Tao, L., Wang, Q. et al. (2023). Potential health impact of microplastics: a review of environmental distribution, human exposure, and toxic effects. *Environment & Health* 1 (4): 249–257.

51 Li, Q., Yuan, M., Chen, Y. et al. (2023). The neglected potential source of microplastics from daily necessities: a study on protective mobile phone cases. *Journal of Hazardous Materials* 441: 129911.

52 Ng, E.-L., Huerta Lwanga, E., Eldridge, S.M. et al. (2018). An overview of microplastic and nanoplastic pollution in agroecosystems. *Science of the Total Environment* 627: 1377–1388.

53 Cheng, Y., Yang, Y., Bai, L., and Cui, J. (2024). Microplastics: an often-overlooked issue in the transition from chronic inflammation to cancer. *Journal of Translational Medicine* 22 (1): 959.

54 Banerjee, A., Billey, L.O., McGarvey, A.M., and Shelver, W.L. (2022). Effects of polystyrene micro/nanoplastics on liver cells based on particle size, surface functionalization, concentration and exposure period. *Science of the Total Environment* 836: 155621.

55 Greven, A.-C., Merk, T., Karagöz, F. et al. (2016). Polycarbonate and polystyrene nanoplastic particles act as stressors to the innate immune system of fathead minnow Pimephales promelas. *Environmental Toxicology and Chemistry* 35 (12): 3093–3100.

56 Joksimovic, N., Selakovic, D., Jovicic, N. et al. (2022). Nanoplastics as an invisible threat to humans and the environment. *Journal of Nanomaterials* 2022 (1): 6707819.

57 Greten, F.R. and Grivennikov, S.I. (2019). Inflammation and cancer: triggers, mechanisms, and consequences. *Immunity* 51 (1): 27–41.

58 Tamargo, A., Molinero, N., Reinosa, J.J. et al. (2022). PET microplastics affect human gut microbiota communities during simulated gastrointestinal digestion, first evidence of plausible polymer biodegradation during human digestion. *Scientific Reports* 12 (1): 528.

59 Lu, L., Wan, Z., Luo, T. et al. (2018). Polystyrene microplastics induce gut microbiota dysbiosis and hepatic lipid metabolism disorder in mice. *Science of the Total Environment* 631–632: 449–458.

60 Li, B., Ding, Y., Cheng, X. et al. (2020). Polyethylene microplastics affect the distribution of gut microbiota and inflammation development in mice. *Chemosphere* 244: 125492.

61 Yin, K., Wang, D., Zhang, Y. et al. (2023). Dose-effect of polystyrene microplastics on digestive toxicity in chickens (Gallus gallus): multi-omics reveals critical role of gut-liver axis. *Journal of Advanced Research* 52: 3–18.

62 Liu, Z., Yu, P., Cai, M. et al. (2019). Effects of microplastics on the innate immunity and intestinal microflora of juvenile Eriocheir sinensis. *Science of the Total Environment* 685: 836–846.

63 Wright, S.L. and Kelly, F.J. (2017). Plastic and human health: a micro issue? *Environmental Science & Technology* 51 (12): 6634–6647.

64 Lim, D., Jeong, J., Song, K.S. et al. (2021). Inhalation toxicity of polystyrene micro(nano)plastics using modified OECD TG 412. *Chemosphere* 262: 128330.

65 Dong, C.-D., Chen, C.-W., Chen, Y.-C. et al. (2020). Polystyrene microplastic particles: in vitro pulmonary toxicity assessment. *Journal of Hazardous Materials* 385: 121575.

66 Pitt, J.A., Kozal, J.S., Jayasundara, N. et al. (2018). Uptake, tissue distribution, and toxicity of polystyrene nanoparticles in developing zebrafish (Danio rerio). *Aquatic Toxicology* 194: 185–194.

67 Wei, J., Wang, X., Liu, Q. et al. (2021). The impact of polystyrene microplastics on cardiomyocytes pyroptosis through NLRP3/Caspase-1 signaling pathway and oxidative stress in Wistar rats. *Environmental Toxicology* 36 (5): 935–944.

68 Zhang, Y., Yin, K., Wang, D. et al. (2022). Polystyrene microplastics-induced cardiotoxicity in chickens via the ROS-driven NF-κB-NLRP3-GSDMD and AMPK-PGC-1α axes. *Science of the Total Environment* 840: 156727.

69 Kwon, W., Kim, D., Kim, H.-Y. et al. (2022). Microglial phagocytosis of polystyrene microplastics results in immune alteration and apoptosis in vitro and in vivo. *Science of the Total Environment* 807: 150817.

70 Liang, B., Huang, Y., Zhong, Y. et al. (2022). Brain single-nucleus transcriptomics highlights that polystyrene nanoplastics potentially induce Parkinson's disease-like neurodegeneration by causing energy metabolism disorders in mice. *Journal of Hazardous Materials* 430: 128459.

71 Kashiwada, S. (2006). Distribution of nanoparticles in the see-through medaka (oryzias latipes). *Environmental Health Perspectives* 114 (11): 1697–1702.

72 Zhao, Q., Zhu, L., Weng, J. et al. (2023). Detection and characterization of microplastics in the human testis and semen. *Science of the Total Environment* 877: 162713.

73 Amereh, F., Babaei, M., Eslami, A. et al. (2020). The emerging risk of exposure to nano(micro)plastics on endocrine disturbance and reproductive toxicity: from a hypothetical scenario to a global public health challenge. *Environmental Pollution* 261: 114158.

74 Sun, L., Liao, K., and Wang, D. (2021). Comparison of transgenerational reproductive toxicity induced by pristine and amino modified nanoplastics in Caenorhabditis elegans. *Science of the Total Environment* 768: 144362.

75 Hirt, N. and Body-Malapel, M. (2020). Immunotoxicity and intestinal effects of nano- and microplastics: a review of the literature. *Particle and Fibre Toxicology* 17 (1): 57.

76 Li, Y., Xu, M., Zhang, Z. et al. (2022). in vitro study on the toxicity of nanoplastics with different charges to murine splenic lymphocytes. *Journal of Hazardous Materials* 424: 127508.

77 Rodrigues, J.P., Duarte, A.C., Santos-Echeandía, J., and Rocha-Santos, T. (2019). Significance of interactions between microplastics and POPs in the marine environment: a critical overview. *TrAC Trends in Analytical Chemistry* 111: 252–260.

78 Mishra, G. and Rahi, S. (2022). Need of toxicity studies for cosmetic products and their approaches. *Progress in Nuclear Energy, Volume 6, Biological Sciences* 2 (1): 105–109.

79 Sharma, M.D., Elanjickal, A.I., Mankar, J.S., and Krupadam, R.J. (2020). Assessment of cancer risk of microplastics enriched with polycyclic aromatic hydrocarbons. *Journal of Hazardous Materials* 398: 122994.

80 Kim, H., Zaheer, J., Choi, E.-J., and Kim, J.S. (2022). Enhanced ASGR2 by microplastic exposure leads to resistance to therapy in gastric cancer. *Theranostics* 12 (7): 3217–3236.

81 Wang, Z., Lin, T., and Chen, W. (2020). Occurrence and removal of microplastics in an advanced drinking water treatment plant (ADWTP). *Science of the Total Environment* 700: 134520.

82 Wang, Q., Bai, J., Ning, B. et al. (2020). Effects of bisphenol A and nanoscale and microscale polystyrene plastic exposure on particle uptake and toxicity in human Caco-2 cells. *Chemosphere* 254: 126788.

83 Goodman, K.E., Hare, J.T., Khamis, Z.I. et al. (2021). Exposure of human lung cells to polystyrene microplastics significantly retards cell proliferation and triggers morphological changes. *Chemical Research in Toxicology* 34 (4): 1069–1081.

84 Palaniappan, S., Sadacharan, C.M., and Rostama, B. (2022). Polystyrene and polyethylene microplastics decrease cell viability and dysregulate inflammatory and oxidative stress markers of MDCK and L929 cells in vitro. *Exposure and Health* 14 (1): 75–85.

85 Saklayen, M.G. (2018). The global epidemic of the metabolic syndrome. *Current Hypertension Reports* 20 (2): 12.

86 Shi, C., Han, X., Guo, W. et al. (2022). Disturbed gut-liver axis indicating oral exposure to polystyrene microplastic potentially increases the risk of insulin resistance. *Environment International* 164: 107273.

87 Liang, B., Zhong, Y., Huang, Y. et al. (2021). Underestimated health risks: polystyrene micro- and nanoplastics jointly induce intestinal barrier dysfunction by ROS-mediated epithelial cell apoptosis. *Particle and Fibre Toxicology* 18 (1): 20.

88 Umamaheswari, S., Priyadarshinee, S., Kadirvelu, K., and Ramesh, M. (2021). Polystyrene microplastics induce apoptosis via ROS-mediated p53 signaling pathway in zebrafish. *Chemico-Biological Interactions* 345: 109550.

89 Çobanoğlu, H., Belivermiş, M., Sıkdokur, E. et al. (2021). Genotoxic and cytotoxic effects of polyethylene microplastics on human peripheral blood lymphocytes. *Chemosphere* 272: 129805.

90 Brandts, I., Cánovas, M., Tvarijonaviciute, A. et al. (2022). Nanoplastics are bioaccumulated in fish liver and muscle and cause DNA damage after a chronic exposure. *Environmental Research* 212: 113433.

5

Characterization and Quantification of Microplastics

Jiří Militký[1], Mohanapriya Venkataraman[1], Jakub Wiener[1], and Aravin Prince Periyasamy[2]

[1] *Department of Material Engineering, Faculty of Textile Engineering, Technical University of Liberec, Czech Republic*
[2] *Textile and Nonwoven Materials, VTT Technical Research Center of Finland, Espoo, Finland*

5.1 Introduction

Microplastics are already found all over the planet in air, soil, surface, and underground waters, fresh and salty and their presence was also proven in Arctic snow [1]. Due to this, they appeared also in the food chain. Release of primary microplastics occur at various stages of the lifecycle of plastic and plastic products. The main sources of primary microplastics are tires, synthetic textiles, marine coatings, road marking, personal care products, plastic pellets and city dust. Secondary microplastics releases due to mismanaged waste during their disposal and environmental degradation [2]. Some polymers such as polyvinyl chloride, polystyrene and polycarbonates are released during degradation toxic monomers [3]. Microplastics obey high sorption capacity for hydrophobic organic chemicals as polycyclic aromatic hydrocarbons which can be important for their transport in the seawater [4]. For microplastics as results of fragmentation of macroplastics due to environmental degradation are surface properties different from bulk material (e.g. increased hydrophilicity and adsorption ability e.g. for pollutants) [5, 6].

Microplastics are partially eliminated by the digestive tract, but some of them accumulate in the body, not only in the human body, but also in the body of animals, which are often part of the food chain, and which are subsequently consumed by humans. Due to their chemical–physical properties (i.e. their size and chemical composition), microplastics penetrate between cells and tissues and accumulate in the human body (lungs, gastrointestinal tract, blood and lymphatic vessels). Microplastics also cause technological problems in industrial plants,

Textile Microplastics in the Environment: Sources, Impacts, and Solutions,
First Edition. Aravin Prince Periyasamy.

where, for example, valves become clogged in circuits with large water flows such as power plants and heating plants. In equipment using service water, for example, for cooling, microplastics are deposited and their subsequently caused malfunctions [7]. It is therefore important to estimate portion of microplastics in real situations, separation of them from mixtures, identifying their chemical nature and avoiding their effect on environment and life. Recent advances in areas connected with microplastics are discussed in the book [8]. This book is a reference source of information about microplastics behavior in the environment, their analysis characterization and remediation. Basic sources of information about microplastics characterization including detailed description of major techniques are presented in the book [9]. This chapter is oriented as well to fibrous microplastics which are unique due to their physical properties as results of supramolecular structure of used polymers and due to different textile processing.

5.2 Fibrous Microplastics

Conventional microplastics (i.e. Particles from synthetic polymers in the size range 1 μm to 5 mm) are from mixture of polymer with different chemical nature, different polymeric chain length and different morphologies which have major influence on their properties such as density, hydrophility, electric charge and adhesion to other materials. Big parts of microplastics are formed from partially degraded polyolefins and their derivates. Some of them have low-density and are floating on water surface. Other are due to higher density in dispersion and are deposited on bottom in calm conditions. Common wastewater treatment is here not very efficient, and huge amount of microplastics appears in the effluent [7].

Some technologies as biobased degradation by enzymes (esterase for polyethylene terephthalate) are also inefficient for degrading mixtures of different kind of microplastics [10]. In real polluted water there are as well some particles classified according the size as microplastics which are from natural polymers (cellulose and proteins mainly) degraded often very slowly. One of the important sources of micro and nanoplastics is packing materials including drinking bottles. PET nanoplastics collected from commercially bottled drinking water had average mean size of ~88.2 nm and their concentration were about 10^8 particles mL^{-1} [11]. Main degradation products in PET bottles (presence of CO_2 and increased temperature) are acetaldehyde, formaldehyde, 2,4-di-*tert*-butyl-phenol (2,4-d*t*BP) and bis(2-hydroxyethyl) terephthalate migrated into the water [12]. It was found that presence of PET, the proliferation and migration of some human cells were suppressed [13].

One of the largest sources of primary and secondary microplastics are undegraded and partially damaged or degraded fibrous textile products. These fibrous microplastics, commonly colored, also contain chemicals used for dyeing

Table 5.1 Densities of fibrous polymers and typical fibers.

Fiber type	PA 6	PA 6.6	PES	PP	PAN
Density [kg m^{-3}] of amorphous polymer at 25 °C	1084	1070	1330	850	—
Density [kg m^{-3}] of fully crystalline polymer at 25 °C	1230	1240	1480	950	1280
Density of typical fibres [kg m^{-3}]	1140	1140	1380	940	1130

chemical finishing and functionalization of textiles. They are usually composed of higher-density polymers containing some functional end groups (acid, amine, hydroxyl), making these particles not easy to separate as shown in Table 5.1. Due to the more varied chemical composition and possible higher toxicity, the complicated elimination from the environment can be expected.

Thus, fibrous microplastics have an extremely high degree of danger, occur in the air and water, including drinking water, and immediately threaten practically the entire human population. In cases, when textile microplastics are transported by air, they usually accumulate mainly in household and office dust and support the development of allergies, asthma, and many respiratory problems. Major sources of microplastic emission from textile materials and their reduction techniques are discussed in article [14]. Textile microplastics released during washing pollute all kinds of water (i.e. rivers, lakes, dams, seas, wastewater treatment plants) including drinking water. Major portion of microplastics is arising from polyethylene terephthalate (PET) fibers released (Fig. 5.1) mainly during the washing of garments [15–17]. Approximately 35% of worldwide ocean primary microplastics may come from synthetic textile laundry [2]. Fibrous microplastics are potentially more dangerous than other types of microplastics for the following reasons:

Figure 5.1 Structure of PET microfibril.

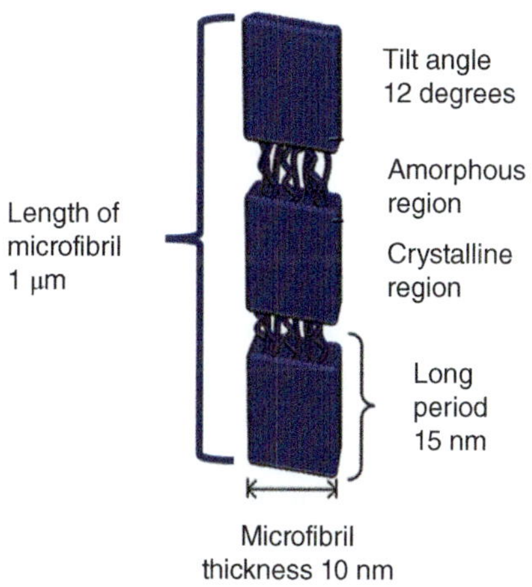

- They are often fiber fragments and are usually needle-shaped [18]. The reason is the internal fibrillar structure of the fibers and the geometric shape of the textile fibers, which are long and thin (the typical length/diameter ratio is 10^3). The consequence of the needle shape can be easier mechanical damage to tissues and cells, mechanical disruption of cell walls, etc.
- Textile fibers are usually contaminated with non-polymeric potentially toxic compounds, which can be textile dyes, finishing agents, matting particles and, more recently, metal and metal oxide particles.
- Fiber microplastics are generated because of the everyday use and maintenance of textiles, so their adverse cumulative effects are considered serious.
- Fibrous microplastics can be easily simulated by suspensions of microfilaments of predetermined length of e.g. 0.2 mm prepared by simple way [19].

With the development of physical and chemical modification methods, the designation of fibers according to the predominant chemical composition (polyester, polyamide, etc.) gives practically no information about their behavior and properties. The typical feature of fibrous microplastics is their thickness (diameter) which is in narrow interval (from few micrometers till 20–30 μm) only. Therefore, fragments of fibers are usually in the range of nanoplastics (less than 1 μm). The length of standard synthetic fibers is over 3 cm and therefore the length of their fragments are often in the range of mesoplastics (5–25 mm) or macroplastics (over 25 mm). The basic structural element of fibers is the microfibril, which is characterized by regular alternation of amorphous and crystalline regions. The microfibrils are clustered into higher formations of "fibrils," where they are interconnected by tie chains. PET microfibrils have a diameter of 10–15 nm and length $l = 1$ μm and are thin, long elements of elliptical cross-section. Fibrils have a diameter of 30–45 nm and fibril length is depended on the size of macromolecular chains and fibers annealing in wide range 200–900 nm [14]. Forces between fibrils are weaker and therefore, they can be separated during the fracture of fibers.

Fibrils can be organized to fibrillar bundles as well. At the individual levels, there are always structural elements in the elongated spindle-shaped form. The fiber length of microplastics released during washing and drying varied from the shortest fragment length of 37,000 nm to the longest one of approximately 4667 μm [17] which is indication of fibrillar bundles. The geometry of nanoparticles after washing and abrasion was comprehensively studied in work [17]. They used special term "microplastic fiber" as thick cylinders of diameters 16 ± 7 μm, fibrils as long tapes of diameters of 3 ± 1 μm and nanoplastics (hydrodynamic diameter 173–188 nm). These dimensions indicate that fragments of PET were over to fibrils dimensions. It is known that after axial splitting of high strength fibers is the diameter of the fiber at the break point is 2–4 μm. The fibrous structure is typical of both natural and synthetic fibers (see Fig. 5.2).

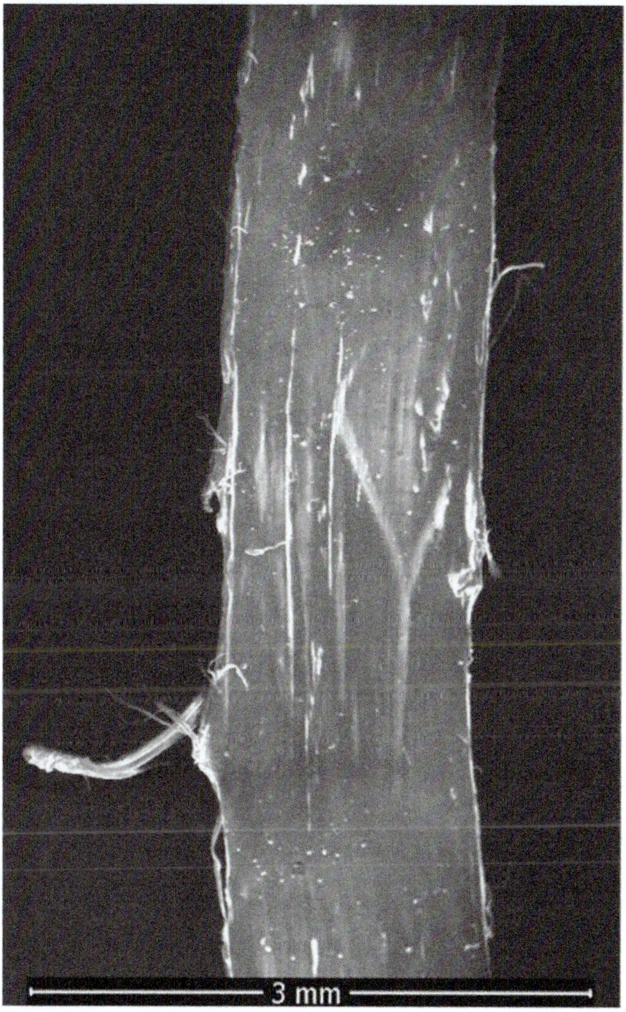

Figure 5.2 Evidence of fibrillar structure of fibres. (*Source:* Gultekin [20] / MDPI / CC BY 4.0.)

Many properties of synthetic fibers and therefore influenced by the partially crystalline structure (crystallinity of PET is about 40–70%). Most synthetic fibers are made from linear semi crystalline polymers. Some of their properties depend on the amount of tie (binding) polymeric chains and their parts, i.e. taut tie chains connecting microfibrils and fibrils. Physical (e.g. birefringence and swelling) and mechanical (e.g. moduli) anisotropy of fiber and fragment properties are caused by the fibrous structure. In the direction of the fiber axis,

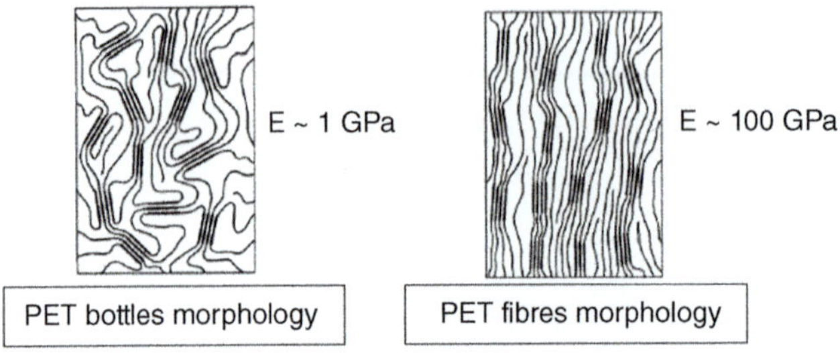

Figure 5.3 Influence of macromolecular chain orientation on the size of the initial modulus.

it is an oriented system, where the individual chains with backbone connected by covalent bonds share several secondary (electrostatic) bonds preventing their deformation. The modulus of the crystalline regions in this direction is typically $EK_p = 100–150\,GPa$. In the direction perpendicular to the axis of the fiber, it acts much fewer secondary bonds, so that the resistance to deformation is significantly lower (this is typical for not oriented PET as are bottles or packing foils). The module of crystalline region in this direction is typically $EK_K = 4\,GPa$. Finally, in amorphous areas where it is not significantly parallel arrangement of chains is the modulus typically $E_a = 1–0.6\,GPa$ (Fig. 5.3).

High initial modulus indicates also high initial tensile stiffness of fibers and fibrous microplastics (fragments). The acidity of PET fibers surface is caused by the carboxylic end groups ranges from $2\,10^{-2}$ to $4\,10^{-2}$ mol kg^{-1} [21]. These groups are responsible for additional degradation because they catalyze the hydrolysis of

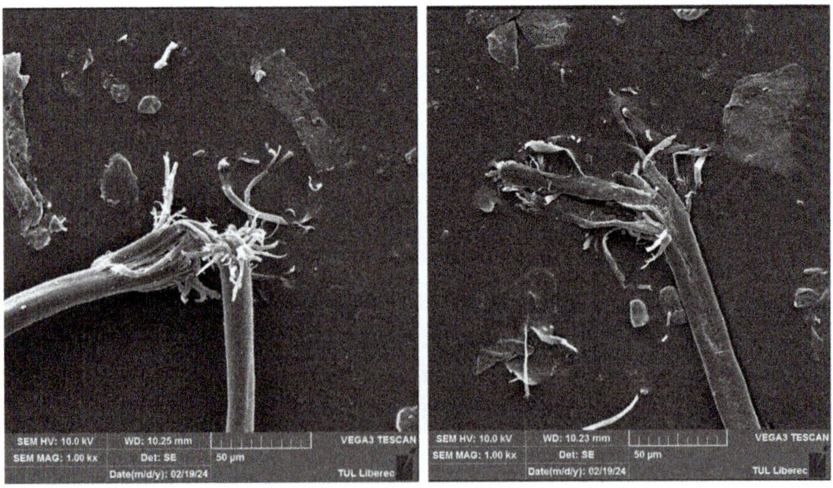

Figure 5.4 Failure of PET fibers by dry abrasion.

Figure 5.5 PET fibrous fragments released after washing and drying. (*Source:* Militký et al. [18] / MDPI / CC BY 4.0.)

ester bonds. The abrasion of PET fibers leads generally to various fibrous fragments with very wide distribution of size (Fig. 5.4).

Washing and drying effect on fragmentation of PET are not so strong as dry mechanical abrasion but local broken area looks similar (see Fig. 5.5).

Usually, it is assumed that textile structures from staple yarns release more microplastics in comparison with structures from endless multifilament [22] (see Fig. 5.6). Majority of multifilament yarns are used for clothing purposes after texturing, which is technology (mainly BCF – bulked continuous filaments) able to break some fibrils (see Fig. 5.6). Our unpublished results showed that the release of microplastics from PET textured yarns fabrics was significantly higher that microplastics released from staple PET yarns fabrics of the same construction. Breaking of PET fibers is closely connected with phenomena of pilling. The most significant effect on fiber strength, a thus also on dynamic fatigue in bending has the relative molecular weight of the polymer. Decrease

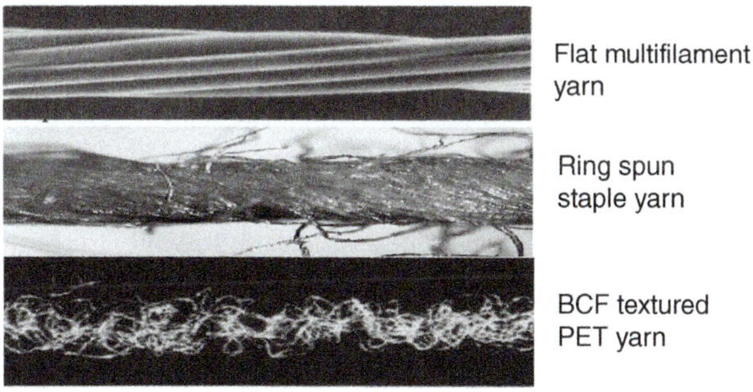

Figure 5.6 Structure of staple, flat and textured filaments yarns.

specific strength by 30%, which occurs due to the reduction of the degree of polymerization, will cause a decrease resistance in bending by 91% – see fiber TREVIRA 350. So-called PET fibers with suppressed pilling are fabricated from polymeric chains with lower length suppressing the strength in repeated bending. These fibers are more prone to fragmentation and creation of microplastics. Realization of pilling suppression (pills will be generated but simply removed) by reducing molecular mass is limited by the necessity of ensuring sufficient viscosity of melts during spinning. There are some common techniques such as use of compounds of boron or silicon, small stable cross-linking (below 1%) by using of pentaerythritol and use of polyester modified by small percentage (till 3%) of sulfoisophthalic acid (or its sodium salt). Generally, all treatment reducing pilling will decrease bending strength and then increase of microplastics generation. It is therefore necessary to investigate deeply starting PET materials according to molecular and supramolecular structure to explain their tendency to create microplastics.

5.3 Microplastics Collection

Microplastics is a very broad assembly of materials with prescribed size intervals, but with very different chemical nature, shapes, densities, surface morphology and electrical charge. Some of them are less dense than water and are flowing on its surface. Others are quickly settling to the ground forming part of the sediments. Smaller particles like nanoplastics are transported by chaotic Brownian motion and bigger ones are influenced by gravitational forces mainly. Selection of proper sampling strategy is dependent on the aim of microplastics investigation. In some cases, e.g. for evaluation of microplastic removal efficiency by using one pressure membrane, it is relatively simple to select sampling strategy ensuring the same conditions for statistical treatment of data [23] by using simulated (artificial) microplastics of known composition and concentration. More complicated is the situation when the real mix of microplastics should be used but in one location and period only (analysis real wastewater plant releasing efficiency). Most complicated are situations when it is necessary to select randomly location (space), type of medium (soil, air, water etc.), and season (time scale) for respond of general question of type “intensity of microplastic in arctic snow”. It is necessary to specify what is population from which is creating “random” sample. This population is often very huge, and creation of random sample is impossible. In these cases, is often created protocol specifying conditions like “it is sample of fibrous microplastics from surface of Ganga River in Varanasi ghat No 3, on 12 April 2023 sampled at noon”. It looks too detailed but the level of microplastic is changed very frequently in different locations and time if their main source is, e.g. domestic washing [22].

There exist currently a lot of different methods for microplastics collection [24, 25]. The medium type investigated (e.g. air, soil, running water, still water, sewage sludge, sediments, wastewater, biota) has major influence on the sampling and preparation process [26]. Primary samples are often concentrated (the amount of microplastics is low) and their cleaning (removal of organic materials, mud, sand, etc.) should be realized. For separation process the physical and chemical properties of the investigated substances (as is density, electrostatic force, size, shape and dimension) are often used [27]. Contaminated samples will require more preparation, while treated drinking water and bottled water are relatively clean media, therefore all the steps as for other samples will not be necessary [26]. The distribution of extracted microplastics is critically dependent on the sampling method, pretreatment of collected samples and technique used for quantification of microplastics presence. There are three major types of sampling: bulk (the entire volume of the sample is used), volume-reduced (use only the portion of the sample that is important for further processing), or selective (extraction of particles that are identified visually e.g. on the surface of sediments) [28]. Results of 50 studies on microplastics in fresh water, drinking-water and wastewater concluded that methods for microplastics sampling and analysis need to be improved for obtaining reproducible and reliable results [29]. The nine quality control criteria adapted from criteria developed for biota samples were specified in work [30].

5.3.1 Water Sampling

Sampling microplastics in the water (from surface till bed) can be realized by various devices, such as nets, pumping systems and discrete devices [31, 32]. For surface waters sampling there are used nets such as Manta nets, Neuston nets (see Fig. 5.7), or zooplankton net [24, 34] and discrete devices as buckets, jerry cans, and containers. For deeper water (below surface) the bongo nets or vessels collecting water samples can be used and for bed are constructed benthic trawls [24]. These trawls are processing large volumes of water and plastic particles are directly concentrated during sampling.

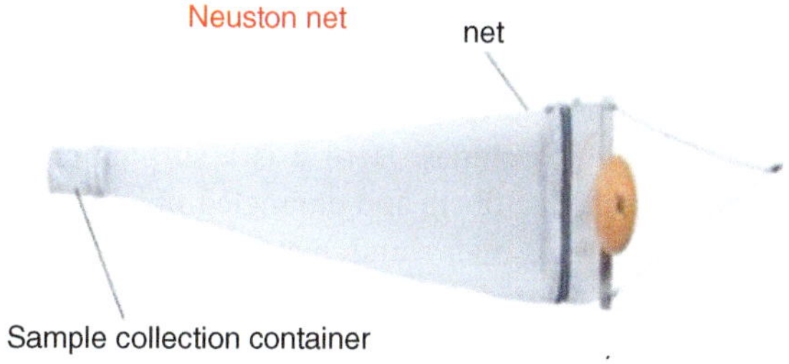

Figure 5.7 Neuston net example. (*Source:* Peng et al. [33] / MDPI / CC BY 4.0.)

Sampling nets mesh size is obviously in wide range from 20 to 5000 μm [31]. Filtered water volume is measured usually by attached flowmeter or acoustic Doppler current profilers [24]. The rotating glass cylinder collecting 50–60 mm on the water surface (surface microlayer sampler) was also created [28]. If the density ρ_0 of the microplastic particle with radius R_0 is smaller than the water density ρ_w (for seawater is $\rho_w \sim 1007\,\text{kg m}^{-3}$ due to presence of salts), it will float at the surface. Due to biofouling assumed as creation of surface film of thickness $d(d \ll R_0)$ with larger density ρ_f that density of water it is possible to predict final mean density of biofouled particle ρ_b from approximate relation [35].

$$\rho_b \approx \rho_0 \frac{R_0^3}{(R_0 + d)^3} + \rho_f \left[1 - \frac{R_0^3}{(R_0 + d)^3} \right] \tag{5.1}$$

For fibrous microplastics with diameter r is Eq. 5.1 replaced by relation

$$\rho_{bf} \approx \rho_0 \frac{r^2}{(r + d)^2} + \rho_f \left[1 - \frac{r^2}{(r + d)^2} \right] \tag{5.2}$$

where ρ_{bf} is the final density of biofouled fibrous microparticle. For microbial biomass $\rho_f = 1500\,\text{kg m}^{-3}$ [36], Thickness of biofouled film and particles mean density are growing and, in some thickness, d_w will reach density of water, which is for circular particle equal to relation [35].

$$d_w = R_0 \left[\left(\frac{\rho_f - \rho_0}{\rho_f - \rho_w} \right)^{1/3} - 1 \right] \tag{5.3}$$

and for fibrous microplastics is these d_{wf} thickness equal to relation

$$d_{wf} = r \left[\left(\frac{\rho_f - \rho_0}{\rho_f - \rho_w} \right)^{1/2} - 1 \right] \tag{5.4}$$

The result will start to sink from surface. Under the assumption of the constant rate of fouling it is possible to predict that for smaller particles, there will be faster biofouling and quicker starting of sinking. In this situation the microparticles will not be fully collected by surface sampling and on the bed will be smaller particles. It is supported by results of field research where it was found that microplastics size in open-sea sediments is smaller than that in water column and at the water surface [28]. A special category of water sampling is wastewater from washing machines. Here it is a surplus of debris composed of particles, fibrous fragments, fibers and entangled fibers, often in globular form, classified as macro, meso, micro and nanoplastics. Other components are impurities from washed materials and product of interaction of components of washing solution and washed (often precipitated). The number of microfibers released from polyester and cotton textiles after first wash varied in the range 2.1×10^5 to 1.3×10^7 [37].

For right sampling it is here necessary to specify washing process (amount of water, temperature, time, rotations) and washing bath (surfactants, washing powder and other chemicals as softening agents) [14, 37–40]. Specification of washed textile material and dirt types is recommended. It was proven that some materials (constructions) are producing huge amounts of fibrous waste after more cycles of washing. Common example is knitted fleece [41]. For standard structures the first two washings are sufficient for release majority of fibrous materials including microplastics [14]. Majority of washing machines have special filter on the machine output for removal of bigger particles which is capturing a part of microplastics in filtration cake. Effluents are therefore without microplastics generated by washing or presented in washed textile material.

5.3.2 Sediment Sampling

Very frequent is investigation of microplastics in sediments from sandy beaches (mostly in high tide line) in variable depth below sediment surface (obviously 0–32 cm). The standard depth of sampling is 5 cm below the surface of sediment [28]. In permeable sands, microplastics are accumulated together with bio based organic particles in deeper sediment layers. For sampling there are here used different stainless-steel spoons or spatula and box corer [42]. The seabed sediments are sampled by bottom trawl [43]. The concentration of microplastics is expressed by the number of particles per square meter (particles m^{-2}); per cubic meter (particles m^{-3}); and per kg of dry sediment (particles kg^{-1}) [24].

5.3.3 Biological Sampling

Microplastics are present as well in biota and therefore it is necessary to have a proper biological sampling scheme. Microplastics have a similar size range as planktonic organisms and other suspended particles and are available to marine invertebrates [44]. The two species of marine invertebrates (blue mussel *Mytilus edulis* and the lugworm *Arenicola marina*) were analyzed [45]. It was found that microplastics were present in all organisms, i.e. 0.2 ± 0.3 g/1 for *M. edulis* and 1.2 ± 2.8 g/1 for *A. marina*. Organisms can be catches in grasps, traps, creels by bottom crawling, by trawls at different water levels, by hand or by electrofishing [25]. Biota samples are firstly conserved (e.g. by formalin) or frozen (at −20 °C) and later are dissected to collect microplastics from their various tissues and organs including gut content or digestive system. It is suggested to analyze microplastics and other samplings of 50 specimens [25].

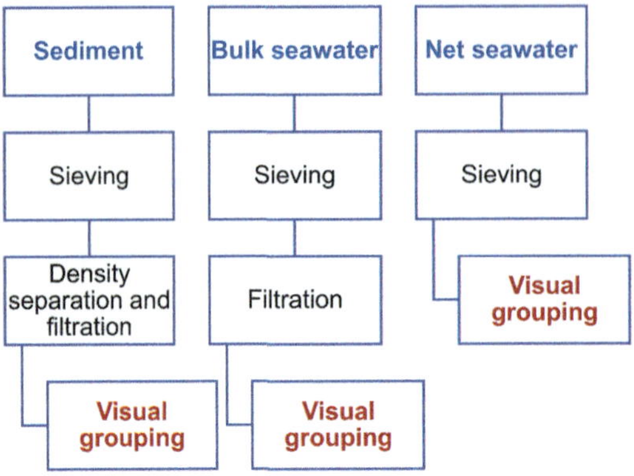

Figure 5.8 Suggested treatment for different types of samples.

5.3.4 Sampling Methods

Sampling methods can be divided into two basic categories: Grab sampling of smaller volume and on-site filtration sampling (using nets or sieves) of comparatively large volume. Grab sampling mostly focuses on sediment and sludge. On-site filtration sampling using nets or sieves for floating and suspended microplastics can be realized by dragging a net or pumping water through a net (or a sieve). Nets with mesh opening sizes of about 100 and 300 μm are conventionally used for sampling microplastics in marine or freshwater environments. All kinds of samples should be sieved over a standard mesh of 5 mm or 1 mm for removal of bigger particles (debris) [28]. The mesh size of 0.5 mm is also often used because 500 μm corresponds to the threshold for visual identification. Suggested treatment for different types of samples is shown in Fig. 5.8.

5.4 Samples Pretreatment

Collected samples contain instead of microplastics a huge portion of impurities and other wastes like sand, other organic material, biota (flora and fauna of different size) suspended oily matter etc. Separation of microplastics from samples is usually realized by density flotation, filtration and sieving. Since sand or sediments have densities around 2700 kg m^{-3}, the difference in density can be used to separate the lighter microplastics from the heavier sand or sediments. On the other hand, soil organic matter has typically a density of 1000–1400 kg m^{-3} and therefore will not be effectively separated from microplastics by density extraction [46]. Larger sediments are often removed by different types of separators such as microplastic separator (MPSS—working on principle of different densities) [47, 48] or electrostatic separators (30 kV by using of corona electrode) [49].

For removal of organic debris, oxidation by some substances or acidic, alkaline and enzymatic degradation are used [25]. Main problem here is possible degradation of microplastics as well [50].

Oxidation is realized by using acids, alkalis or other oxidizing compounds (hydrogen peroxide). Very good results for removal of mussel soft tissue were obtained by using HNO_3 heated for 1 h at 60 °C and then for 1 h at 100 °C [51], but for some polymers partial dissolution occur. It was found that most polymers are resistant against the use of KOH but some of them (including PET) are partially degraded by using of NaOH [52]. Hydrogen peroxide is an efficient oxidizing compound but concentrated H_2O_2 solutions have negative influence on microplastics extraction and therefore it is recommended to use 10% H_2O_2 for exposure time of 18 h [53]. Much better results were obtained by using Fenton's reaction for advanced oxidation process by H_2O_2 in the presence of a catalyst (Fe^{2+}) at pH 3–5, by reaction scheme ($Fe^{2+} + H_2O_2 \rightarrow Fe^{3+} + OH + OH^-$) [46]. Enzymatic degradation by cellulases lipases and proteases is generally very selective and has not negative influence on common microplastics degradation [54]. Main limitation is due to the time needed for processing samples. A very effective methodology uses firstly filtration on an aluminum filter instead of the plastic filter for capturing microplastics. The second step is the use of oxidation of organic particles (i.e. the remains of plant and animal shells) using hydrogen peroxide [10] and then drying the filter at 30 °C to prevent signal interference of infrared spectroscopy with water.

5.5 Extraction of Microplastics

An extraction process based on density differences of denser solution and lighter particles is known as density flotation. The samples with microplastics are mixed with a liquid of a defined density allowing lighter particles to float. Further cleaning requires the use of chemicals or enzymes for removing organic and inorganic contaminants. Microplastics are then collected from the supernatant and filtered [26]. As liquids with known density are often used concentrated or saturated salt solution. e.g. sodium chloride (NaCl), where density is in range 1150–1300 kg m^{-3}, sodium iodide (NaI) where density is in range 1550–1800 kg m^{-3} or zinc chloride ($ZnCl_2$) where density is in range 1500–1800 kg m^{-3} [55]. NaCl solution is one of the most common solutions used in extraction of light microplastics composed of polyethylene, polypropylene and polyamide by density flotation. The particles smaller than 1 mm were separated by only 40%. For microplastics with higher density that 1200 kg m^{-3} as are majority of fibrous microplastics is necessary to use NaI or with benefit $ZnCl_2$ which can be simply reused [56]. Both NaI and $ZnCl_2$ are classified as hazardous materials, and this can be motivation for using more safe chemicals such as magnesium chloride ($MgCl_2$) which can have a density of 1310 kg m^{-3} and is more suitable for microplastic density separation than NaCl. Another possibility is to use sodium dodecylsulfate solution [57].

Two step procedures for sediment sampling based on microplastic concentration and extraction by flotation in dense solution, were realized by using of the elutriation column [51] or the air-induced overflow system [58]. Elutriation realize separation of organic matter (biota) from sediment by injecting water at the bottom of the elutriation device. If the water velocity is well adjusted, the organisms are carried from the bottom to the top of the system and then are extracted, whereas sediment remains in the device [51]. The aim of air-induced overflow method is pre-extraction to decrease the sediment sample mass for the second density separation step by flotation. The method is based on fluidization in a saturated NaCl (26% solution). The original sediment mass is reduced by up to 80% [58]. The "sieved" NaCl solution can be stored for reuse for a maximum of six subsequent runs. For effective extraction of light microplastics from sediment the moderate bubbling was generated by a constant air flow (approximately 0.1 L s^{-1}) in NaCl solution. The air bubbles are forcing lighter particles (microplastics with density below 1200 k m^{-3}) to move more frequently and quickly to the top layer. The total supernatant is then transferred to subsequent treatment and the remaining sediments are discarded [58].

The air-induced overflow method is like standard air flotation used to separate solids and liquids phases. Impurities including microplastics and microorganisms from water sampling mainly are (after addition of the coagulants) precipitated into flakes. They are transported then by the air current to the surface where they form a layer which can be then transferred to subsequent treatment. There exists also technique of microplastics separation from samples by high-pressure solvent extraction using an organic solvent under subcritical pressure. However, this method is changing microplastics morphology.

5.6 Separation of Microplastics

Microplastics can be simply separated using sieves of several mesh fineness, which enables their classification into size groups. A series of stainless-steel mesh sieves (according to ISO 3310-1) with mesh sizes from 1 mm, 250 μm, 125 μm, 90 μm, to 32 μm were used in research project [59]. A series of copper sieves of diminishing pore size: 250 μm, 125 μm, and 62.5 μm were used in study [28, 60–63].

The simple and widely used separation method of microplastics is filtration. It is also frequently used as a preliminary step for density separation. It is a physical separation process, in which the size of the captured solids is dependent on the pore size. Pore size of the filter paper varies, most generally between 1 and 2 μm.

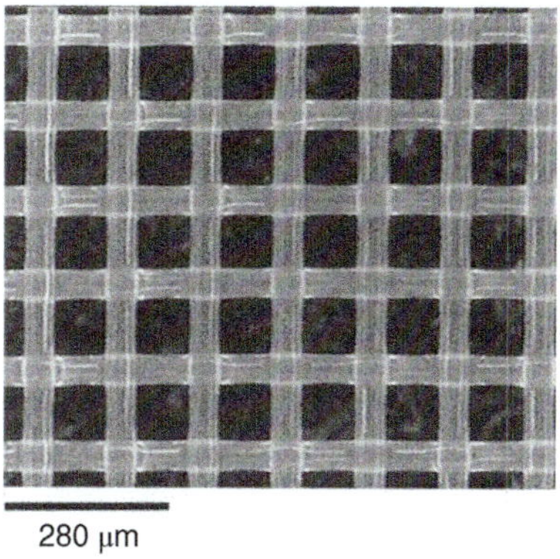

Figure 5.9 Example of polyamide monofilament filters.

Widely used materials for filters are glass fiber, polyamide fibers and cellulose ester fibers. Generally, materials having no adverse effect on subsequent spectral analysis as aluminum-coated polycarbonate membrane filters [61] or copper coated polyester nanoassembly are beneficial [62]. Frequently used filters from polymeric materials can be potential source cross-contamination of the sample. Polyamide filters are frequently used for treatment of wastewater from washing machines and airborne particles from tumble dryers (see Fig. 5.9).

Microplastics can be removed from surface of filters very easily by using translucent adhesive tape (for microscopic analysis) or collected to Petri dish (for weighting). The microplastics adhered to tape after filtration of wastewater from home washing of polyester fleece is shown Fig. 5.10 [63]. It is also possible to use spray-coated glass slide by pressure sensitive adhesive (PSA resin containing poly [2-ethylhexyl acrylate]). Thickness of PSA films at about 7 μm [64].

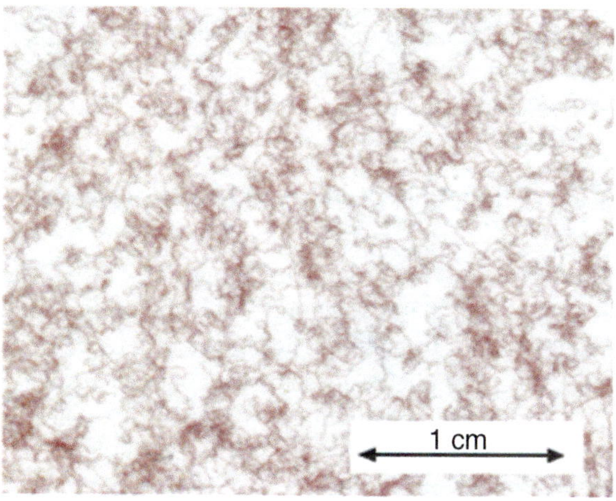

Figure 5.10 Microplastics adhered to adhesive tape after filtration of wastewater from home washing of polyester fleece.

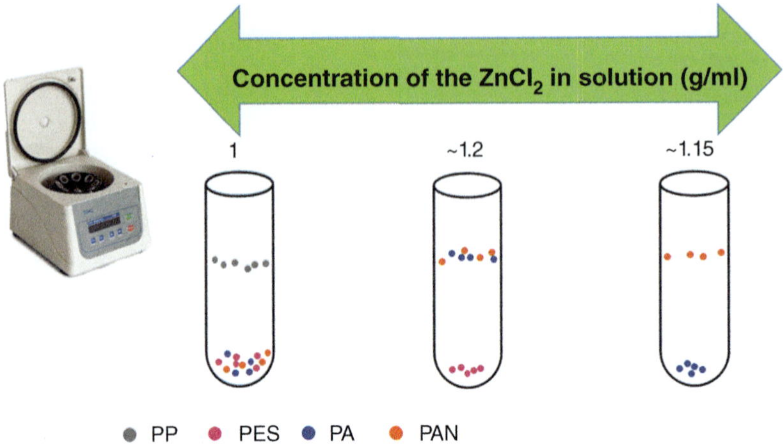

Figure 5.11 Use a centrifuge to accelerate the extraction process with the speed of 6000 rpm for 5 min.

For polymeric mixture separation in the phase of microplastics extraction, the methods based on density differences are commonly used. Densities of microplastics in water, sediments and air are in relatively narrow range from 50 kg m^{-3} for polystyrene foam to 2100–2300 kg m^{-3} for polytetrafluoroethylene (Teflon) [35]. Densities of fibrous microplastic are usually in 1 narrower interval (see Table 5.1) but are slightly dependent on the degree of crystalline and supramolecular structure mainly.

For acceleration of purified fibrous microplastics mixture separation from sediment and supernatant, the density centrifugation can be simply used (Fig. 5.11).

The biota action, UV degradation processes and weathering can change density of some polymeric particles, which can be reason for their wrong identification [35]. It was found that high-density polyethylene particles decreased in seawater their specific density from 950 to 810–850 kg m^{-3}, and polystyrene particles increased their density from 1040–1080 to 1240–1410 kg m^{-3} [65].

5.7 Characterization of Microplastics

In general, the analysis of microplastics can be separated into morphological and physical characterization; and chemical characterization includes often quantification. In consequence of the different principles of used methods, it is often very complicate to compare obtained results. There are also differences in determination of the range of detected particles for selected characterization technique [27].

5.7.1 Morphological Characterization

The basic morphological characteristics of microplastics as the dimension, shape and color are often evaluated visually by the naked eye (limited by resolution about 0.1 mm), by using of optical microscopy or advanced microscopical methods as scanning electron microscopy (SEM) and laser confocal microscopy with resolution about 1 μm which can be used for evaluation of surface morphology [24, 27, 66]. Optical microscope magnification suitable for microplastics identification is in range from 4.5× to 400× [31]. To suppress misidentification of microplastics by optical investigation some criteria were proposed [31],

- The particles have no observable organic or cellular structures.
- In the case of fibers, the diameter should be consistent along their length.
- Particles should present clear and homogeneous colors.
- The further high magnification should be used in the case of transparent or white particles.

If these criteria are fulfilled, particles can be defined as plastic. Scanning electron microscopy (SEM) enabling determination of the micro/nanoplastics surface chemical composition when it is completed with energy dispersive X-ray spectroscopy EDS suitable for elemental qualitative/quantitative analysis of surface [67]. SEM can be used for characterization of microplastic surface changes due to degradation processes such as aging or oxidative and mechanical weathering. SEM–EDX analysis was also used for identification of organic and inorganic black particles in waters [68]. On the other hand, SEM may cause damage of microplastics and nanoplastics surfaces due to use of electron beam. SEM combined with dynamic image analysis device was also used for determination of microplastic fibers released by textile washing [69]. Particle size distributions and concentrations of spherical polyethylene microplastics in solutions can be measured by electrochemical analysis using carbon electrode. Basic assumption is that microplastic are electrochemically inert (pristine) [70].

Another method for particle size distribution is nephelometric procedures. The principle is to measurement of diffusely scattered light on dispersed particles by a helium-neon or argon laser. Measured dispersion is illuminated by a laser beam and the scattered light is monitored by a detector [23]. These and other methods can be used in combination with the digital image processing technique to find particle size distribution or equivalent area diameter of the particle profiles [71]. For fibrous microplastics morphology characterization it can be used similar methods as for fiber identification [72]. Microplastic can be directly identified by using optical microscopy without or with labeling using dyes. It is an easy, fast and cheap method. For staining of microparticles the fluorescent dyestuffs [70] and optical brightening agents can be used. Most often, for staining of various microplastics used dyestuff Nile Red, (see Fig. 5.12) with

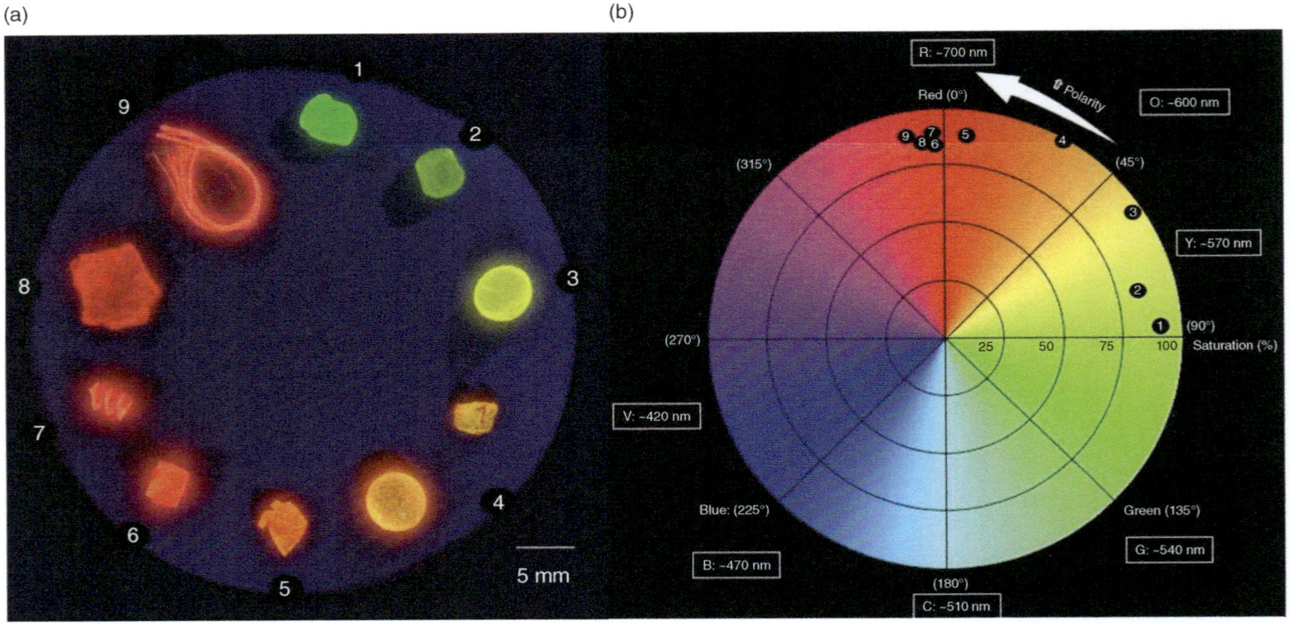

Figure 5.12 Staining with Nile Red in acetone. (*Source:* Ho and Masura [73] / MDPI / CC BY 4.0.)

intense fluorescence properties in the apolar medium, i.e. fluorescence connected with the polarity of the solvent used for its dissolution. In water, the solubility is low and there is no fluorescence. The different solvents, such as methanol, ethanol, and acetone are used for the preparation of Nile Red solutions [74]. Staining is usually realized at elevated temperatures (60–70 °C) [75, 76].

Nile Red can be used for staining microplastics in a range of shades from yellow to deep red depending on their surface hydrophobicity. The emitted color of Nile Red can be also affected by surface contamination from the environment [76, 77]. Influence of Nile Red concentration in methanolic solution on staining of polyester fibrous particles in drinking water allowed to select suitable concentration of 1 mg L^{-1} [78].

[79] The Nile Red can be replaced by selected azo dyes containing aromatic amines [75] or commercially available fluorescent disperse dyes. Three disperse dyestuffs selected according to their fluorescence in microscope fluorescence filter ranges in three emission regions: far red (685 nm), red (593 nm), and green (525 nm) were successfully applied [77]. There are a lot of factors that have influence on the accuracy of the visual evaluation result, and the right identification of particle nature (plastic or not) is often impossible [26]. It was experimentally determined that visual identification alone is not appropriate for identification of small microplastics [57]. The percentage of right visual identification of microplastics is dependent on their type, size and color. Fibrous microplastics had a higher percentage of correct identification (75%) than particles (64%) [57]. One simple technique for testing the nature of potential microplastics is the hot point test. This test is using hot needle to contact with particle which will be in case of microplastic sticky [80]. The same problems with right identification arise by using of scanning electron microscopy. It is a direct method that is very accurate and does not require calibration but is not possible to use identification based on color and should be accompanied by chemical identification of their nature. For geometrical characterization of microplastics in water dispersion the particle size analyzers can be simply used. Especially for nanoparticles geometry and size distribution the nanoparticle tracking analysis was developed [81]. This system is based on light scattering and Brownian motion for evaluation of the particle size distribution of samples in liquid suspension. The diffusion of particles in a liquid suspension by random Brownian motion modeled by the Stokes-Einstein equation where the diffusion coefficient D of a spherical particle is inversely proportional to their size

$$D = \frac{k_b\, T}{3\, \pi\, \eta\, d} \tag{5.5}$$

where k_b is the Boltzmann constant, T is temperature, η is dynamic viscosity of the liquid and d is the particle diameter. Diffusion coefficient can be evaluated from mean squared displacement of particles $\overleftrightarrow{(\Delta x^2)}$ (estimated by tracking analysis),

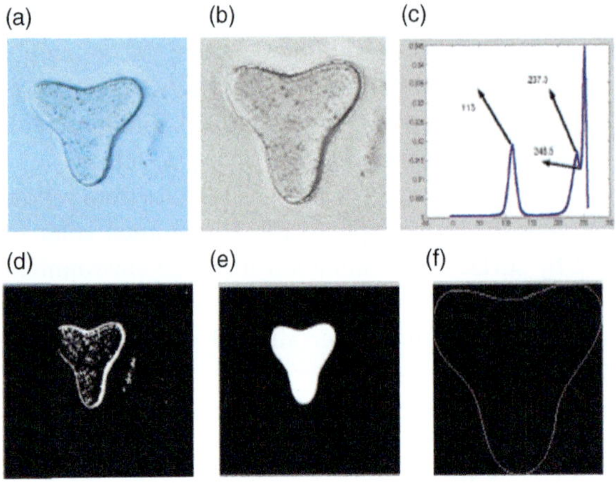

Figure 5.13 Basic steps of image processing: (a) original color image, (b) grey scale image, (c) evaluation of threshold (d) thresholding, (e) image cleaning, (f) contouring.

which is equal to 2 $D\,t$, where t is elapsed time. Nanoparticle tracking analysis was used to characterize nanoplastics with sizes between 10 nm and 1000 nm [81].

In majority of cases (as optical microscopy, scanning electron microscopy (SEM), and fluorescence microscopy) is the final stage of microplastics morphological characterization the data analysis of images used for quantifying colors, shapes, sizes, and surface topologies by using proper techniques and software tools [79]. The color can be simply defined by RGB color coordinates used for digitalization of color images. Their combination was used for calculation of index for fluorescence values (R + G)/R [82]. The basic steps of image treatment are shown in Fig. 5.13.

Transformation from original color image to grey scale image (range 0–256 levels) is realized routinely by all kinds of software for image processing [83]. The first critical step is thresholding where images in gray scale transformed into black and white scale images, black is background and white is object. Common thresholding is based on Otsu procedure which is a discrete analogue of Fisher's discriminant analysis [84]. The mathematical morphology (usually combination of erosion and dilatation) is then used for objects cleaning and extracting information from images (e.g. granulometry, skeletonization and watershed) based on spatial structure and connections [83]. Boundaries, contours, skeletons, and convex hulls can be found. Particle size is often simply characterized by diameter of circular object d_c having the same area as particle projected surface area A i.e. $d_c = \sqrt{4A/\pi}$. The length of fibrous microplastics is one of difficult task and can be traced using straight or segmented lines [82].

For automatic interpretation of images and microplastics identification the machine learning and in particular deep learning [85] and deep convolutional neural networks (CNNs) [79] can be used. Their application is based on the extension of traditional neural networks [86]. Neural network models are very flexible, and they can improve their performance by learning from experience. It is simple to tune the degree of fit by, e.g. adjusting the number of nodes. Neural networks are strictly data dependent, and their performance will be improved as sample size increases. On the other hand, the use of neural network models has serious disadvantages. The main problem is the impossibility of preserving shape and limiting behavior. They are critically dependent on the quality and space covering of training data sets and they are valid in subspace covered by this data set without extrapolation out of this subspace. Training data set is usually so big to be presented in papers and are stored in computer memory only, which reduced portability of these models [86].

5.7.1.1 Chemical Characterization

There are several approaches to determine chemical composition of microplastics [87]. The mostly used are spectroscopic, thermo-analytical and other chemical techniques described in detail in the books [63, 88–90].

5.7.1.2 Spectroscopic Methods

Electromagnetic waves, such as light, scatter upon contact with a sample. Most of this scattering is elastic (Raleigh or Mie), retaining the same energy and color. The very small part is following inelastic scattering (Raman) with different energy and colors. Elastic scattering is responsible for reflection and inelastic scattering for emission (see Fig. 5.14).

Inelastic Stokes scattering appears due to absorbing energy i.e. frequency of the scattered photon is lower than that of incident photon. Anti-Stokes scattering appears due to emitting energy that is frequency of the scattered photon is higher than of the incident photon. Spectroscopic techniques investigate the absorption, reflection and emission of matter molecules subjected to contact with photons (light). The impact of these photons produces excitement of the vibrational and rotational states of the molecules which are characteristic of different materials.

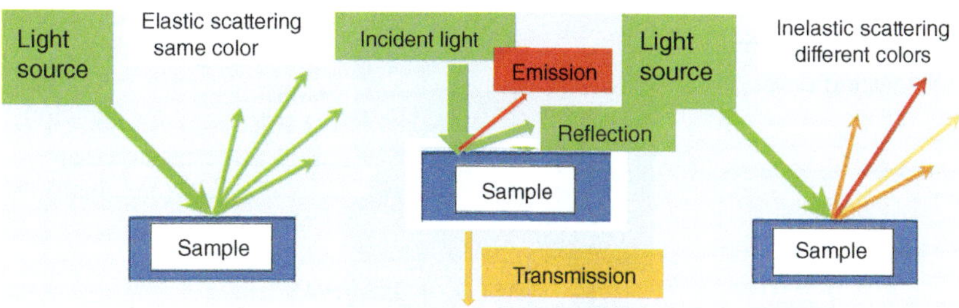

Figure 5.14 Schematic interaction of photons with matter. (*Source:* Bhargava et al. [91] / MDPI / CC BY 4.0.)

The applicability of spectroscopic techniques alone or in combination with microscopes for microplastics identification and quantification is widely proposed, but as well limited due to complicated identification of unknown matters and difficult quantitation of multicomponent spectra [90]. A raw vibrational spectrum of microplastics is usually disturbed by many factors connected with matrix effect, baseline effects and random noise of measurements effects which interfere with spectroscopic signal. The workflow of microplastics identification based on vibrational spectra pretreatment can be comprised into follows [92]:

- Recording vibrational spectra of microplastics (under suitable measurement conditions).
- Increasing signal-to-noise ratio (usually by nonparametric data smoothing), if necessary.
- Performing the baseline correction (often removal of blank spectra e.g. pure water).
- Normalizing data into proper intervals.
- Comparing data with a suitable spectra library (commonly by library searching).

Suitable methods for the realization of these tasks are deeply discussed in [92]. General scheme of spectral data analysis is shown in Fig. 5.15.

Electromagnetic waves are split into different regions based on their frequencies/wavelengths. There are gamma rays (γ, highest frequency/lowest

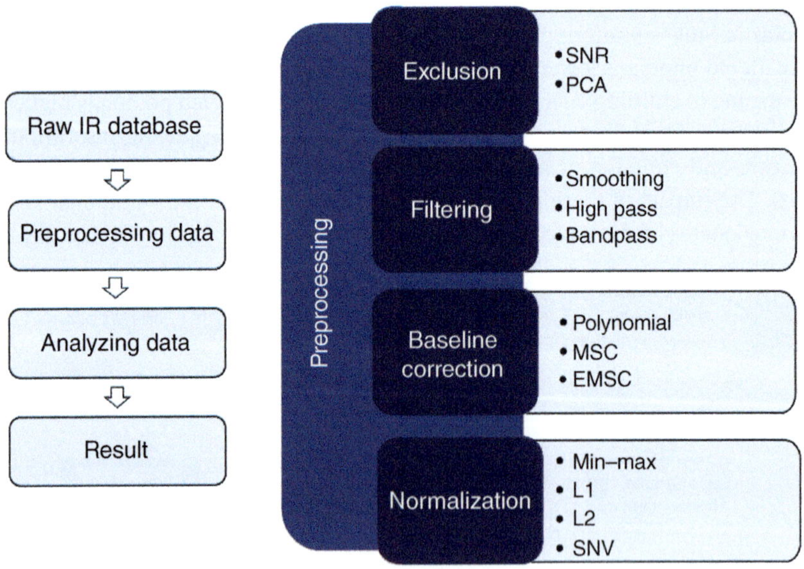

Figure 5.15 Spectral data analysis. (*Source:* Mokari et al. [93] / MDPI / CC BY 4.0.)

wavelengths), X-rays, ultraviolet–visible waves, infrared waves, microwaves, and radio waves (lowest frequency/longest wavelength). The strength of electromagnetic waves in different regions are completely different due to their wave nature characteristics as frequencies ν [Hz or s^{-1}] or wavelengths λ [m] or wavenumbers W [m^{-1}], where $W = 1/\lambda$. In practice are often used wavelength λ [μm] and wavenumber W [cm^{-1}]. For this case is $W = 10^4/\lambda$. These characteristics are related to electromagnetic waves energy E [J] through Planck equation $E = h\,\nu$, where $h = 6.626\ 10^{-34}$ [J s] is Planck constant. Generally, the vibrational frequencies of a molecule depend on their nature as the nature of the motion, mass of the atoms and nature of the chemical bonds.

For quantitative analysis of organic and inorganic molecules, the near-infrared (12,500–4000 cm^{-1} i.e. 0.8–2.5 μm), mid-infrared (4000–400 cm^{-1} i.e. 2.5–25 μm), and Raman spectroscopy (characterized by excitation laser wavelength in visible of near infrared region) are commonly used [90]. The frequencies at which the molecule absorbs light depends on the matter of atoms, their geometrical arrangements and the strength of the bonds [89]. These vibrations have a characteristic frequency. Typically, group frequency analysis is used to detect the presence and absence of various functional groups. Raman spectra of polypropylene and polyethylene are shown in Fig. 5.16.

Absorption, transmission, reflectance or Raman scattering spectra are analyzed by vibrational spectroscopy for the presence or absence of characteristic frequency bands that result from excited vibrational states in the tested sample. The energy difference between an excited and ground state is characteristic for a particular molecular structure and is presented in the spectral signal. Vibrational spectroscopic methods are used to identify the chemical structure of polymers (and microplastics) by comparison of their spectrum with the reference spectrum. FTIR is a very well established, relatively fast, and reliable IR spectroscopic method, allowing clearer identification of polar groups [91]. Raman spectroscopy is a slower method, signal is weaker (only 10^{-8} of the photons impacting the sample are translated to signal) and it allows identification of non-polar, symmetric bonds [81]. FTIR and Raman spectroscopies are based on complementary principles: molecular vibrations that are active in IR are inactive in Raman and vice versa [63]. For microplastics identification the Raman microspectroscopy called Raman microscopy (μ-Raman) and IR microspectroscopy called as IR microscopy (μ-FTIR) are commonly used [95]. It was found that for particles larger than 20 μm microscopical techniques are equally capable. The μ-Raman can detect particles till size 5 μm, but the μ-FTIR is unable to detect particles in the range of size 5–10 μm and only 40% of particles in range of size 11–20 μm [95]. Instead of analysis of each single particle (standard case) it is possible to use automated analysis as full point mapping, where the whole area of interest is virtually split to grid and analyzed by collecting of spectra at points

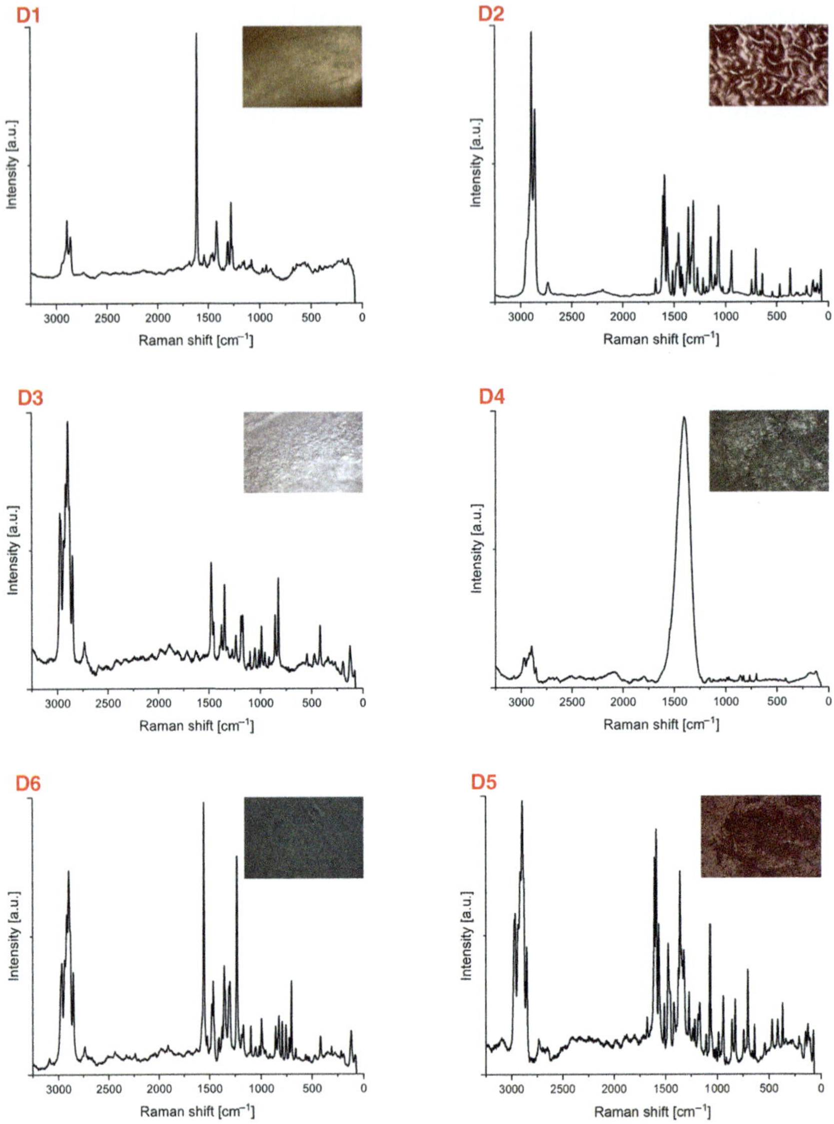

Figure 5.16 Raman spectrum of polypropylene and polyethylene. (*Source:* Rytelewska and Dąbrowska [94] / MDPI / CC BY 4.0.)

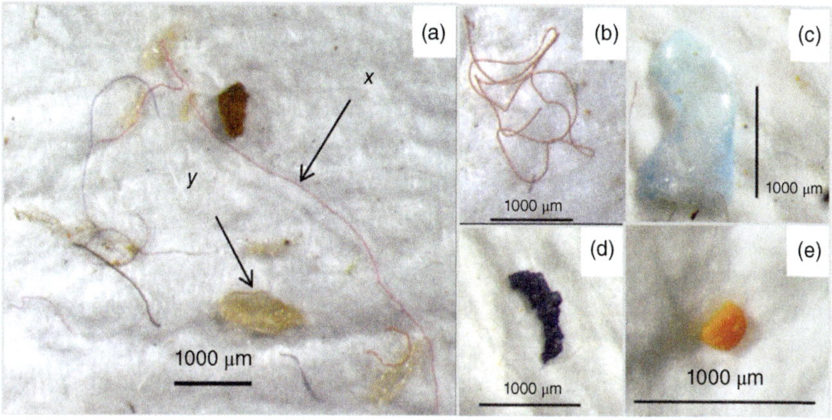

Figure 5.17 Microscopic views of the different types of microplastics. (a) Photograph of a typical filter paper under a stereomicroscope, where x = microplastic fibre and y = biogenic material paper; (b) synthetic microplastic fibre, (c, d) synthetic fragments; and (e) synthetic bead. (*Source:* Song et al. [98] / MDPI / CC BY 4.0.)

of this grid. More sophisticated techniques of spectra mapping are proposed as well [95, 96]. The microscopic image of different microplastics are shown in Fig. 5.17.

Microplastics analysis by IR and Raman microscope can be realized by manual analysis, semi-automated method, and fully automatic identification [98]. The fully automated identification was very quick, but the corresponding false positive identification rate was $80 \pm 15\%$. The semi-automated analysis using spectrum profiling approximately two times quicker than the manual analysis (manual: 6.1 ± 0.8 h, semi-automated: 4.0 ± 0.6 h), and 22% more microplastic particles were detected compared to the manual analysis [98].

Main limitations of IR and Raman methods is presence of molecules other than polymeric ones as result of special treatment or results of random deposition on the sample surface. Practically all polyesters based on fibrous microplastics contain at least traces of rutile TiO_2 used as mating agent in fibers. The presence of these molecules creates an additional peak without changing the rest of Raman spectra [57]. Other additives containing Si-O bonds, such as clay and silica, are changing seriously IR spectra but have practically no effect on Raman spectra [57]. Wery huge problems dealing with application of vibrational spectroscopy relate to molecules strongly absorbing light (as black dyes or carbon black particles) and molecules (as organic colorants, pigments and dyes) strongly fluorescent in visible light which seriously deteriorate spectra. These chemicals can lead to masking or a decrease of the signal to noise ratio leading to a lower degree of spectral resolution and are not part of Raman libraries, due to unsatisfactory reproducibility [99].

Some surfactants used for textiles washing can have strong Raman scattering and should be removed before microplastics analysis simply by rinsing [57, 90].

For measurements in practical conditions and environmental monitoring, various portable and handheld vibrational spectrometers are available on the market. Optical sensors for these devices are based on FTIR, Raman, and other methods [100]. These small devices (e.g. portable Raman spectrometers are around $10 \times 10 \times 5$ cm, weighing less than 1 kg) have limited performance (spectral range, resolution, signal-to-noise ratio and sampling versatility) but are aimed to be used for identification and screening purposes [101]. Handheld devices, where both FTIR and Raman sensors are coupled together, will be probably used for microplastics monitoring in the environment [100]. One real application was detection and identification of meso and microplastics by handheld FTIR spectrometer [102]. The in-situ analysis of microplastics by these devices is advantageous because concentrations of microplastics in the environment are usually very limited. Generally, FTIR is less suitable for monitoring microplastics in water environments because it absorbs infrared radiation. Inexpensive portable Raman sensor for the detection of microplastics in water in a quartz cuvette was developed [103].

5.7.2 Infrared Spectroscopy

Infrared (IR) spectroscopy is based on the transmission, reflection, absorption or attenuation of electromagnetic radiation due to a specified motion of chemical bonds in matter [63]. The absorbances at a different frequency are directly connected with the concentration of the chemical species in the sample (analogy with well-known Lambert Beer law). The molecules absorb photons at frequencies of their natural vibrational frequencies. The resulting energy increases due to an increase in the amplitude of the vibration. To interact with the external infrared light the vibrational motion of a bond is generated a dipole moment (see Fig. 5.18).

Infrared spectroscopy is frequently used for identification of the polymeric composition of microplastics because the IR spectra of an unknown microplastics sample can be compared with the IR spectra of known polymers available from IR spectra libraries [24]. To detect microplastics till 10 μm size, it is possible to connect an IR spectrometer with a microscope [105]. Two types of spectrometers which are commonly used are dispersive (IR) and interferometric Fourier transform (FTIR) instruments. The original IR instruments were of dispersive type, where normal vibrational mode in a molecule induces resonant IR absorption (or emission) of electromagnetic radiation by the interaction of the external electric vector, E, of the incident electromagnetic waves with the electric dipole moment of molecules [63]. These instruments then separate the individual frequencies of energy emitted from the IR source. The introduction of

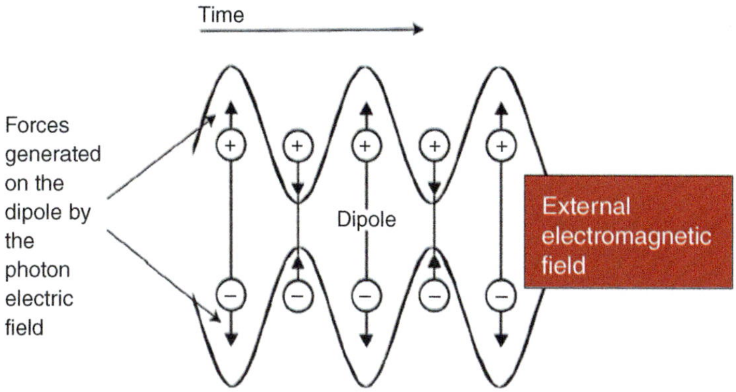

Figure 5.18 The oscillations of the dipole and the external electromagnetic field.

Fourier transform methodology greatly improved the quality of IR spectroscopy. The important component of these spectrometers is a Michelson interferometer, which is splitting incoming light to one beam going to an internal fixed mirror and the beam to the moving mirror. After reflection, the beams recombine inside the interferometer producing the interferogram. This interferogram should be processed by a Fourier transform to convert it into a conventional spectrum [63].

FTIR can be used mainly in reflection or transmission mode [106]. Generally, the transmission mode analyzes the IR radiation after full penetration through the sample and is more suitable for smaller particles, because larger and thicker particles absorb the incident radiation. Reflection mode, mainly investigating the surface of the sample, cannot be hindered by total absorption, and suffers from variations in morphologies. The surface needs to be able to reflect the radiation. If this mode is used on reflective surface as a copper or gold-coated filter, a measurement in reflectance–absorbance mode yields spectra like transmission mode [106]. FTIR can be combined with microscope and focal plane array (FPA) detectors, which allow the spectrochemical molecular imaging [107]. To eliminate the need for special sample preparation and avoid the loss of subsequent spectral reproducibility, the attenuated total reflectance (ATR) can be used. Generally, ATR spectroscopy is used for characterization of surface and sub-surface properties of a material [108]. The main part of ATR is the crystalline material of high refractive index in which the infrared beam is internally reflected. The penetration depth is usually between 0.5–5 μm [104]. The evanescent wave will be altered generating an energy difference of the reflected IR beam which is indicated by the special detector. FTIR transmission spectra of typical microplastics separated from seawater are shown in Fig. 5.19. Charateristical IR bands of typical textiel fibers is shown in Table 5.2.

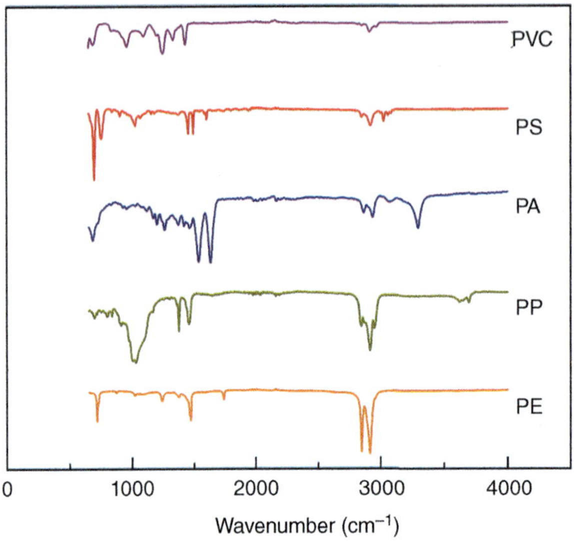

Figure 5.19 FT-IR spectra of microplastics. (*Source:* Ye et al. [109] / MDPI / CC BY 4.0.)

Table 5.2 Typical IR band frequencies for common textile fibers.

Band frequency (cm^{-1})	Type of vibration	Fiber type
3700–3200	O—H stretch (free and bonded)	Cellulosics
3500–3200	N—H stretch	Polyamides, elastane
2930–2840	C—H stretch	Most fibrous polymers
2260–2240	C≡N stretch (saturated nitrile)	Acrylics
1740–1715	C=0 stretch (ester)	Polyesters, acrylics
1670–1630	C=0 stretch (amide)	Polyamide, wool
1650–1590	N—H deform (primary amine)	Wool, polyamide
1570–1515	N—H deform (secondary amine)	Wool, polyamide
1250–1150	C—0 stretch (ester)	Polyester, acrylic
1100–1000	C—0 stretch (alcohols)	Cellulosics
730–650	C—H (aromatic)	Polyester

Source: [104] / Chalmers University of Technology.

For prediction of IR spectra changes due to weathering, some virgin microplastics were artificially weathered in a climate room equipped with UVA lamps, calibrated at 340 nm, and programmed at a temperature of 22 °C, an irradiance of 12 h, and humidity at 60%, for a total of 20 days. Afterwards, samples were subjected to thermally ageing at 45 °C in an air-circulated oven, for another 20 days. In the spectra of aged PET particles, 40 days after the artificial weathering, a new peak at 1614 cm^{-1}, non-existent in the same particles before the ageing process, appears [110]. IR spectroscopy allows the identification of microplastics composition by comparison with the reference spectra library. These reference spectra represent often very clean and ideal polymeric materials, not typically found in the environment. FTIR spectra of microplastics are affected by various, degradation processes. It was found that 8 weeks of treatment of polyethylene microplastics in artificial seawater was significantly more degradative than UV illumination [111]. It is therefore necessary to build a library of nontypical reference polymeric materials (as microplastics) from various sources, such as beach debris, recycled waste, and microplastics released after washing of colored textiles for comparison with unknown environmentally relevant samples. The reference spectra of virgin or weathered plastics and microparticles are commercially available, custom-made, or freely available [79].

5.7.3 Raman Spectroscopy

The Raman effect occurs when a sample is irradiated by monochromatic (laser) light, causing a small fraction of the scattered radiation have shifted frequencies that correspond to the matter vibrational transitions. Raman spectroscopy is detecting the light scattered inelastically at a frequency different from the incident light corresponding to a molecular or lattice vibration in the tested sample [108]. Raman scattering is the process of reradiation of scattered light by dipoles induced in the molecules by the incident light and modulated by the vibrations of the molecules [63]. In normal Raman scattering are the dipoles results of interaction of the electric field component, E, of the incident light and the molecules [63]. Raman scattering is therefore result of the oscillating polarizability in a molecule as a function of vibration and not from the permanent dipole of the molecule. Ground-state molecules create lines shifted to energies lower than the incident light, and weaker lines at higher frequency are due to molecules in excited vibrational states. These lines are result of the inelastic scattering of light by the matter. Elastic collisions result in Rayleigh scattering and appear more intense that unshifted component of the scattered light [63]. Raman spectroscopy, and especially micro-Raman spectroscopy, have better spatial resolution, wider spectral coverage, narrower spectral bands, and lower water interference than FTIR spectroscopy [96]. The typical characteristic peaks on classical Raman spectra of some microplastics are shown in Fig. 5.20.

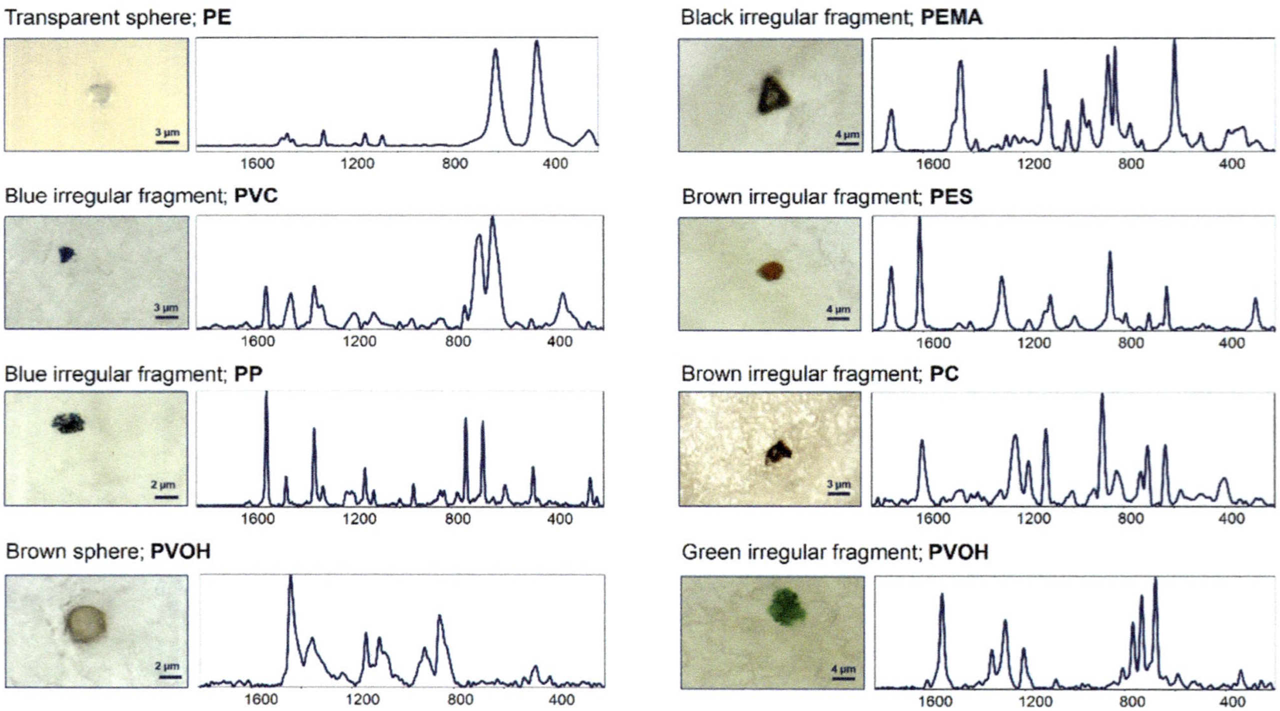

Figure 5.20 Microphotographs and Raman spectra (wavenumbers, cm^{-1}) of some selected MPs. PE: polyethylene; PVC: polyvinyl chloride; PP: polypropylene; PVOH: polyvinyl alcohol; PEVA: poly(ethylene-co-vinyl acetate); PEMA: poly(ethyl methacrylate); PES: polyester, and PC: polycarbonate. (*Source:* Ragusa et al. [112] / MDPI / CC BY 4.0.)

One of the main problems in Raman spectroscopy is fluorescence, which is the emission of a photon of longer wavelengths than the absorbed one. This effect is responsible for creation of very broad peaks interfering with the spectrum. The simple way to suppress the effect of fluorescence is to use exciting lasers of longer wavelengths. Red light laser of 785 nm is the standard but now there are systems with light in the near infra-red region at 1064 nm. The problem with using lower energy lasers is that the intensity of Raman scattering decreases. Excitation lasers are in some cases generating an excess of heat responsible for deterioration of polymeric materials, which can be limitation for identification of microplastics. Therefore, for microplastics identification the laser source wavelength can be proposed as a maximum of e.g. 633 nm [113]. One portable microplastic detection systems based on Raman spectroscopy is composed of collimated laser (5 mW), a sample holder, a notch filter, a diffraction grating, and a CCD sensor all integrated in a 3D printed case (see Fig. 5.21) [103].

Some dyestuffs and organic pigments obey strong fluorescence in visible part of light spectra and can be one of the major reasons of poor Raman signal quality [115]. Raman spectra of textile fibers are shown in Fig. 5.22 [115].

Usually, red and yellow particles had the lowest rates of Raman confirmation [115]. Black color polymers containing carbon type fillers or black particles as rubber tire wear are strongly absorbing the scattered light and show a carbon-like spectrum and are frequently not confirmed by Raman spectroscopy [115]. Other additives can cause partial signal masking to a lower degree

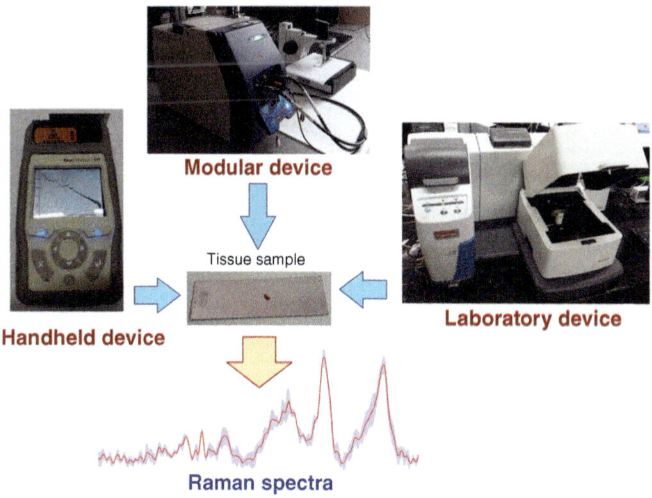

Figure 5.21 Raman spectroscopes. (*Source:* Synytsya et al. [114] / MDPI / CC BY 4.0.)

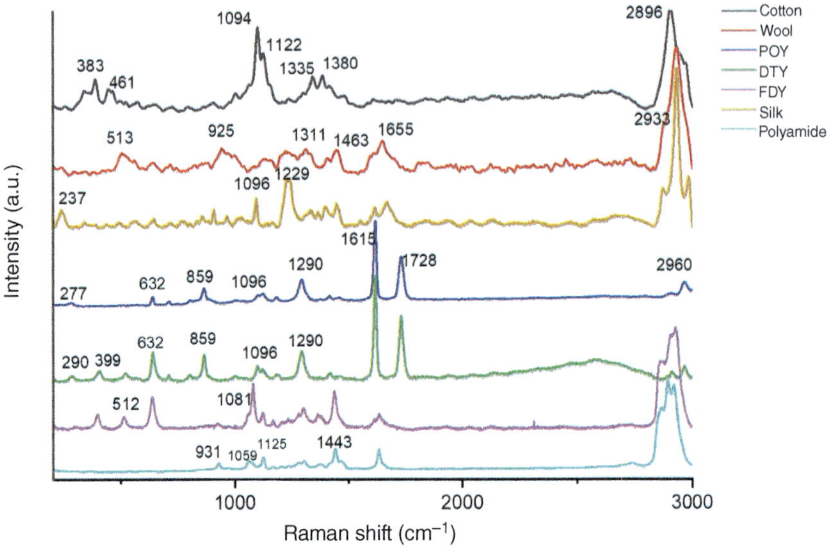

Figure 5.22 Raman spectra of textile fibers. (*Source:* Liu and Chen [116] / MDPI / CC BY 4.0.)

of Raman spectral matching. It was found, that some microplastic analyzed by Raman spectroscopy had moderate match scores with PET, but without the characteristic peak at 1616 cm^{-1} [115]. Raman spectra of polyethylene terephthalate-type microplastics are shown in Fig. 5.23 and therefore the proper samples pretreatment and microplastics extraction should be used and replicate measurements of different spots on a particle should be realized [115].

Especially the microplastics particles <20 μm and small total microplastics masses (<1.6 μg) are challenges for Raman microscopy [113]. The calculation of the minimum number of particles that need to be analyzed by Raman spectroscopy based on the random sampling approach was also proposed [113]. Raman spectra have usually very good reproducibility [107]. In Raman spectroscopy, there appears randomly very narrow strong peaks (called spikes or cosmic rays) as result of interaction of high energy particles with detectors [79].

[84] For correcting of this phenomenon, it is simple to compare multiple spectra of the same materials because Raman bands should be consistent and cosmic rays appear randomly. Improvements of detection (increase signal-to-noise ratios) of Raman spectroscopy are obtained by using special techniques such as Resonance Raman, Surface Enhanced Raman (SERS), Coherent-Anti-Stokes-Raman, Swept-source Raman [118] or Fourier-Transform-Raman (FT-Raman) which are costlier and require complicated sample preparation [57].

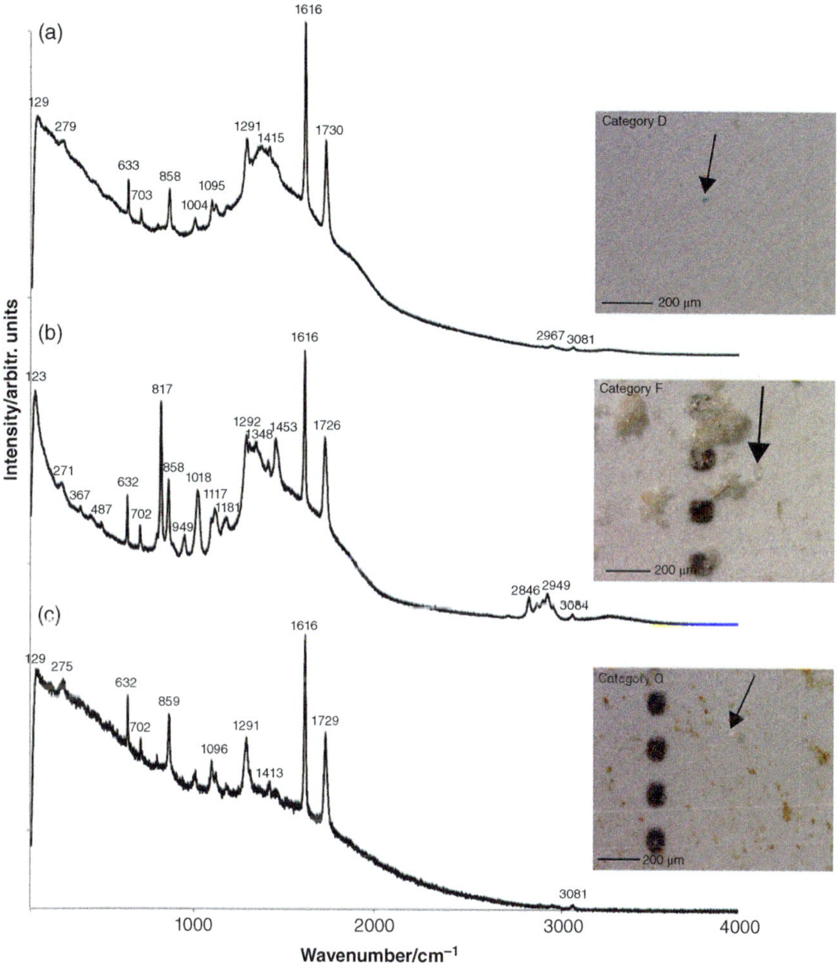

Figure 5.23 Raman spectra of polyethylene terephthalate-type microplastics. (*Source:* Perraki et al. [117] / MDPI / CC BY 4.0.)

5.7.3.1 Thermal Methods

Thermal analysis (TA) covers the methods in which bulk physical property changes of a material, or a mixture of substances are measured in response to programmed changes in temperature in a specified atmosphere. The physical properties measured are transition temperatures or enthalpy (DSC), dimensional changes, viscoelastic properties, dielectric properties and mass changes (TG) [90]. Instead of vibrational spectroscopy oriented to presence, composition, size and shape of particles are these methods based on the identification of mass of the polymers according to its degradation products [113]. Thermal decomposition of the polymers is realized by using TGA or standard pyrolysis. For the simultaneous analysis of PE and PP the TGA coupled to differential scanning calorimetry (TGA-DSC) was used [119]. This technique is widely used for the purity of synthetic materials, thermal degradation or phase transitions examination. The DSC thermograms of PE-LD and PP are shown in Fig. 5.24 [117].

The endothermic phase transition heat flows and peak temperatures obtained by DSC were clearly different for PE and PP. The limit of detection was 2.5 wt.% (for PE) and 5 wt.% (for PP). DSC can be used in combination with optical microscopy for the identification and quantification of microplastics related to environmental problems such as LDPE, HDPE, PP and PET (see Fig. 5.24) [118]. For the identification of microplastics, the characteristic onset temperature (starting temperature of melting) is proposed. For mass quantitation, the melt enthalpy (based on the area of the melting peak) is used. Both characteristics are affected by particle size [118].

TGA decomposition profiles (loss of mass) for common polymers present in microplastics are shown in [121]. All polymers showed only one decomposition step without the formation of residue, except for PET, which formed a residue of about 11%. [118] Due to the overlapping of the decomposition ranges of the different polymers, it was not possible to identify single kinds of polymers by using only the decomposition profile of the TGA in inert atmosphere. A special technique for avoiding this problem is use of thermal mass losses in 10 °C intervals from TGA spectra and sparse principal component regression for the qualitative and quantitative determination of the most common microplastics (PE, PS, PVC, and PET) in an agricultural soil [122]. The amount of mass used for the TGA analysis was between 0.5 and 1 g, which is acceptable for the reproducibility, but the type of used soil will probably affect the degradation dynamics and the evaluation of microplastics composition. The decomposition products of TGA are generally dependent on polymer type [121].

The polyethylene (PE) predominantly decomposes in 1,12-tridecadiene; 1,13-tetradecadiene; 1,14-pentadecadiene; and 1,15-hexadecadiene. The polypropylene (PP) predominantly decomposes in alkanes, alkenes and alkadienes with a ratio of 1:17:4, respectively. Six compounds revealed relatively high abundance. The 2,4-dimethylhept-1-ene; followed by two diastereoisomers of 2,4,6-trimethylnon-1-ene; and four diastereoisomers of 2,4,6,8-tetramethylundec-1-ene.

The polystyrene (PS) thermally decomposes mainly by chain end scission, leading to the formation of secondary benzyl and resonantly stabilized allyl benzene radicals, which are able to abstract further protons from the polymer backbone over a five-, six-, or seven-membered intermediate state. Three main thermal degradations products are: styrene; 2,4-diphenyl-1-butene (dimer); and 2,4,6-triphenyl-1-hexen (trimer).

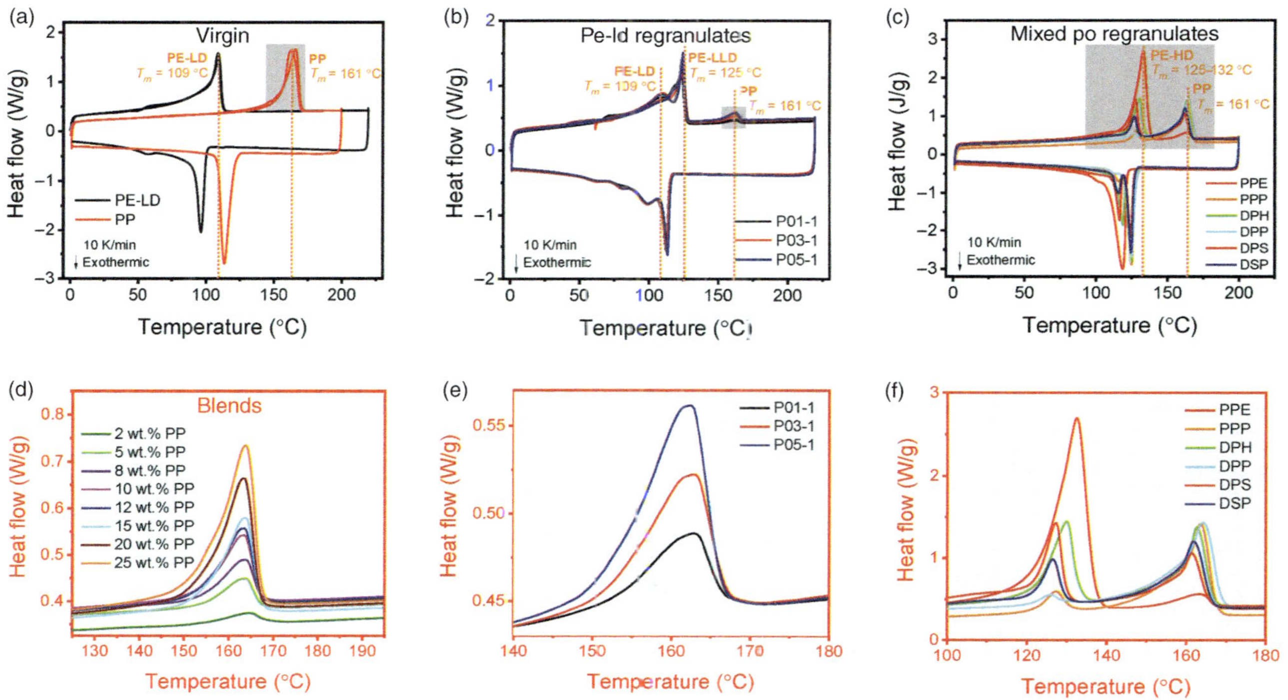

Figure 5.24 DSC thermograms of (a) virgin PE-LD and polypropylene (PP) with (d) magnification of PP melting peak (*Tm* = 161 °C) illustrating melt enthalpy differences by blend PP content (2, 5, 8, 10, 12, 15, 20, and 25 wt.%), (b) P01,03,05 PE-LD regranulates with (e) magnification of PP melting peak (*Tm* = 161 °C) and (c) PPE, PPP, DPH, DPP, DPS and DSP mixed PO regranulates with (*f*) magnification of PE-HD (*Tm* = 125–132 °C) and PP (*Tm* = 161 °C) melting peaks. (*Source:* Karaagac et al. [120] / MDPI / CC BY 4.0.)

The polyethylene terephthalate (PET) main thermal degradation is the transfer of a b-hydrogen that leads to the formation of vinyl esters and acid end groups over a six-membered transition state. Main degradation compounds are benzoic acid; ethyl benzoate; vinyl benzoate and biphenyl. The polyamide 6 (PA 6) thermally decomposes by homolytic chain scission of the N-alkylamide or peptide bond. The most predominant volatile degradation product is b-caprolactam. Other, less frequent degradation products are nitriles; aldehydes and 3,4,5,6-tetrahydro-2H-azepine derivatives. Based on knowledge of these compounds it is possible to identify microplastics (strictly all organic materials) types in unknown polluted matter (solid soil or water dispersion) from their thermal degradation volatiles by proper analytical method [121]. The simple way is direct connection of mass spectrofotometr and TGA instrument. The mass loss measured by the thermo-balance in the TGA system is related to the signal in the mass spectroscopy detector [123]. Thermoanalytical methods of type Py-GC-MS start with pyrolysis (PY) usually by TG in an inert atmosphere. Thermal degradation compounds are analyzed by gas chromatography (GC) and characterized by mass spectrometry (MS). These methods are suitable for single particles and can be applied for sample weights of 0.1–0.5 mg. The Py-GC-MS does not depend on the shape, size or the presence of organic or inorganic contaminants in the plastic particles and samples do not require pretreatment. On the other hand it is not suitable for analysis of complex sample mixtures like environmental samples. This methodology is limited by complexity of data interpretation for cases of many contaminants or plastics with similar masses and degradation temperatures. The evolved gases are not separated before MS detection [123]. The other method (samples about 20 mg) also analyzing the degradation products of the microplastics. After completing thermal decomposition coupled to solid-phase extraction (TGA-SPE) (called thermos extraction) are volatiles adsorbed on a solid-phase absorber with polydimethylsiloxane layer, and absorbers are analyzed by thermally desorbed gas chromatography mass spectrometry (TED-GC-MS) [121, 123]. The current limit of detection is around 1 wt.% (for PE), 0.5 wt.% (for PP), and 0.1 wt.% (for PS) [124]. By using these methods, the rapid identification and quantitative determination of most thermoplastic polymers and elastomers is possible [23–25]. For quantification of common microplastic the combining pressurized liquid extraction (PLE) and pyrolysis GC-MS was developed [125]. The extraction is composed of a pre-extraction step by methanol and a subsequent pressurized liquid extraction at 185 °C and 100 bar using tetrahydrofuran. Limits of quantification for polyethylene, polypropylene, and polystyrene was 0.007 mg g^{-1}. Comparison of the different techniques for microplastics detection and analysis in environmental samples is shown in the Table 5.3 [121].

Table 5.3 Different techniques for microplastics detection and analysis.

Results	Raman	FTIR	TED-CG-MS	Extraction
Identification	PE, PP, PS, PET	PE, PP, PS, PET	PE, PP, PS, PET	PS, PET
Quantification	No	No	Yes[c]	Yes
Particle size distribution/visual image	Yes	Yes	No	No

5.7.3.2 Chemical Methods

Especially for evaluation of polyolefins based microplastics the combination of density separation and a subsequent quantification of microplastics content with high temperature gel-permeation chromatography was described [126]. Twofold density separation in a separation funnel was used. In the first stage, the impurities heavier than polyolefins were removed by using proper liquids with higher density than water (as solution of NaCl). In the second stage a solvent with a lower density than polyolefins (as acetone) were added, to settle the microplastics particles to the bottom of the separation funnel. Liquid Chromatography is based on the different solubility of polymers [121]. For high performance liquid chromatography (HPLC) analysis, a 1,2,4-trichlorobenzene was used as mobile phase, at a flow of 1 mL min^{-1} at 150 °C. Heptane was added to the dried microplastics as a flow marker. Dissolution was performed using preheated 1,2,4-trichlorobenzene, using a dissolution temperature of 160 °C 120 min [126]. Liquid extraction with subsequent size exclusion chromatography (SEC) can be also used [121]. For extraction of polystyrene the tetrahydrofurane (THF) and for PET the hexafluoro isopropanol (HFIP) were proposed. In both cases the temperature of dissolution was 45 °C for 24 h [121].

5.8 Conclusion

This chapter is focused on the very broad class of methods connected with characterization of microplastics in different media (as water and soil) from microplastics collection (sampling techniques), sample preprocessing (removal of organic pollutants, biomaterials, inorganic materials, biomaterials, biota etc.) and extraction (separation into classes according to their size) to their morphological, physical and chemical characterization. The fibrous microplastics released from fibrous structures are discussed in detail because they are unique in comparison with plastics of the same chemical nature.

The basic techniques for morphological characterization of microplastics (as the dimension, shape and color) are briefly discussed. The qualitative and quantitative methods of microplastics analysis based on optical spectroscopy in IR region, thermal methods and other chemical methods are critically described. Besides basic principles of individual methods, the problems arise from presence of impurities, water content, and different foreign materials (TiO_2, dyestuffs, and other agents) are explained. The most reliable techniques and requirements for further improvements are mentioned. The sorting/identification of microplastics are often subjected to high rates of false results and limited range of micro/nanoparticles size. For chemical identification of microplastics are commonly used vibrational optical methods as ATR-FTIR spectroscopy (for particles larger than 500 μm), micro-FTIR spectroscopy (for particles in the size range of 20–500 μm) and Raman microspectroscopy (for particles larger than 1 μm). Raman spectroscopy is seriously distorted by fluorescence caused by colored microplastics (usual for microplastics released from textile structures) and IR techniques are not suitable for water containing samples due to strong absorption of IR radiation by water and overlap of fibrous polymers IR spectra major peaks. NMR spectroscopy combined with thermal decomposition, dissolution or chromatography can be used to characterize microplastics released from textiles composed from pure fibers and their blends. Selected chemometric techniques such as principal component analysis or partial least squares may be coupled to each technique to provide quantitative results. For automatic sorting purposes it is necessary to select techniques performing qualitative as well as quantitative characterization. Till yet the rate of false decision is higher in comparison with lengthy manual sorting based on the same method.

References

1 Bergmann, M., Mützel, S., Primpke, S. et al. (2019). White and wonderful? Microplastics prevail in snow from the Alps to the Arctic. *Science Advances* 5 (8): eaax1157.

2 Boucher, J., and Friot, D. (2017) Primary microplastics in the oceans: a global evaluation of sources, IUCN International Union for Conservation of Nature.

3 Wang, J., Tan, Z., Peng, J. et al. (2016). The behaviors of microplastics in the marine environment. *Marine Environmental Research* 113: 7–17.

4 Lee, H., Shim, W.J., and Kwon, J.-H. (2014). Sorption capacity of plastic debris for hydrophobic organic chemicals. *Science of the Total Environment* 470–471: 1545–1552.

5 Al Harraq, A. and Bharti, B. (2022). Microplastics through the lens of colloid science. *ACS Environmental Au* 2 (1): 3–10.

6 Periyasamy, A.P. (2025). Functionalized textile microplastics: a closer look at the issues, strategy, and legislation on the microplastic reduction. *Kuwait Journal of Science* 100395.

7 Marek, J. (2021). State-of-the-art water treatment in Czech power sector: industry-proven case studies showing economic and technical benefits of membrane and other novel technologies for each particular water cycle. *Membranes (Basel)* 11 (2): 98.

8 Rocha-Santos, T., Costa, M.F., and Mouneyrac, C. (ed.) (2022). *Handbook of Microplastics in the Environment*. Cham: Springer International Publishing.

9 Rocha-Santos, T.A.P. and Duarte, A.C. (ed.) (2017). Characterization and analysis of microplastics. In: *Comprehensive Analytical Chemistry*, vol. 75, i. Elsevier.

10 Liu, S.Y., Leung, M.M.-L., Fang, J.K.-H., and Chua, S.L. (2021). Engineering a microbial "trap and release" mechanism for microplastics removal. *Chemical Engineering Journal* 404: 127079.

11 Zhang, J., Peng, M., Lian, E. et al. (2023). Identification of poly(ethylene terephthalate) nanoplastics in commercially bottled drinking water using surface-enhanced raman spectroscopy. *Environmental Science & Technology* 57 (22): 8365–8372.

12 Bach, C., Dauchy, X., Chagnon, M.-C., and Etienne, S. (2012). Chemical compounds and toxicological assessments of drinking water stored in polyethylene terephthalate (PET) bottles: a source of controversy reviewed. *Water Research* 46 (3): 571–583.

13 Dhaka, V., Singh, S., Anil, A.G. et al. (2022). Occurrence, toxicity and remediation of polyethylene terephthalate plastics. A review. *Environmental Chemistry Letters* 20 (3): 1777–1800.

14 Periyasamy, A.P. and Tehrani-Bagha, A. (2022). A review on microplastic emission from textile materials and its reduction techniques. *Polymer Degradation and Stability* 199: 109901.

15 Šaravanja, A., Pušić, T., and Dekanić, T. (2022). Microplastics in wastewater by washing polyester fabrics. *Materials* 15 (7): 2683.

16 Yang, T., Luo, J., and Nowack, B. (2021). Characterization of nanoplastics, fibrils, and microplastics released during washing and abrasion of polyester textiles. *Environmental Science & Technology* 55 (23): 15873–15881.

17 Choi, S., Kwon, M., Park, M.-J., and Kim, J. (2021). Characterization of microplastics released based on polyester fabric construction during washing and drying. *Polymers (Basel)* 13 (24): 4277.

18 Militký, J., Kremenakova, D., and Venkataraman, M. (2024). Microplastics and fibrous fragments generated during the production and maintenance of textiles. *Fibers* 12 (7): 51.

19 Mossotti, R., Dalla Fontana, G., Anceschi, A. et al. (2021). Preparation and analysis of standards containing microfilaments/microplastic with fibre shape. *Chemosphere* 270: 129410.

20 Gultekin, A. (2024). Effect of fiber type and length on strength, fracture energy, and durability properties of microwave-cured fiber-reinforced geopolymer mortars. *Buildings* 14 (12): 3723.

21 D'Allo, C.A. (1977) The chemical and related characteristics of polyester fibres. 9th Shirley Institute Seminar.

22 De Falco, F., Cocca, M., Avella, M., and Thompson, R.C. (2020). Microfiber release to water, via laundering, and to air, via everyday use: a comparison between polyester clothing with differing textile parameters. *Environmental Science & Technology* 54 (6): 3288–3296.

23 Meloun, M. and Militký, J. (2011). 3 – Statistical analysis of univariate data. In: *Statistical Data Analysis* (ed. M. Meloun and J. Militký), 73–150. Woodhead Publishing India.

24 Rocha-Santos, T. and Duarte, A.C. (2015). A critical overview of the analytical approaches to the occurrence, the fate and the behavior of microplastics in the environment. *TrAC Trends in Analytical Chemistry* 65: 47–53.

25 Stock, F., Kochleus, C., Bänsch-Baltruschat, B. et al. (2019). Sampling techniques and preparation methods for microplastic analyses in the aquatic environment – a review. *TrAC Trends in Analytical Chemistry* 113: 84–92.

26 WHO (2019) Microplastics in drinking-water. Licence: CC BY-NC-SA 3.0 IGO.

27 Murphy, F., Ewins, C., Carbonnier, F., and Quinn, B. (2016). Wastewater treatment works (WwTW) as a source of microplastics in the aquatic environment. *Environmental Science & Technology* 50 (11): 5800–5808.

28 Hidalgo-Ruz, V., Gutow, L., Thompson, R.C., and Thiel, M. (2012). Microplastics in the marine environment: a review of the methods used for identification and quantification. *Environmental Science & Technology* 46 (6): 3060–3075.

29 Koelmans, A.A., Mohamed Nor, N.H., Hermsen, E. et al. (2019).v Microplastics in freshwaters and drinking water: critical review and assessment of data quality. *Water Research* 155: 410–422.

30 Hermsen, E., Mintenig, S.M., Besseling, E., and Koelmans, A.A. (2018). Quality criteria for the analysis of microplastic in biota samples: a critical review. *Environmental Science & Technology* 52 (18): 10230–10240.

31 Cutroneo, L., Reboa, A., Besio, G. et al. (2020). Microplastics in seawater: sampling strategies, laboratory methodologies, and identification techniques applied to port environment. *Environmental Science and Pollution Research* 27 (9): 8938–8952.

32 Razeghi, N., Hamidian, A.H., Wu, C. et al. (2021). Microplastic sampling techniques in freshwaters and sediments: a review. *Environmental Chemistry Letters* 19 (6): 4225–4252.

33 Peng, L., Mehmood, T., Bao, R. et al. (2022). An overview of micro(nano) plastics in the environment: sampling, identification, risk assessment and control. *Sustainability* 14 (21): 14338.

34 Pelamatti, T., Fonseca-Ponce, I.A., Rios-Mendoza, L.M. et al. (2019). Seasonal variation in the abundance of marine plastic debris in Banderas Bay, Mexico. *Marine Pollution Bulletin* 145: 604–610.

35 Chubarenko, I., Bagaev, A., Zobkov, M., and Esiukova, E. (2016). On some physical and dynamical properties of microplastic particles in marine environment. *Marine Pollution Bulletin* 108 (1–2): 105–112.

36 Bratbak, G. and Dundas, I. (1984). Bacterial dry matter content and biomass estimations. *Applied and Environmental Microbiology* 48 (4): 755–757.

37 Sillanpää, M. and Sainio, P. (2017). Release of polyester and cotton fibers from textiles in machine washings. *Environmental Science and Pollution Research* 24 (23): 19313–19321.

38 Hernandez, E., Nowack, B., and Mitrano, D.M. (2017). Polyester textiles as a source of microplastics from households: a mechanistic study to understand microfiber release during washing. *Environmental Science & Technology* 51 (12): 7036–7046.

39 Napper, I.E. and Thompson, R.C. (2016). Release of synthetic microplastic plastic fibres from domestic washing machines: effects of fabric type and washing conditions. *Marine Pollution Bulletin* 112 (1–2): 39–45.

40 Hartline, N.L., Bruce, N.J., Karba, S.N. et al. (2016). Microfiber masses recovered from conventional machine washing of new or aged garments. *Environmental Science & Technology* 50 (21): 11532–11538.

41 Kaikobad, M., Mahmud, F., Sayam, A. et al. (2023). Evaluation of physical and mechanical characteristics of three-thread fleece knit fabric for their structural changes. *Heliyon* 9 (11): e21588.

42 Vianello, A., Boldrin, A., Guerriero, P. et al. (2013). Microplastic particles in sediments of Lagoon of Venice, Italy: first observations on occurrence, spatial patterns and identification. *Estuarine, Coastal and Shelf Science* 130: 54–61.

43 Van Cauwenberghe, L., Vanreusel, A., Mees, J., and Janssen, C.R. (2013). Microplastic pollution in deep-sea sediments. *Environmental Pollution* 182: 495–499.

44 Wright, S.L., Thompson, R.C., and Galloway, T.S. (2013). The physical impacts of microplastics on marine organisms: a review. *Environmental Pollution* 178: 483–492.

45 Van Cauwenberghe, L., Claessens, M., Vandegehuchte, M.B., and Janssen, C.R. (2015). Microplastics are taken up by mussels (Mytilus edulis) and lugworms (Arenicola marina) living in natural habitats. *Environmental Pollution* 199: 10–17.

46 Hurley, R.R., Lusher, A.L., Olsen, M., and Nizzetto, L. (2018). Validation of a method for extracting microplastics from complex, organic-rich, environmental matrices. *Environmental Science & Technology* 52 (13): 7409–7417.

47 Imhof, H.K., Schmid, J., Niessner, R. et al. (2012). A novel, highly efficient method for the separation and quantification of plastic particles in sediments of aquatic environments. *Limnology and Oceanography: Methods* 10 (7): 524–537.

48 Zobkov, M.B. and Esiukova, E.E. (2017). Evaluation of the Munich plastic sediment separator efficiency in extraction of microplastics from natural marine bottom sediments. *Limnology and Oceanography: Methods* 15 (11): 967–978.

49 Felsing, S., Kochleus, C., Buchinger, S. et al. (2018). A new approach in separating microplastics from environmental samples based on their electrostatic behavior. *Environmental Pollution* 234: 20–28.

50 Lusher, A.L., Welden, N.A., Sobral, P., and Cole, M. (2017). Sampling, isolating and identifying microplastics ingested by fish and invertebrates. *Analytical Methods* 9 (9): 1346–1360.

51 Claessens, M., Van Cauwenberghe, L., Vandegehuchte, M.B., and Janssen, C.R. (2013). New techniques for the detection of microplastics in sediments and field collected organisms. *Marine Pollution Bulletin* 70 (1–2): 227–233.

52 Kühn, S., van Werven, B., van Oyen, A. et al. (2017). The use of potassium hydroxide (KOH) solution as a suitable approach to isolate plastics ingested by marine organisms. *Marine Pollution Bulletin* 115 (1–2): 86–90.

53 Frias, J., et al. (2018) Standardised protocol for monitoring microplastics in sediments. Deliverable 4.2. JPI-Oceans BASEMAN Project.

54 Mintenig, S.M., Int-Veen, I., Löder, M.G.J. et al. (2017). Identification of microplastic in effluents of waste water treatment plants using focal plane array-based micro-Fourier-transform infrared imaging. *Water Research* 108: 365–372.

55 Cutroneo, L., Reboa, A., Geneselli, I., and Capello, M. (2021). Considerations on salts used for density separation in the extraction of microplastics from sediments. *Marine Pollution Bulletin* 166: 112216.

56 Rodrigues, M.O., Gonçalves, A.M.M., Gonçalves, F.J.M., and Abrantes, N. (2020). Improving cost-efficiency for MPs density separation by zinc chloride reuse. *MethodsX* 7: 100785.

57 Lenz, R., Enders, K., Stedmon, C.A. et al. (2015). A critical assessment of visual identification of marine microplastic using Raman spectroscopy for analysis improvement. *Marine Pollution Bulletin* 100 (1): 82–91.

58 Nuelle, M.-T., Dekiff, J.H., Remy, D., and Fries, E. (2014). A new analytical approach for monitoring microplastics in marine sediments. *Environmental Pollution* 184: 161–169.

59 Cecchi, T., Poletto, D., Berbecaru, A.C. et al. (2024). Assessing microplastics and nanoparticles in the surface seawater of Venice lagoon – part I: methodology of research. *Materials* 17 (8): 1759.

60 Desforges, J.-P.W., Galbraith, M., Dangerfield, N., and Ross, P.S. (2014). Widespread distribution of microplastics in subsurface seawater in the NE Pacific Ocean. *Marine Pollution Bulletin* 79 (1–2): 94–99.

61 Oßmann, B.E., Sarau, G., Schmitt, S.W. et al. (2017). Development of an optimal filter substrate for the identification of small microplastic particles in food by micro-Raman spectroscopy. *Analytical and Bioanalytical Chemistry* 409 (16): 4099–4109.

62 Hýbl, V. (2022) Vývoj nanovlákenného filtru vhodného pro základní stanovení mikroplastů a opakovanou dead-end filtraci.

63 Koenig, J.L. (2001). *Infrared and Raman Spectroscopy of Polymers*. Shropshire: iSmithers Rapra Publishing.

64 Rieland, J.M., Hu, Z., Deese, J.S., and Love, B.J. (2023). Pressure sensitive adhesives for quantifying microplastic isolation. *Separation and Purification Technology* 307: 122819.

65 Morét Ferguson, S., Law, K.L., Proskurowski, G. et al. (2010). The size, mass, and composition of plastic debris in the western North Atlantic Ocean. *Marine Pollution Bulletin* 60 (10): 1873–1878.

66 Pivokonsky, M., Cermakova, L., Novotna, K. et al. (2018). Occurrence of microplastics in raw and treated drinking water. *Science of the Total Environment* 643: 1644–1651.

67 Reimer, L. (2000). Scanning electron microscopy: physics of image formation and microanalysis, second edition. *Measurement Science and Technology* 11 (12): 1826–1826.

68 Norén, F. and Naustvoll, L.J. (2010) Survey of microscopic anthropogenic particles in skagerrak, pilot study Report commissioned by Klima-og Forurensningsdirektoratet.

69 Haap, J., Classen, E., Beringer, J. et al. (2019). Microplastic fibers released by textile laundry: a new analytical approach for the determination of fibers in effluents. *Water (Basel)* 11 (10): 2088.

70 Shimizu, K., Sokolov, S.V., Kätelhön, E. et al. (2017). *In situ* detection of microplastics: single microparticle-electrode impacts. *Electroanalysis* 29 (10): 2200–2207.

71 Shanthi, C., Porpatham, R., and Pappa, N. (2014). Image analysis for particle size distribution. *International Journal of Engineering and Technology* 6: 1340–1345.

72 Houck, M.M. (2009). 1 – Introduction to textile fiber identification. In: *Identification of Textile Fibers* (ed. M.M. Houck), 3–5. Woodhead Publishing.

73 Ho, D. and Masura, J. (2025). Dyeing to know: harmonizing nile red staining protocols for microplastic identification. *Colorants* 4 (2): 20.

74 Martinez, V. and Henary, M. (2016). Nile red and Nile blue: applications and syntheses of structural analogues. *Chemistry – A European Journal* 22 (39): 13764–13782.

75 Yoshino, J., Kano, N., and Kawashima, T. (2013). Fluorescent azobenzenes and aromatic aldimines featuring an N–B interaction. *Dalton Transactions* 42 (45): 15826–15834.

76 Shim, W.J., Song, Y.K., Hong, S.H., and Jang, M. (2016). Identification and quantification of microplastics using Nile red staining. *Marine Pollution Bulletin* 113 (1–2): 469–476.

77 Karakolis, E.G., Nguyen, B., You, J.B. et al. (2019). Fluorescent dyes for visualizing microplastic particles and fibers in laboratory-based studies. *Environmental Science & Technology Letters* 6 (6): 334–340.

78 Alvim, C.B., Bes-Piá, M.A., Mendoza-Roca, J.A., and Alonso-Molina, J.L. (2023). Identification of microfibers in drinking water with Nile red. Limitations and strengths. *Journal of Environmental Chemical Engineering* 11 (3): 109697.

79 Cowger, W., Gray, A., Christiansen, S.H. et al. (2020). Critical review of processing and classification techniques for images and spectra in microplastic research. *Applied Spectroscopy* 74 (9): 989–1010.

80 Devriese, L.I., van der Meulen, M.D., Maes, T. et al. (2015). Microplastic contamination in brown shrimp (Crangon crangon, Linnaeus 1758) from coastal waters of the Southern North Sea and Channel area. *Marine Pollution Bulletin* 98 (1–2): 179–187.

81 Comfort, N., Cai, K., Bloomquist, T.R. et al. (2021). Nanoparticle Tracking Analysis for the Quantification and Size Determination of Extracellular Vesicles. *Journal of Visualized Experiments* (169): 10-3791.

82 Hermabessiere, L., Himber, C., Boricaud, B. et al. (2018). Optimization, performance, and application of a pyrolysis-GC/MS method for the identification of microplastics. *Analytical and Bioanalytical Chemistry* 410 (25): 6663–6676.

83 Umbaugh, S.E. (2022). *Digital Image Processing and Analysis*. Boca Raton: CRC Press.

84 Otsu, N. (1979). A threshold selection method from gray-level histograms. *IEEE Transactions on Systems, Man, and Cybernetics* 9 (1): 62–66.

85 LeCun, Y., Bengio, Y., and Hinton, G. (2015). Deep learning. *Nature* 521 (7553): 436–444.

86 Militký, J. (2011). Fundamentals of soft models in textiles. In: *Soft Computing in Textile Engineering*, 45–102. Elsevier.

87 Li, J., Liu, H., and Paul Chen, J. (2018). Microplastics in freshwater systems: a review on occurrence, environmental effects, and methods for microplastics detection. *Water Research* 137: 362–374.

88 Kaushik, S. and Kumar, B. (ed.) (2023). *Analytical Methods in Chemical Analysis*. De Gruyter.

89 Larkin, P.J. (2018). Chapter 1 – Introduction: infrared and Raman spectroscopy. In: *Infrared and Raman Spectroscopy*, Seconde (ed. P.J. Larkin), 1–5. Elsevier.

90 Bart, J.C.J. (2006). *Plastics Additives: Advanced Industrial Analysis* (ed. J.C.J. Bart). Amsterdam: IOS Press.

91 Bhargava, R., Wang, S.-Q., and Koenig, J.L. (2003). FTIR microspectroscopy of polymeric systems. In: *Liquid Chromatography/FTIR Microspectroscopy/Microwave Assisted Synthesis*, 137–191. Berlin, Heidelberg: Springer Berlin Heidelberg.

92 Renner, G., Nellessen, A., Schwiers, A. et al. (2019). Data preprocessing & evaluation used in the microplastics identification process: a critical review & practical guide. *TrAC Trends in Analytical Chemistry* 111: 229–238.

93 Mokari, A., Guo, S., and Bocklitz, T. (2023). Exploring the steps of infrared (IR) spectral analysis: pre-processing, (classical) data modelling, and deep learning. *Molecules* 28 (19): 6886.

94 Rytelewska, S. and Dąbrowska, A. (2022). The Raman spectroscopy approach to different freshwater microplastics and quantitative characterization of polyethylene aged in the environment. *Microplastics* 1 (2): 263–281.

95 Araujo, C.F., Nolasco, M.M., Ribeiro, A.M.P., and Ribeiro-Claro, P.J.A. (2018). Identification of microplastics using Raman spectroscopy: latest developments and future prospects. *Water Research* 142: 426–440.

96 Ren, L., Liu, S., Huang, S. et al. (2023). Identification of microplastics using a convolutional neural network based on micro-Raman spectroscopy. *Talanta* 260: 124611.

97 Chan, C.K.-M., Lo, C.K.-Y., and Kan, C.-W. (2024). A systematic literature review for addressing microplastic fibre pollution: urgency and opportunities. *Water (Basel)* 16 (14): 1988.

98 Song, Y.K., Hong, S.H., Eo, S., and Shim, W.J. (2021). A comparison of spectroscopic analysis methods for microplastics: manual, semi-automated, and automated Fourier transform infrared and Raman techniques. *Marine Pollution Bulletin* 173: 113101.

99 Hummel, D.O. (2002). *Atlas of Plastics Additives*. Berlin, Heidelberg: Springer Berlin Heidelberg.

100 Crocombe, R.A. (2018). Portable spectroscopy. *Applied Spectroscopy* 72 (12): 1701–1751.

101 Crocombe, R.A., Kammrath, B.W., and Leary, P.E. (2023). Portable Raman spectrometers: how small can they get? *Spectroscopy* 32–40.

102 Tang, P.L., Mckumskay, R., Rogerson, M. et al. (2019). Handheld portable FTIR spectroscopy for the triage of micro and meso sized plastics in the marine environment incorporating an accelerated weathering study and an aging estimation. *Spectroscopy* 34: 54–60.

103 Iri, A.H., Shahrah, M.H.A., Ali, A.M. et al. (2021). Optical detection of microplastics in water. *Environmental Science and Pollution Research* 28 (45): 63860–63866.

104 Peterson, A. (2015) Towards recycling of textile fibers separation and characterization of textile fibers and blends. Master's thesis in Materials Chemistry and Nanotechnology, Department of Chemistry and Chemical Engineering, Chalmers University of Technology, Gothenburg, Sweden.

105 Asamoah, B.O., Uurasjärvi, E., Räty, J. et al. (2021). Towards the development of portable and in situ optical devices for detection of micro-and nanoplastics in water: a review on the current status. *Polymers (Basel)* 13 (5): 730.

106 Primpke, S., Christiansen, S.H., Cowger, W. et al. (2020). Critical assessment of analytical methods for the harmonized and cost-efficient analysis of microplastics. *Applied Spectroscopy* 74 (9): 1012–1047.

107 Hirschmugl, C.J. and Gough, K.M. (2012). Fourier transform infrared spectrochemical imaging: review of design and applications with a focal plane array and multiple beam synchrotron radiation source. *Applied Spectroscopy* 66 (5): 475–491.

108 Yoshikawa, M. (2023). *Advanced Optical Spectroscopy Techniques for Semiconductors*. Cham: Springer International Publishing.

109 Ye, C., Lin, J., Li, Z. et al. (2024). Characteristics of Microplastic Pollution in Agricultural Soils in Xiangtan, China. *Sustainability* 16 (17): 7254.

110 Savino, I., Campanale, C., Trotti, P. et al. (2022). Effects and impacts of different oxidative digestion treatments on virgin and aged microplastic particles. *Polymers (Basel)* 14 (10): 1958.

111 Alrbaihat, M.R. and Abu-Afifeh, Q. (2023). Eco-friendly microplastic removal through physical and chemical techniques: a review. *Annals of Advances in Chemistry* 7 (1): 14–24.

112 Ragusa, A., Notarstefano, V., Svelato, A. et al. (2022). Raman microspectroscopy detection and characterisation of microplastics in human breastmilk. *Polymers (Basel)* 14 (13): 2700.

113 Anger, P.M., von der Esch, E., Baumann, T. et al. (2018). Raman microspectroscopy as a tool for microplastic particle analysis. *TrAC Trends in Analytical Chemistry* 109: 214–226.

114 Synytsya, A., Kováčová, Z., Janstová, D. et al. (2024). Comparative study on handheld, modular, and laboratory Raman instruments for the analysis of colon tissues and colorectal polyps. *Applied Sciences* 14 (2): 791.

115 Jiang, Y., Zhao, Y., Wang, X. et al. (2020). Characterization of microplastics in the surface seawater of the South Yellow Sea as affected by season. *Science of the Total Environment* 724: 138375.

116 Liu, K. and Chen, H. (2025). Identification and visualization textile fibers by Raman imaging. *Materials* 18 (7): 1682.

117 Perraki, M., Skliros, V., Mecaj, P. et al. (2024). Identification of microplastics using μ-Raman spectroscopy in surface and groundwater bodies of SE Attica, Greece. *Water (Basel)* 16 (6): 843.

118 Song, J., So, P.T.C., Yoo, H., and Kang, J.W. (2024). Swept-source Raman spectroscopy of chemical and biological materials. *Journal of Biomedical Optics* 29 (S2): S22703–S22703.

119 Majewsky, M., Bitter, H., Eiche, E., and Horn, H. (2016). Determination of microplastic polyethylene (PE) and polypropylene (PP) in environmental samples using thermal analysis (TGA-DSC). *Science of the Total Environment* 568: 507–511.

120 Karaagac, E., Jones, M.P., Koch, T., and Archodoulaki, V.-M. (2021). Polypropylene contamination in post-consumer polyolefin waste: characterisation, consequences and compatibilisation. *Polymers (Basel)* 13 (16): 2618.

121 Elert, A.M., Becker, R., Duemichen, E. et al. (2017). Comparison of different methods for MP detection: what can we learn from them, and why asking the right question before measurements matters? *Environmental Pollution* 231: 1256–1264.

122 David, J., Weissmannová, H.D., Steinmetz, Z. et al. (2019). Introducing a soil universal model method (SUMM) and its application for qualitative and quantitative determination of poly(ethylene), poly(styrene), poly(vinyl chloride) and poly(ethylene terephthalate) microplastics in a model soil. *Chemosphere* 225: 810–819.

123 Peñalver, R., Arroyo-Manzanares, N., López-García, I., and Hernández-Córdoba, M. (2020). An overview of microplastics characterization by thermal analysis. *Chemosphere* 242: 125170.

124 Ivleva, N.P., Wiesheu, A.C., and Niessner, R. (2017). Microplastic in aquatic ecosystems. *Angewandte Chemie International Edition* 56 (7): 1720–1739.

125 Dierkes, G., Lauschke, T., Becher, S. et al. (2019). Quantification of microplastics in environmental samples via pressurized liquid extraction and pyrolysis-gas chromatography. *Analytical and Bioanalytical Chemistry* 411 (26): 6959–6968.

126 Hintersteiner, I., Himmelsbach, M., and Buchberger, W.W. (2015). Characterization and quantitation of polyolefin microplastics in personal-care products using high-temperature gel-permeation chromatography. *Analytical and Bioanalytical Chemistry* 407 (4): 1253–1259.

6

Role of Machine Learning and Machine Vision on Microplastics

6.1 Introduction

It is essential to monitor microplastic contamination to get insights into their origins and the effects they have on ecosystems and human well-being. Nevertheless, the task of collecting these minute particles is arduous because of their little presence in wastewater. There is no universally applicable methodology for analyzing microplastics, with optical identification methods being the prevailing approach. Initially, the visual identification technique for quantifying and categorizing microplastics, which relies on assessing their color, size, and form, is widely employed. Subsequently, microplastics are identified based on their density [1]. Vibrational spectroscopic techniques such as Fourier transform infrared spectroscopy (FTIR) [2], Raman spectroscopy [3], and energy-dispersive X-ray spectroscopy (EDS) [4] are employed for the identification of microplastics. However, it is important to note that these procedures are not automated and need extra expertise. FTIR, Raman spectroscopy, and Py-GC-MS are commonly used techniques to characterize microplastics [2]. However, these approaches have limitations in terms of lengthy characterization time, large sample needs, and low efficiency. In addition, unforeseen variables such as the impact of aging might generate spikes in the spectra of microplastics, leading to discrepancies in their identification [5]. Although physical characterization is a direct approach, it might result in imprecise quantification results and difficulties in discerning chemical specifics [4]. Despite its low cost and rapid implementation, this method is vulnerable to operator biases and mistakes, frequently resulting in misidentification [6]. Esch et al. [7] showed that visual counts conducted by expert operators can vary by up to 30%. Microplastics generated during the domestic washing of white cloths can sometimes absorb staining dyes, like Nile Red, which are frequently employed to enhance the visual detection of microplastics by giving

Textile Microplastics in the Environment: Sources, Impacts, and Solutions,
First Edition. Aravin Prince Periyasamy.

them color or fluorescence [6]. However, due to the intricate nature of domestic washing, it is practically impossible to exclusively wash white cloths [6].

Incorporating significant amounts of experimental data into the numerical model for microplastic analysis presents challenges due to the limited availability of long-term research on temporal fluctuations in microplastic pollution, as well as the labor-intensive nature of microplastic monitoring methods. The process of collecting, identifying, quantifying, and categorizing microplastic particles is a very labor-intensive stage in the analytical procedure. The use of visual identification in the classification of microplastics, mostly relying on their morphology and dimensions, has significant worth. Nevertheless, it is imperative to integrate chemical verification methods to fully grasp the magnitude of microplastic pollution and pinpoint certain plastic types that require further examination via numerical modeling. Meanwhile, to identify the chemistry of microplastics, it is possible to use vibrational spectroscopy techniques, such as FTIR and Raman spectroscopy. Nevertheless, it is important to acknowledge that these methodologies need thorough and labor-intensive evaluation protocols. The application of reliable computer-vision–based technologies in microplastic research offers substantial benefits by enhancing the efficiency of microplastic particle monitoring. The areas of machine learning and computer vision have become more intertwined, as machine learning provides innovative approaches to tackle issues in computer vision, while computer vision acts as a valuable tool to extract meaningful insights from visual data. The utilization of computer vision techniques and machine learning algorithms holds promise for enabling the automated monitoring and data collection of microplastics in their natural environment. A significant benefit of computer vision is its ability to objectively extract hidden information from visual inputs and accurately assess the geometric characteristics of objects, including eccentricity, sphericity, elongation, and other relevant features, across many applications [8]. To enhance the precision of microplastic generation predictions, it is imperative to develop more accurate models based on the collected microplastics. The utilization of digital technology has shown substantial potential in the realm of microplastic investigation, attracting considerable interest among researchers [9]. Machine learning can be used to design a control system that effectively releases microfibers and identifies their individual properties, such as length and width. Machine vision (MV) technology is currently utilized in the textile sector for inspection purposes because of its rapid, real-time functionality, accuracy, and efficiency. MV inspection equipment supplants human visual capabilities in duties such as surveillance, measurement, and judgment. This technology has several benefits, including the ability to operate without physical contact, consistent performance, dependability, exceptional precision, uninterrupted functionality, optimal efficiency, and adaptability [10]. This chapter examines the application of machine learning (ML) methods for measuring microplastics in the textile origin.

6.2 Machine Learning for Microplastics

Machine learning (ML) is an influential component of artificial intelligence (AI) that can swiftly analyze intricate data, identify characteristics, and accomplish automatic categorization or prediction. ML is a dependable automated approach for accurately detecting and forecasting the toxicity and adsorption capability of microplastics. By integrating hyperspectral imagery with machine learning, the process of identifying microplastics may be expedited. ML models have the capability to forecast pollution caused by microplastics and determine the significant factors that affect toxicity, adsorption capacity, and microbial abundance. This can help alleviate the workload involved in research. Nevertheless, doing research on microplastics has several problems, including the requirement for a substantial level of prior knowledge and a scarcity of available references. Notwithstanding these difficulties, ML has the capacity to enhance the detection and quantification of microplastics. Although ML has promising possibilities, there are still several obstacles in microplastics research regarding the tasks of identifying, predicting, and extracting features. Environmental, textile researchers require a substantial amount of prerequisite information before they can effectively utilize ML technologies. Furthermore, the progress of ML applications, other than identification, is still in its early stages, and there are few sources available for studying microplastics [11]. Figure 6.1 provides a comprehensive explanation of supervised and unsupervised machine learning approaches.

Machine learning algorithms are utilized in many applications, including the avoidance of water pollution, evaluation of water quality, and detection of pollutants. They offer important assistance in the analysis, categorization,

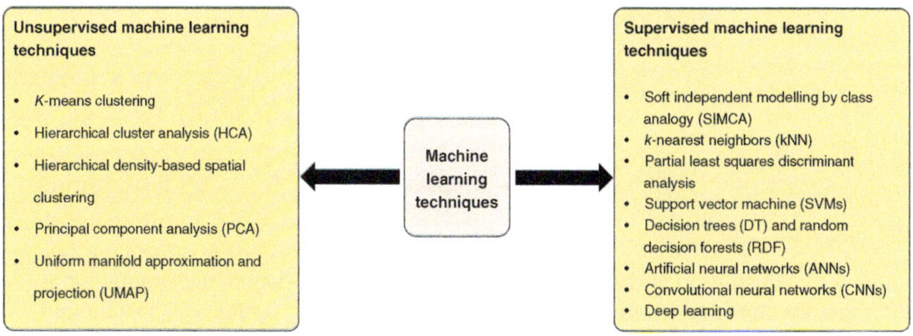

Figure 6.1 Supervised and unsupervised machine learning techniques.

and forecasting of data. Nevertheless, the success of machine learning models cannot be accurately predicted before training. Consequently, researchers must train datasets using various methods concurrently. Machine learning algorithms acquire the ability to make judgments by analyzing and processing training data, much like how students learn from lessons in school. Supervised machine learning uses class labels to train the model, whereas unsupervised machine learning utilizes clustering techniques to group unlabeled data points based on their similarity [12]. Most machine learning models need the inclusion of supplementary parameters, known as hyperparameters, which encompass architecture and training parameters. Optimization of hyperparameters may be necessary through cross-validation [13], as suggested by the pipeline described by Back et al. [14]. Machine learning is a highly effective approach for automatically enhancing the quality of spectra [15]. Federated learning is a machine learning methodology in which several devices or nodes collaboratively train a model while maintaining the localization of their data, hence safeguarding privacy and security. It encompasses localized data, wherein each device or node possesses its own dataset, does local computations, and aggregates models. Locally trained models are regularly sent to a central server, which consolidates them to formulate a global model. The procedure encompasses initialization, local training, model updating, and aggregation. The central server consolidates the changes to enhance the global model. The procedure continues until the model attains acceptable performance. Federated learning has benefits like privacy, efficiency, and scalability; nevertheless, it encounters problems such as communication overhead and heterogeneity, since disparate devices may own differing quantities and types of data, which can impact model performance. The illustration of the concept of local learning, including data and computation at several unconnected sites, is effectively elucidated in Figure 6.2.

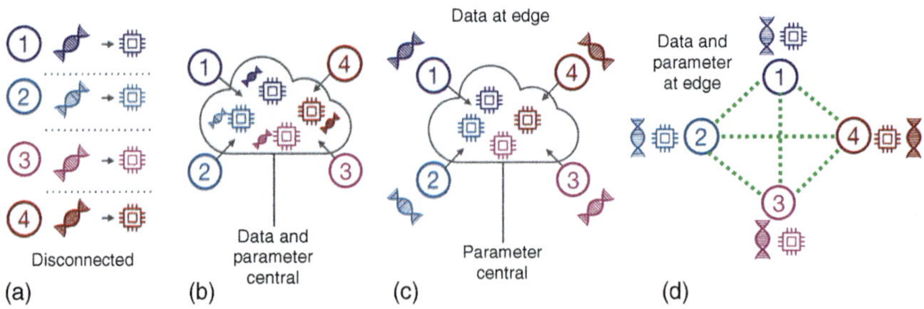

Figure 6.2 Illustration of the concept of local learning with data and computation at different, disconnected locations. (a) Local learning; (b) Central learning; (c) Federated learning; (d) Swarm learning. *Source:* Reproduced with permission from [16] / Springer Nature / CC BY 4.0.

6.2.1 Unsupervised Machine Learning Techniques

Unsupervised ML is a potent method for analyzing information without the need for human experience in spectroscopy. Its primary objective is to discern shared characteristics among distinct subsets of a given dataset. Clustering algorithms are often used as unsupervised machine learning approaches that group data together based on shared characteristics. Their performance is impacted by the selected distance measure. Nevertheless, the intricate nature of microplastic spectra can often hinder the ability to achieve distinct cluster distinction. Some common forms of clustering methods include hierarchical clustering, *k*-means clustering, and principal component analysis (PCA) [17].

6.2.1.1 *k*-Means Clustering

k-Means is an iterative clustering technique that aims to maximize the objective function by partitioning data points into *k* clusters. A higher value of *k* corresponds to smaller groups with greater granularity, whereas a lower value of *k* corresponds to bigger groups with less granularity. The method categorizes data points into *k* groups by establishing a centroid for each category. The locations of these centroids in the data space are continuously optimized through iterative processes, considering the distances between data points and clusters. The performance is greatly impacted by the choice of *k*. The flowchart of *k*-means clustering is well shown in Figure 6.3a. The *k*-means clustering algorithm's stages are delineated in the input. It entails the selection of several clusters, the random selection of cluster centroids, the assignment of data points to the nearest centroid, the updating of centroid positions, and the repetition of these steps until convergence. When all data points are near their respective centroids and no data point switches clusters, the algorithm terminates. Wander et al. [18] used *k*-means clustering on the particle FTIR spectra in an ambient microplastic sample. Their findings showed that the eight groups produced were not able to accurately represent the unique properties of each polymer. Conversely, these clusters represented more general classifications of materials, such as organic fibers (i.e. plant-based cellulosic fibers), synthetic polymers (i.e. polyamide, polyester, polyacrylic, and polypropylene), or animal-derived fibers (i.e. wool and silk). Unfortunately, their dataset lacked the necessary information to differentiate between certain types of polyester, such as polyethylene terephthalate (PET) fiber, polybutylene terephthalate (PBT) fiber and polytrimethylene terephthalate (PTT), using just the *k*-means method [18].

6.2.1.2 Hierarchical Cluster Analysis

Hierarchical cluster analysis (HCA) is a method that constructs a hierarchical structure of clusters, beginning with all data being allocated to individual clusters and culminating in the formation of a single surviving cluster. Hierarchical

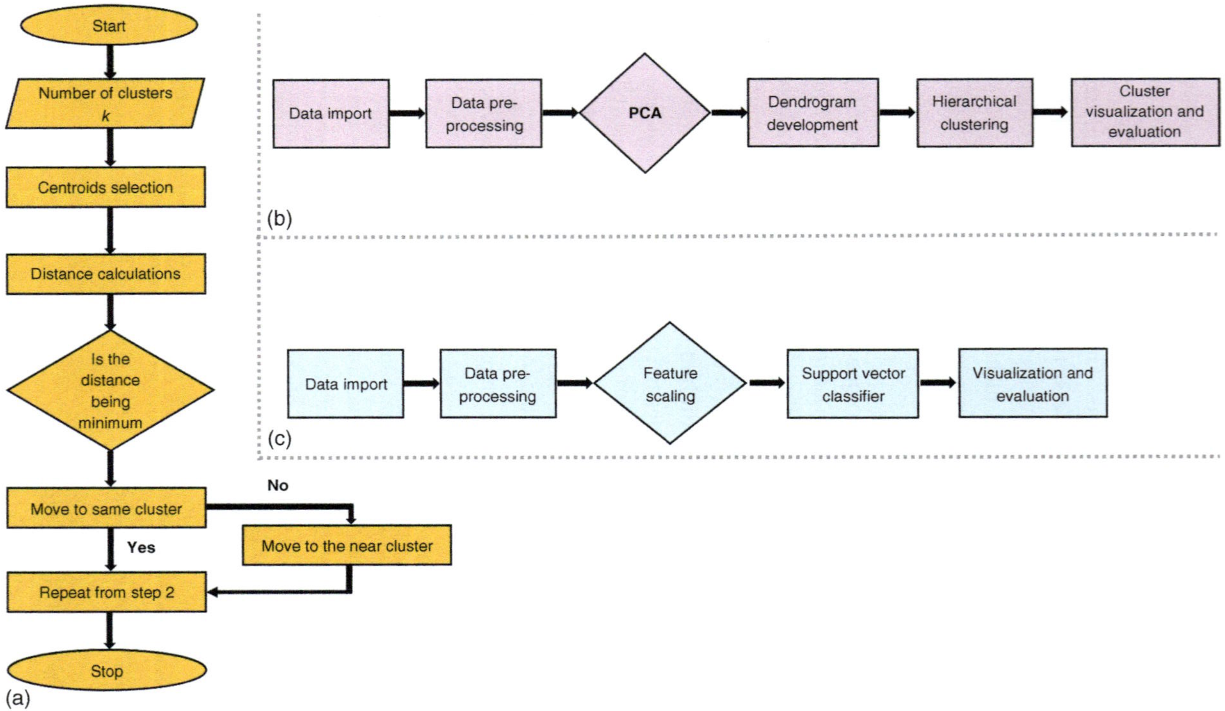

Figure 6.3 Flowchart of *k*-means clustering (a), workflow for hierarchical clustering (b), and workflow for support vector machine (SVM) (c).

clustering algorithm gathers and consolidates comparable data elements. Manual clustering is the preferred method for organizing an FTIR library into clusters since it enables refining depending on the significance of substances and the impact of interference effects. Iterative assessment demonstrates that the level of specificity within clusters is significant. The utilization of a fusion of HCA and PCA may be employed to evaluate classification outcomes and identify any anomalous data points [19]. HCA has been employed to differentiate between cellulosic fibers of plant origin (i.e. cotton, linen, jute) and those that are artificially produced cellulosic fibers like viscose, lyocell, and modal [20]. It can be speculated to use such tools for differentiating the microfibers. However, it is impractical to determine the specific kinds of polymers using HCA (Figure 6.3b).

6.2.1.3 Hierarchical Density-Based Spatial Clustering

Hierarchical density-based spatial clustering of applications with noise clustering, a clustering algorithm, prioritizes the identification of enduring clusters at all levels of the hierarchy, enabling more precise differentiation between classes [21]. Nevertheless, this method may provide a surplus of clusters (i.e. 21 clusters) [18], raising questions over its suitability for distinguishing microplastics, as it leaves more than half of the spectra unassigned [21].

6.2.1.4 Principal Component Analysis

Principal component analysis (PCA) is now widely employed in diverse disciplines such as chemistry and environmental research to decipher Raman spectrum matrices for the purpose of particle, material, or cell identification [22]. PCA is an appropriate method for analyzing microplastics by deciphering their Raman spectra matrix. It has been utilized for the analysis of spectral images to differentiate potential microplastics from background spectra and produce dimensionality-reduced data for machine learning models. In research conducted by Wander et al. [18], they were able to significantly decrease the number of spectra that needed to be analyzed in later stages by 94%. They did this by utilizing PCA to extract potential microplastics μFTIR spectra [23]. Nevertheless, PCA alone is insufficient for accurately categorizing microplastics in real data, since the score plot reveals the presence of clusters [24]. To isolate microplastics from sample matrices using bulk sample spectra and PCA to determine microplastics concentration and types. While the first principal components can reflect microplastics' presence or absence, their applicability to real-world samples with low microplastic contents is questionable. PCA can help optimize method parameters and train ML models faster on PCA-treated training data. It can provide a qualitative overview of datasets, distinguish potential microplastics from filter substrate or matrix spectra, and speed up subsequent analysis steps. However, a full discrimination of multiple plastic types in real-world microplastics data cannot be achieved using PCA alone [14] (see Figure 6.4a).

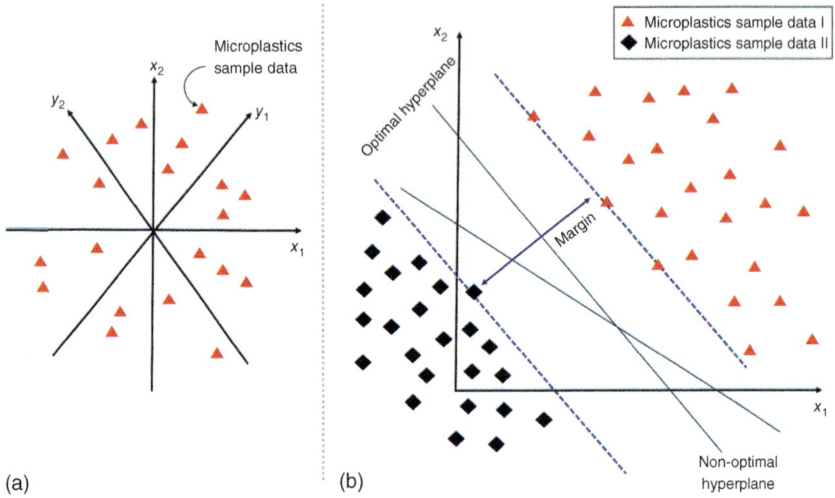

Figure 6.4 The logic flowchart for microplastics identification using principal component analysis (a); and the schemes of a support vector machine (SVM) that separates microplastic sample data I and II (b).

6.2.1.5 Uniform Manifold Approximation and Projection

Uniform manifold approximation and projection (UMAP) is a technique that transforms data from high-dimensional datasets into lower dimensions while maintaining the structure observed in the original higher-dimensional environment. It could identify complex connections among data points [25]. UMAP successfully grouped real-world μFTIR microplastic data according to fiber types (i.e. polymer), and filter substrate with noise or undetermined composition [18]. Wander et al. [18] employed false-color maps to graphically depict the class assignment of individual particles.

6.2.2 Supervised Machine Learning Techniques

ML algorithms are employed to acquire knowledge about characteristics of a given set of data and make predictions about characteristics of unknown data. These algorithms may be classified into two distinct groups: unsupervised learning, which solely relies on input data, and supervised algorithms, which need pairings of inputs and desired outputs. Classification tasks employ machine learning models to forecast results of unseen data, where the distinct output is referred to as a class label and is chosen from a predetermined set of potential classes [26]. Binary classification involves the differentiation of two distinct classes, whereas multiclass classification involves the categorization of more than two classes. Diverse learning methods may be employed to address classification challenges, whereby hyperparameters are defined during the training phase. The choice of a learning algorithm and its hyperparameters is contingent upon the specific characteristics of the data on which it is being applied. Optimal hyperparameter values must be determined through tuning to get the most effective configuration for the learning algorithm with the given dataset. Learning methods such as *k*-nearest neighbors (kNN), support vector machines (SVM), and random decision forests (RDF) are commonly used.

6.2.2.1 Soft Independent Modeling of Class Analogies

Soft independent modeling by class analogy (SIMCA) is a statistical method for supervised data classification. It assigns a sample to a class if its PCA projection is close enough to the class's cluster. A study by Da Silva et al. [27] trained SIMCA models for microplastic classification in μFTIR images and found that *partial least squares discriminant analysis* (PLS-DA) outperformed SIMCA models in terms of sensitivity, sensibility, and misclassification error. PLS-DA was less sensitive to edge effects and poorly focused regions of particles. The proposed method offers an efficient automated approach for microplastic polymer characterization, abundance numeration, and size distribution, with substantial benefits for method standardization.

6.2.2.2 *k*-Nearest Neighbors

The *k*-nearest neighbors (kNN) method is a widely employed supervised learning technique that categorizes unfamiliar samples by comparing them to the closest training examples [28]. The quantity of closest neighbors is a predetermined hyperparameter, and the classification may be determined by conducting a majority vote, where the class that appears most frequently among the nearest neighbors is allocated. The kNN algorithm selects the *k* library items that bear the closest resemblance to a query spectrum. However, it does not include any genuine learning process. When the value of *k*s is big, it is probable that many classes are recognized as near neighbors. The Plastic pOllutionS ExtractIon, DetectiOn and aNalysis (POSEIDON) program, which is based on kNN, was developed to identify microplastics in ATR-FTIR spectra [19]. Chen et al. [29] constructed a kNN database consisting of 475 ATR-FTIR spectra of four categories of microplastics found on beaches. They employed a least trimmed squares estimator as a distance measure. Michel et al. [30] obtained the second-highest level of accuracy by using ATR-FTIR and NIR spectra. Their results were somewhat lower than those reached by linear discriminant analysis, but better than those achieved by an SVM. With the assistance of Figure 6.5, it is easy to understand the kNN classification. Suppose that we have a new data point x_1, and there are two categories: microplastic sample data I and microplastic data II. It is imperative that we ascertain the category to which this data point belongs. The k-NN algorithm can be employed to resolve this issue. By considering the data point's closest companions, k-NN assists in determining its category or class.

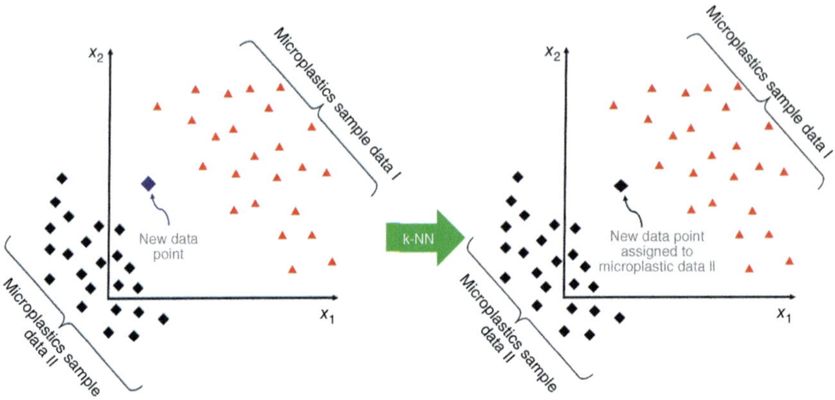

Figure 6.5 Description of *k*-nearest neighbors (k-NN) algorithm.

6.2.2.3 Partial Least Squares Discriminant Analysis

Partial least squares (PLS) regression is a statistical technique used to decrease the number of variables in a dataset by converting them into latent variables. It has the capability to establish connections between input and output data by using linear or nonlinear transformations. PLS regression may provide scores and loading graphs, like PCA [31, 32]. PLS-DA, a discriminant analysis method based on thresholds, may be used to query spectra using PLS [27]. It has effectively been employed to train models for the purpose of recognizing microplastic spectra in sediment samples from μFTIR images [27]. PLS-DA demonstrated a notable level of sensitivity (>95%) and a minimal rate of misclassification error (<1%) when distinguishing between several categories of microplastics. PLS-DA was employed in a separate investigation to categorize large microplastics in short-wave infrared images, resulting in notable accuracy rates for specific microplastic varieties. Serranti et al. [33] applied PLS-DA model to distinguish polyethylene (PE), polystyrene (PS), and polypropylene (PP). The accuracy rates for properly categorizing PE and PP were 99.68% and 97.68%, respectively; however, the accuracy rate for correctly classifying PS was lower at 81.30%. The authors attribute this phenomenon to the sample's low concentration of PS. However, the authenticity of this argument cannot be judged due to insufficient modeling technique documentation, notably the uncertainty surrounding the execution of a thorough segregation of test and training data.

6.2.2.4 Support Vector Machine

Support vector machines (SVMs) are very efficiently supervised learning algorithms used for classification of data. They aim to identify the ideal hyperplane that maximizes the margins between different classes (Figure 6.3c). SVM model, which was trained using PCA preprocessed hyperspectral imaging (HSI) spectra of six plastic references, achieved an average sensitivity of up to 94%. The classification of dark and tiny particles (<1 mm) was shown to be more challenging compared to bigger, light-colored, and colorful particles. Michel et al. [30] demonstrate that ML classifiers successfully identified consumer plastic types using Attenuated total reflectance (ATR)–FTIR, NIR reflectance spectroscopy, laser-induced breakdown spectroscopy (LIBS), and X-ray fluorescence (XRF) spectroscopy techniques with success rates of 99%, 91%, 97%, and 70%, respectively. The success rates of microplastic classification were similar or lower, possibly due to plastic alterations caused by environmental weathering processes. The success rates for ATR-FTIR, NIR reflectance spectroscopy, LIBS, and XRF were 99%, 81%, 76%, and 66% respectively. In another study, SVM has been utilized for the identification of spiked-in microplastic in NIR spectra of seawater samples. The study also found that SVM was the most suitable technique for identifying white PE in soil. The precision rates for detecting PE particles in the size ranges of 1–5 mm and 0.5–1 mm were 84% and 77%, respectively. Bifano et al. [34] demonstrated that the SVM model has the ability to accurately differentiate between the three classes, namely "no plastic," "PP," and "polyolefin." Additionally, it successfully enables the quantitative estimation of microplastics concentration (see Fig. 6.4b). The detection of microplastics in FTIR images using a two-step approach. An SVM model was trained in pre-processed spectra to differentiate between microplastic and matrix spectrum. Additionally, a set of binary classifiers was trained using a one-versus-all strategy to distinguish between different microplastic classes and it shows the accuracy of 95% raises concerns about its real-world applicability due to the limited size of the test set and the lack of differentiation between training and test data [35]. de Medeiros Back et al. [14] found that SVM models exhibit high accuracy and simplicity, requiring minimal tuning of hyperparameters.

6.2.2.5 Decision Trees and Random Decision Forests (RDF)

Decision Trees (DT) are a classification algorithm that utilizes training data to build base classifiers, allowing machine learning models to make predictions about target variable values. As the sample size increases, decision trees tend to produce complex and extensive trees, which can result in overfitting [36]. RDF models have demonstrated strong performance in the classification of microplastics in FTIR images, achieving sensitivity between 94% and 100% [37]. FTIR analysis of plankton, sediment, compost, and sea salt samples successfully identified microplastics. RDF models achieve a 95.45% accuracy in distinguishing 17 classes of synthetic and natural polymers, including background signal [38]. ML models are not infallible due to data variance. A recent study demonstrated the successful classification of Raman spectra using RDF, with no classification errors observed in a limited test set. Although a larger test set would increase

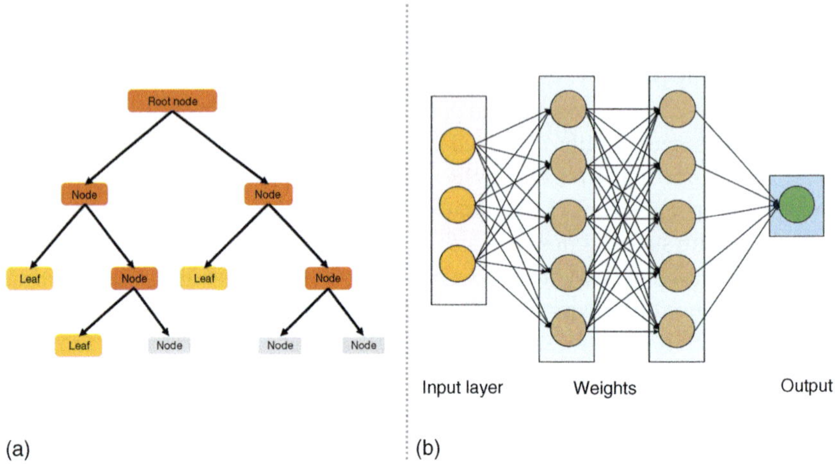

Figure 6.6 The decision tree, orange and yellow boxes represent the features and classes, respectively (a); artificial neural network passing information through weighted connections between neurons (b).

the statistical significance of the results, the study demonstrates the suitability of RDF models for analyzing Raman spectra [39]. Figure 6.6a shows the decision tree processing modules in relation to microplastics identification.

6.2.2.6 Artificial Neural Networks

Artificial neural networks (ANNs) play a significant role in environmental engineering, specifically in drainage systems, and rainfall and runoff prediction. ANNs imitate the functioning of biological neural systems by transmitting information among interconnected neurons to generate predictions regarding class membership. Training ANNs can be computationally expensive and prone to overfitting. Neural networks frequently exhibit overfitting behavior, resulting in a lack of generalizability in their predictions. To mitigate this issue, it is necessary to employ a carefully designed neural architecture, conduct thorough hyperparameter selection, and rigorously validate the model using independent datasets [40]. The checklist has undergone a thorough review and is essential for precise prediction. Convolutional neural networks (CNNs) are designed for image analysis and operate by processing input data in small pixel patches, resulting in improved efficiency. CNNs have demonstrated their utility in various tasks within a given domain after their initial training [41]. Back et al. [14] compared seven ML algorithms for classifying ATR-FTIR spectra and found that ANNs were among the top models. However, due to the significant computational resources needed, they opted to proceed with an SVM model for further investigation. CNN usage for microplastic detection in VIS/NIR spectra of bulk soil samples resulted in a moderate accuracy of 78.5%. Further, an autoencoding neural net was recently shown to substantially improve spectral quality to ease subsequent classification steps [15]. CNNs show greater specificity and sensitivity than traditional methods like multilayer perception, due to shared weight filters and pooling layers. They have analyzed Raman and surface-enhanced Raman scattering (SERS) data, acquiring up to nine molecular gradients simultaneously. For mixture analysis in Raman spectroscopy, CNNs offer better sensitivity and specificity compared to other techniques. They identify tongue squamous cell carcinoma using vibrational signatures of healthy and cancerous cells. CNNs are effective for raw SERS spectra pattern recognition and component identification through transfer learning, creating an open-access spectral library and accelerating training [42]. To achieve this task, artificial neurons compute the sum of a variable number of weighted inputs and then apply a nonlinear function to generate an output. By grouping and connecting these neurons together, neural networks are created (see Figure 6.6b). The application of segmentation techniques is essential for obtaining statistically relevant data on the dimensions, notably length and diameter, of microfibers. This data is derived from images produced by an MV camera. The segmentation algorithms utilized must efficiently differentiate the microfibers from the surrounding backdrop and other objects in the picture. Automated methods frequently utilized in medical image analysis for recognizing and measuring cells are also relevant for detecting microfibers. The method has promise for the detection and measurement of microfibers in future efforts [43]. Nevertheless, these techniques frequently face difficulties in precisely classifying anomalous object patterns and environmental noise [44]. Moreover, several algorithms depend on human-crafted feature selection or form assumptions, thereby limiting their use to the identification of unique items with certain features amidst other contrasting objects in an image [43].

James et al. [45] utilized machine learning methodologies to identify particles through the scattering patterns of microplastics. Numerous developments in neural networks have enabled effective object classification and diffraction imaging. The authors present a sensing method indicating that a neural network can quantify the substance and number of particles from a single scattering pattern, contingent upon receiving suitable training data (as seen in Figure 6.7a). Figure 6.7b depicts the neural network utilized for microspheres and real-time assessments. Before the training data was introduced to the neural network, the photos were preprocessed by cropping them to dimensions of 112 by 112 pixels and normalizing their values to a range of 0–1. After the input layer, both the microspheres neural network and the real-time neural network experience convolutions. The microspheres neural network does 64 convolutions of dimensions 112 by 112, whereas the real-time neural network executes 32 convolutions of dimensions 56 by 56. Both networks perform convolutions of dimensions 112 by

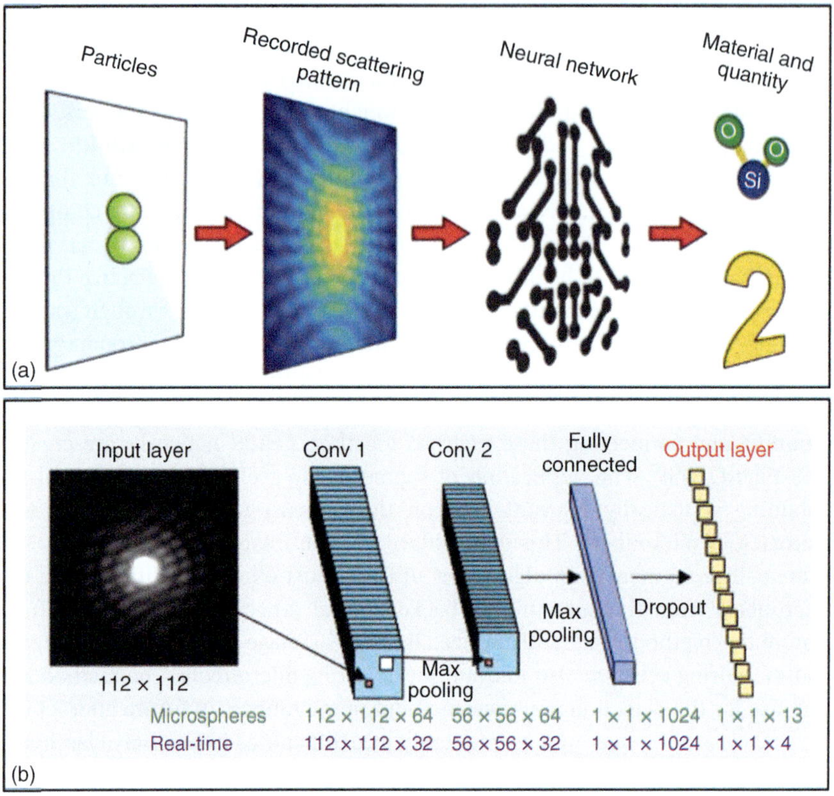

Figure 6.7 Proposed application of a NN to quantify an object directly from the scattering pattern (a); diagram of the NN used for microspheres and real-time experiments (b). (*Source:* Reproduced with permission from [45] / The Optical Society / CC BY 4.0.)

112 and 56 by 56, respectively. The microspheres and real-time neural network layers have kernel sizes of 3 and 15, with a pooling factor of 2 and a stride of 2. The output from the second layer was sent to a layer consisting of 1024 completely linked neurons [45].

6.2.2.7 Deep Learning

Deep learning is a valuable tool for multidisciplinary research on microplastics, necessitating collaboration across diverse fields [46]. Navarro et al. [47] proposed a hybrid approach that integrates U-Net and VGG16 neural networks to effectively count and classify common microplastics in controlled laboratory settings. Navarro et al. [47] employed deep learning architectures to classify microplastic samples from images captured without specialized acquisition equipment. Their findings demonstrate promising results, exhibiting comparable or even superior accuracy when compared to alternative approaches based on traditional computer vision techniques or manual inspection. This research presents a hybrid methodology that integrates bounding box and instance segmentation approaches for the examination of microplastics [47]. The procedure comprises two stages: segmentation and categorization of microplastics. The preliminary phase of instance segmentation is recognizing particles within an image sample, with particular focus on differentiating between particles of various classes and the background. This facilitates the acquisition of particle-associated pixels, which are crucial for microplastic research. In the second step, a refined CNN is utilized to categorize each particle [47].

6.2.2.8 Image and AI or ML

Vitali et al. [48] conducted a critical assessment to quantify microplastics in bottled water using artificial intelligence. This evaluation was based on Nile Red staining, fluorescence microscopy, and automated image processing for image analysis. A Random Forest classifier was trained to provide a strong semantic segmentation of fluorescence pictures of Nile Red–stained microplastics, which is crucial for accurate automated microplastic quantification. An evaluation was conducted to determine the precision of the technique in measuring and quantifying microplastics. Reproducibility refers to the capacity to repeat a process or experiment with consistent results. Inside-laboratory reproducibility specifically pertains to the consistency of results obtained inside the same laboratory. An efficient, high-capacity, and straightforward method for quantitatively analyzing microplastics is already accessible for further advancement in the realm of microplastic analysis in food.

6.3 Challenges

Unfortunately, machine learning in microplastic analysis and classification necessitates supervised training on extensive datasets, and many of these methodologies may exhibit suboptimal performance on real-world images that significantly differ from the training data, particularly when attempting to detect and classify diminutive and low-resolution objects against a vast, uncorrelated background.

6.3.1 Cameras and Their Sensibility

Cameras are employed to capture photos of microfibers in domestic laundering under diverse settings, including detergent type, water quality, turbidity, and foam. Camera power calibration for enhanced microfiber recognition relies on resolution, pixel rate, and frame rate. Camera sensitivity is contingent upon the intensity of light from the source being captured, and it may be adjusted in a controlled setting. Consequently, a camera with enhanced sensitivity and reassessed color sensitivity is essential. The picture processing software is utilized for algorithmic applications, picture segmentation, and feature extraction. Image segmentation delineates objects for quantification or verification, removing background or residual interference. Feature extraction uses image processing software to get measures like the existence, diameter, and length of microfibers, subsequently leading to authentication and recognition. Images are assessed by several methodologies, and fiber edges are calculated to ascertain their length, breadth, and additional properties. Microplastics are differentiated from non-synthetic microparticles using visual classification, which is constrained in its capacity to precisely define polymer structures. It can ascertain the presence or absence of plastic components inside a particle [49]. The categorization of microfibers cannot rely exclusively on visual assessment, as this approach is not deemed a valid method. Digital discoveries need input data; nevertheless, this procedure may lead to discrepancies in results and may not be appropriate for sharing data throughout the full supply chain [50]. The fluorescence of fluorescent microparticles is affected by the luminous emittance of an LED lantern; hence, a UV LED light was continuously operated for detection and photography. Errors may arise from temperature, camera sensor, visible light, contrast, focus, and camera stabilization [51]. To guarantee dependable outcomes, the area of the filter membrane in the final image is optimized. A regulated dark-room setting was established to reduce temperature fluctuations and visible light interference. Filters were photographed using a high-definition SLR camera at a 45° angle, utilizing UV light, in a controlled environment.

6.3.2 Influence of Textile Functionalization on Image Capturing and Processing

The coloring process, including dyeing and printing, is an essential element of the textile value chain. It significantly influences the identification and distinction of microfibers and their sources. Diverse categories of dyes are employed in coloring procedures, including reactive, direct, and acid dyes, which can penetrate the fiber structure [52, 53]. Moreover, pigments and vat dyes are utilized, demonstrating physical adsorption on the fiber's surface. Fibers like silk and viscose rayon have a glossy characteristic that might influence picture capture and subsequent image processing [54, 55]. Some fibers exhibit distinctive properties, including the incorporation of photochromic and thermochromic pigments, which greatly influence their identification [56, 57]. Furthermore, it is customary to utilize optical brighteners, referred to as optical brightening agents (OBAs), fluorescent brightening agents (FBAs), or fluorescent whitening agents (FWAs), to augment the brightness and whiteness of white fabrics [52]. Moreover, FWAs are often used in domestic laundry detergents to improve the color and brightness attributes of fabrics post-wash. Finishing procedures like flame retardancy and water repellency modify the surface qualities of a material, significantly influencing the picture capture process. The elements significantly impact on the picture capture process, followed by the quantitative analysis of microfibers.

6.3.3 Reference Libraries and Matching Software

Reference libraries facilitate the expeditious acquisition of information pertaining to the morphological characteristics, surface topographies, and chromatic properties of microfibers, thereby enabling the validation of their classification. The proper functioning of any software for spectral matching is highly dependent on its reference libraries [58]. At present, there exists no established repository of resources that can be utilized for the discernment of microfibers. Nonetheless, there exist certain libraries, such as commercial libraries like the National Institute of Standards and Technology, KnowItAll by Bio-Rad, and Polymer Spectral Libraries by Thermo Fisher Scientific. The software systematic identification of microplastics in the environment integrates two pipelines [59], and microplastic automated particle/fiber analysis (MPAPP), for analyzing FTIR images [60]. The Open Specy platform is available for access both online at www.openspecy.org and through an R package. The user can preprocess Raman or FTIR spectra before comparing them to on-board databases using Pearson's correlation. To facilitate the creation of varied and accurate databases specifically designed for microplastic analysis, users have the option to contribute their own spectra to be included in the databases [61]. The POSEIDON program uses a kNN model to detect ATR-FTIR spectra of marine microplastics, but not yet it developed for microplastics from the textile laundry process (i.e. pre-stage). It uses unspoiled and aged spectrum from prevalent MP categories and uses an iterative procedure to determine the number of neighbors. The program can match up to 50,000 spectra in a single operation [19].

However, these libraries pertain to generalized domains such as microplastics and do not cater to microfibers. The exorbitant expenses associated with these libraries render them accessible solely to individuals or entities with considerable financial resources, such as governmental or corporate organizations. Environmental particles often contain additives such as dyes and plasticizers and are susceptible to weathering and/or degradation. In contrast, commercial libraries are typically composed predominantly of spectra derived from reference materials of pure polymers. As of present, the dissemination of plastic libraries accessible to the public remains restricted, with only a few examples in existence, including a library designed for employment in Raman spectroscopy [62] and another intended for employment in IR. Moreover, in the context of visual classification, the available reference images are frequently limited in both quantity and variety, and their acquisition from scholarly sources can pose a challenge. Numerous publications have been released within the microplastic field that could prove advantageous in identifying microfibers. Despite the availability of polymer libraries for precise material categorization, challenges persist in the process of matching. Microfibers originating from textile sources exist in various forms and comprise diverse constituents such as dyes, printing agents, finishing agents such as phosphorous (for flame retardancy), fluorocarbon and silane (for water repellency), formaldehyde (for crease recovery), and an assortment of nanoparticles. The spectra of a particular polymer type may exhibit significant variations depending on the specific dyes, printing, and finishing chemicals that are incorporated into the fiber structure. The spectra obtained from multiple locations on identical microfibers may exhibit significant variations. Moreover, polymers derived from environmental specimens are often exposed to various conditions such as photodegradation, thermal degradation, and biodegradation, which may result in leaching or denaturation of chemical constituents. This, in turn, can alter the initial polymer composition and its chemical signature [63].

Several software programs are used for the analysis of microplastic pollution in spectroscopy data. The simple software, compatible with techniques including single-point FTIR, Raman, ATR-FTIR, and μFTIR imaging, utilizes on-board and adaptable databases for weighted Pearson correlation matching. It groups neighboring same-substance pixels and generates spectral data in formats such as .dmd, .fsm, .dx, .jdx, and .txt. Its performance offers over 90% accuracy in real-world environmental samples. The MPAPP software processes false-color images for particle/fiber analysis. Open Specy, compatible with single-point μFTIR, ATR-FTIR, and Raman, also uses on-board databases for Pearson correlation matching, providing similar performance to commercial databases and supporting the addition of shared spectra to its FTIR and Raman databases. The μMAP software, designed for single-point μFTIR, uses a weighted bivariate logistic distance for database matching, analyzing optical images with an online spectrum monitor. Its performance ensures maximum accuracy and minimum false negatives in particle selection. TUM-ParticleTyper evaluates optical images from FTIR, Raman, EDX, and fluorescence microscopes, providing high accuracy and comparable performance to manual evaluation and established routines. GEPARD, compatible with single-point μFTIR and Raman, analyzes optical images and controls measurement points through CSV format coordinates, showing qualitative assessment of false-color images. POSEIDON, designed for single ATR-FTIR spectra, utilizes a kNN model for database matching and provides high sensitivity for specific polymer classes. PlaMAPP, using SWIR images, relies on on-board databases matching with scaled matching penalties, offering substantial accuracy but requiring further clarity on its development specifics. Lastly, uFTIR, tailored for μFTIR-imaging, employs the Spectral Angle

Mapper for database matching and supports scaling, differentiation, and clipping of images. Its performance includes high accuracy in the analysis of spiked soil samples and comparison with simple results.

6.3.4 Data Sharing and Storage with the Supply Chain

In contemporary times, the storage and dissemination of data have become crucial in the highly digitalized world and are now considered indispensable elements of efficient supply chain management methodologies. The safety and availability of data storage, specifically servers, is a crucial consideration. Insufficient data backup procedures can lead to data loss, which can be further compounded by increased complexity. To mitigate the release of microfibers, it is imperative to openly distribute quantified microfiber data across the various stages of the textile value chain. This will improve the overall performance of the textile supply chain and promote sustainability. In the time to come, the utilization of machine learning for microfiber classification will witness the adoption of various techniques and data fusion methods, along with the implementation of advanced machine learning methodologies. Additionally, the availability of a comprehensive reference library will further enhance the efficacy of machine learning-based quantification and data sharing for microfiber identification. This will enable more precise microfiber identifications.

6.4 Conclusion and Future Scope

The utilization of ML and MV for the identification, categorization, and analysis of microplastics represents a notable progression in environmental monitoring and pollution control. This chapter highlighted that unsupervised learning methods, including *k*-means clustering, hierarchical clustering, and PCA, and provide effective tools for uncovering concealed patterns in unlabeled data, facilitating the first categorization and segmentation of microplastic particles. Concurrently, supervised learning models such as SVM, kNN, decision trees, CNNs, deep learning, and ANN exhibit high accuracy in classifying polymer types, identifying particle shapes, and predicting degradation status when trained on well-labeled datasets. Although these technologies provide significant potential for enhancing the speed, precision, and automation of microplastic analysis, they also face substantial hurdles. A significant restriction is the sensitivity and resolution of imaging instruments (i.e. camera), which directly affect model performance, particularly in identifying tiny particles. Inconsistent illumination, inadequate camera calibration, and fluctuating magnifications exacerbate image analysis challenges. The deficiency of high-quality reference libraries and consistent datasets constitutes a significant obstacle, impeding the repeatability and validation of models across various research. Moreover, data-sharing limitations among institutions arising from proprietary formats, inadequate infrastructure, or apprehensions over intellectual property impede collaborative progress and the cross-validation of AI tools.

A universal database is essential due to the data-driven nature of machine learning. Consequently, extensive microplastic data and comprehensive databases are essential for large-scale microplastic research utilizing machine learning. The existing quantity and scope of microplastic datasets significantly trail the scientific and decision-making requirements for machine learning. The existing regional disparity in microplastic studies has resulted in inadequate evidence for forecasting the worldwide distribution pattern of microplastics. Comprehensive initiatives for worldwide research are urgently required to gather reliable and comparative data. Extensive strategies or programs would also diminish the quantity of rudimentary and repeated investigations.

To fully leverage the potential of machine learning and machine vision in microplastic research, it is essential to promote interdisciplinary collaboration across data science, environmental science, and materials science fields. Investments in high-resolution imaging technology, the production of open-source annotated datasets, and the implementation of defined techniques for image capture and preprocessing are crucial advancements. Furthermore, establishing platforms for secure and accessible data exchange can expedite innovation and facilitate the large-scale implementation of AI technologies. In summary, machine learning and machine vision have the capacity to significantly mitigate the worldwide issue of microplastic pollution; nevertheless, achieving this potential necessitates overcoming technological constraints, enhancing data infrastructures, and fostering transparency in scientific inquiry. Through deliberate development, these technologies can become essential to the next generation of sustainable environmental monitoring and textile lifecycle management.

6.4.1 Future Scope

Presently, there are few methodologies for assessing microplastic emissions in the textile sector, including the DIN SPEC 4872 Test Method and the AATCC Test Method 212-2021. Nevertheless, these findings mostly pertain to the final product and are not disseminated across the full supply chain. This results from manual data entry, human mistakes, and redundancy difficulties. The textile value chains encompass the procurement of fibers from one nation, the production of yarn and fabric in a different nation, and the assembly in yet another one. In extended supply chains, input on microfiber release is not sent to the raw material processor, which might improve the qualities of raw materials to mitigate microfiber emissions. A closed-loop system for data exchange in the textile manufacturing chain is required to resolve this issue (see Fig. 6.8).

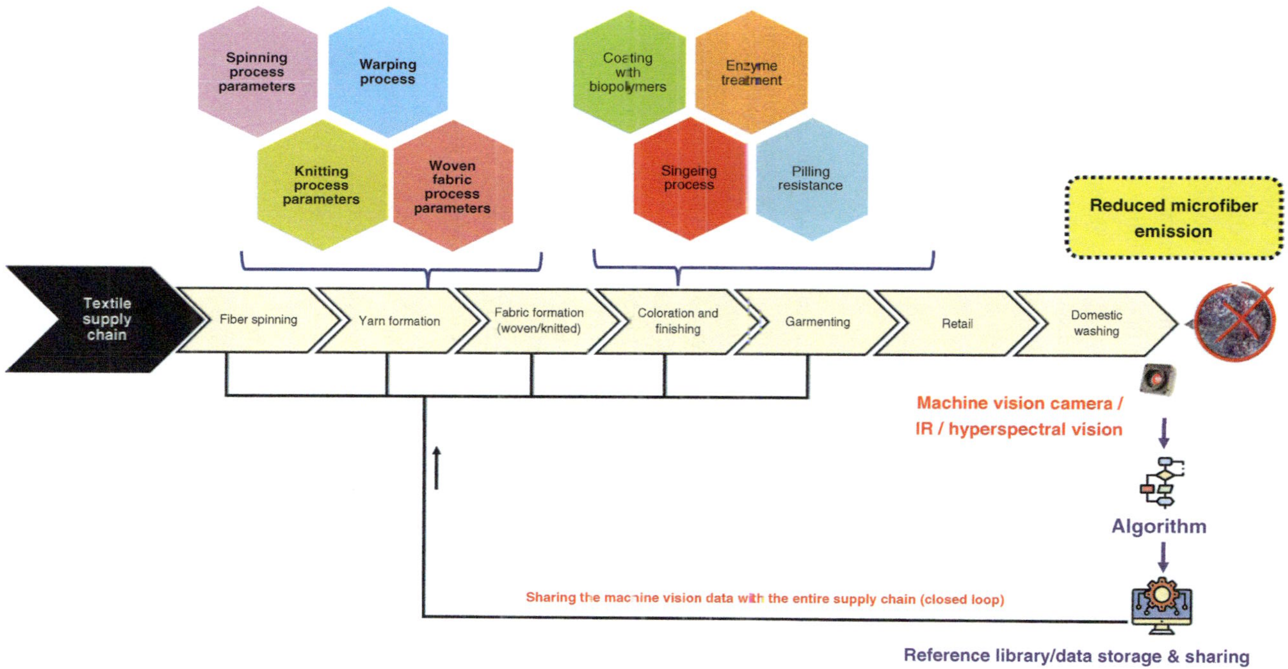

Figure 6.8 Strategy for the integration of machine vision inside the textile production process. (*Source:* Reproduced with permission from [64] / MDPI / CC BY 4.0.)

This system relies on machine vision data gathered during domestic laundering. This technology can create data to identify essential operations and enhance supply chains. MV technology may be utilized in the textile sector for inspection, delivering rapid, real-time, precise, and efficient outcomes. It substitutes the human eye for surveillance, measurement, and assessment, providing benefits such as noncontact operation, repeatability, dependability, high precision, continuity, efficiency, and adaptability. The camera in this system captures real-time fabric imagery and transmits it to the backend system. Upon concluding the deep learning process, the system may ascertain whether the fabric illustrated in the present image exhibits a defect and if a corresponding corrective action should be executed for that defect.

References

1 Strungaru, S.-A., Jijie, R., Nicoara, M. et al. (2019). Micro- (nano) plastics in freshwater ecosystems: abundance, toxicological impact and quantification methodology. *TrAC Trends in Analytical Chemistry* 110: 116–128.

2 Xu, J.-L., Thomas, K.V., Luo, Z., and Gowen, A.A. (2019). FTIR and Raman imaging for microplastics analysis: state of the art, challenges and prospects. *TrAC Trends in Analytical Chemistry* 119: 115629.

3 Sobhani, Z., Al Amin, M., Naidu, R. et al. (2019). Identification and visualisation of microplastics by Raman mapping. *Analytica Chimica Acta* 1077: 191–199.

4 Wang, Z.-M., Wagner, J., Ghosal, S. et al. (2017). SEM/EDS and optical microscopy analyses of microplastics in ocean trawl and fish guts. *Science of the Total Environment* 603–604: 616–626.

5 Phan, S., Padilla-Gamiño, J.L., and Luscombe, C.K. (2022). The effect of weathering environments on microplastic chemical identification with Raman and IR spectroscopy: part I. Polyethylene and polypropylene. *Polymer Testing* 116: 107752.

6 Prata, J.C., Reis, V., Matos, J.T.V. et al. (2019). A new approach for routine quantification of microplastics using Nile red and automated software (MP-VAT). *Science of the Total Environment* 690: 1277–1283.

7 von der Esch, E., Kohles, A.J., Anger, P.M. et al. (2020). TUM-ParticleTyper: a detection and quantification tool for automated analysis of (microplastic) particles and fibers. *PLoS One* 15 (6): e0234766.

8 Zakaria, M.F., Choon, H.S., and Suandi, S.A. (2012). Object shape recognition in image for machine vision application. *International Journal of Computer Theory and Engineering* 4 (1): 76–80.

9 Li, J., Liu, Y., Gao, Y. et al. (2023). Study on the extraction method of microplastic system in textile wastewater. *Polymers (Basel)* 15 (6): 1394.
10 Norton-Wayne, L., Bradshaw, M., and Jewell, A.J. (1992) Machine vision inspection of web textile fabric, in BMVC92, Springer London, London, pp. 217–226.
11 Weisser, J., Pohl, T., Heinzinger, M. et al. (2022). The identification of microplastics based on vibrational spectroscopy data – a critical review of data analysis routines. *TrAC Trends in Analytical Chemistry* 148: 116535.
12 Greener, J.G., Kandathil, S.M., Moffat, L., and Jones, D.T. (2022). A guide to machine learning for biologists. *Nature Reviews. Molecular Cell Biology* 23 (1): 40–55.
13 Xu, Q.-S. and Liang, Y.-Z. (2001). Monte Carlo cross validation. *Chemometrics and Intelligent Laboratory Systems* 56 (1): 1–11.
14 de Medeiros Back, H., Vargas Junior, E.C., Alarcon, O.E., and Pottmaier, D. (2022). Training and evaluating machine learning algorithms for ocean microplastics classification through vibrational spectroscopy. *Chemosphere* 287: 131903.
15 Brandt, J., Mattsson, K., and Hassellöv, M. (2021). Deep learning for reconstructing low-quality FTIR and raman spectra – a case study in microplastic analyses. *Analytical Chemistry* 93 (49): 16360–16368.
16 Warnat Herresthal, S. et al. (2021). Swarm learning for decentralized and confidential clinical machine learning. *Nature* 594 (7862): 265–270.
17 Abbas, K.A., Gharavi, A., Hindi, N.A. et al. (2023). Unsupervised machine learning technique for classifying production zones in unconventional reservoirs. *International Journal of Intelligent Networks* 4: 29–37.
18 Wander, L., Vianello, A., Vollertsen, J. et al. (2020). Exploratory analysis of hyperspectral FTIR data obtained from environmental microplastics samples. *Analytical Methods* 12 (6): 781–791.
19 Kedzierski, M., Falcou-Préfol, M., Kerros, M.E. et al. (2019). A machine learning algorithm for high throughput identification of FTIR spectra: application on microplastics collected in the Mediterranean Sea. *Chemosphere* 234: 242–251.
20 Comnea-Stancu, I.R., Wieland, K., Ramer, G. et al. (2017). On the identification of rayon/viscose as a major fraction of microplastics in the marine environment: discrimination between natural and manmade cellulosic fibers using Fourier transform infrared spectroscopy. *Applied Spectroscopy* 71 (5): 939–950.
21 Campello, R.J.G.B., Moulavi, D., Zimek, A., and Sander, J. (2015). Hierarchical density estimates for data clustering, visualization, and outlier detection. *ACM Transactions on Knowledge Discovery from Data* 10 (1): 1–51.

22 Laptenok, S.P., Martin, C., Genchi, L. et al. (2020). Stimulated Raman microspectroscopy as a new method to classify microfibers from environmental samples. *Environmental Pollution* 267: 115640.

23 Gál, L., Oravec, M., Kiššová, M. et al. (2020). Forensic discrimination of black laser prints by a combination of chemometric methods and μ-ATR-FTIR spectroscopy. *Chemical Papers* 74 (10): 3269–3277.

24 Meng, X., Chen, S., Li, D. et al. (2023). Identification of marine microplastics based on laser-induced fluorescence and principal component analysis. *Journal of Hazardous Materials* 465: 133352.

25 Vermeulen, M., Smith, K., Eremin, K. et al. (2021). Application of uniform manifold approximation and projection (UMAP) in spectral imaging of artworks. *Spectrochimica Acta. Part A, Molecular and Biomolecular Spectroscopy* 252: 119547.

26 Lin, J., Liu, H., and Zhang, J. (2022). Recent advances in the application of machine learning methods to improve identification of the microplastics in environment. *Chemosphere* 307: 136092.

27 da Silva, V.H., Murphy, F., Amigo, J.M. et al. (2020). Classification and quantification of microplastics (<100 μm) using a focal plane array–fourier transform infrared imaging system and machine learning. *Analytical Chemistry* 92 (20): 13724–13733.

28 Ertel, W. (2017). Machine learning and data mining. In: *Introduction to Artificial Intelligence*, 175–243. Cham: Springer International Publishing.

29 Chen, X., Zhou, J., Yuan, L.-M. et al. (2021). Spectroscopic identification of environmental microplastics. *IEEE Access* 9: 47615–47620.

30 Michel, A.P.M., Morrison, A.E., Preston, V.L. et al. (2020). Rapid identification of marine plastic debris via spectroscopic techniques and machine learning classifiers. *Environmental Science & Technology* 54 (17): 10630–10637.

31 Lee, L.C., Liong, C.-Y., and Jemain, A.A. (2018). Partial least squares-discriminant analysis (PLS-DA) for classification of high-dimensional (HD) data: a review of contemporary practice strategies and knowledge gaps. *Analyst* 143 (15): 3526–3539.

32 Wold, S., Sjöström, M., and Eriksson, L. (2001). PLS-regression: a basic tool of chemometrics. *Chemometrics and Intelligent Laboratory Systems* 58 (2): 109–130.

33 Serranti, S., Palmieri, R., Bonifazi, G., and Cózar, A. (2018). Characterization of microplastic litter from oceans by an innovative approach based on hyperspectral imaging. *Waste Management* 76: 117–125.

34 Bifano, L., Meiler, V., Peter, R., and Fischerauer, G. (2022) Detection of microplastics in water using electrical impedance spectroscopy and support vector machines. Sensors and Measuring Systems; 21th ITG/GMA-Symposium, 1–4.

35 Xu, J.-L., Hassellöv, M., Yu, K., and Gowen, A.A. (2020). Microplastic characterization by infrared spectroscopy. In: *Handbook of Microplastics in the Environment*, 1–33. Cham: Springer International Publishing.

36 Deepnarain, N., Nasr, M., Kumari, S. et al. (2019). Decision tree for identification and prediction of filamentous bulking at full-scale activated sludge wastewater treatment plant. *Process Safety and Environmental Protection* 126: 25–34.

37 Hufnagl, B., Stibi, M., Martirosyan, H. et al. (2022). Computer-assisted analysis of microplastics in environmental samples based on μFTIR imaging in combination with machine learning. *Environmental Science & Technology Letters* 9 (1): 90–95.

38 Weisser, J., Beer, I., Hufnagl, B. et al. (2021). From the well to the bottle: identifying sources of microplastics in mineral water. *Water (Basel)* 13 (6): 841.

39 Vinay Kumar, B.N., Löschel, L.A., Imhof, H.K. et al. (2021). Analysis of microplastics of a broad size range in commercially important mussels by combining FTIR and Raman spectroscopy approaches. *Environmental Pollution* 269: 116147.

40 Yang, J., Xu, J., Zhang, X. et al. (2019). Deep learning for vibrational spectral analysis: recent progress and a practical guide. *Analytica Chimica Acta* 1081: 6–17.

41 Ng, W., Minasny, B., and McBratney, A. (2020). Convolutional neural network for soil microplastic contamination screening using infrared spectroscopy. *Science of the Total Environment* 702: 134723.

42 Fan, X., Ming, W., Zeng, H. et al. (2019). Deep learning-based component identification for the Raman spectra of mixtures. *Analyst* 144 (5): 1789–1798.

43 Meng, Y., Zhang, Z., Yin, H., and Ma, T. (2018). Automatic detection of particle size distribution by image analysis based on local adaptive canny edge detection and modified circular Hough transform. *Micron* 106: 34–41.

44 Riccio, D., Brancati, N., Frucci, M., and Gragnaniello, D. (2019). A new unsupervised approach for segmenting and counting cells in high-throughput microscopy image sets. *IEEE Journal of Biomedical and Health Informatics* 23 (1): 437–448.

45 Grant-Jacob, J.A., Mackay, B.S., Baker, J.A.G. et al. (2018). Real-time particle pollution sensing using machine learning. *Optics Express* 26 (21): 27237.

46 Montesinos López, O.A., Montesinos López, A., and Crossa, J. (2022). Fundamentals of artificial neural networks and deep learning. In: *Multivariate Statistical Machine Learning Methods for Genomic Prediction*, 379–425. Cham: Springer International Publishing.

47 Lorenzo-Navarro, J., Castrillón-Santana, M., Sánchez-Nielsen, E. et al. (2021). Deep learning approach for automatic microplastics counting and classification. *Science of the Total Environment* 765: 142728.

48 Vitali, C., Peters, R.J.B., Janssen, H.-G. et al. (2024). Quantitative image analysis of microplastics in bottled water using artificial intelligence. *Talanta* 266: 124965.

49 Lusher, A.L., Bråte, I.L.N., Munno, K. et al. (2020). Is it or isn't it: the importance of visual classification in microplastic characterization. *Applied Spectroscopy* 74 (9): 1139–1153.

50 Isobe, A. et al. (2019). An interlaboratory comparison exercise for the determination of microplastics in standard sample bottles. *Marine Pollution Bulletin* 146: 831–837.

51 Prata, J.C., Alves, J.R., da Costa, J.P. et al. (2020). Major factors influencing the quantification of Nile red stained microplastics and improved automatic quantification (MP-VAT 2.0). *Science of the Total Environment* 719: 137498.

52 Periyasamy, A.P. (2023). Microfiber emissions from functionalized textiles: potential threat for human health and environmental risks. *Toxics* 11 (5): 406.

53 Periyasamy, A.P. (2025). Functionalized textile microplastics: a closer look at the issues, strategy, and legislation on the microplastic reduction. *Kuwait Journal of Science* 52 (3): 100395.

54 Venkatesan, H. and Periyasamy, A.P. (2019). Eco-fibers in the textile industry. In: *Handbook of Ecomaterials*, vol. 3, 1413–1433. Cham: Springer International Publishing.

55 Periyasamy, A.P. and Militky, J. (2020). Sustainability in regenerated textile fibers. In: *Sustainability in the Textile and Apparel Industries*, 63–95. Cham: Springer International Publishing.

56 Periyasamy, A.P., Vikova, M., and Vik, M. (2017). A review of photochromism in textiles and its measurement. *Textile Progress* 49 (2): 53–136.

57 Periyasamy, A.P., Vikova, M., and Vik, M. (2019). Photochromic polypropylene filaments: impacts of mechanical properties on kinetic behaviour. *Fibres & Textiles in Eastern Europe* 27 (3): 19–25.

58 Renner, G., Schmidt, T.C., and Schram, J. (2018). Analytical methodologies for monitoring micro(nano)plastics: which are fit for purpose? *Current Opinion in Environmental Science & Health* 1: 55–61.

59 Primpke, S., Cross, R.K., Mintenig, S.M. et al. (2020). Toward the systematic identification of microplastics in the environment: evaluation of a new independent software tool (siMPle) for spectroscopic analysis. *Applied Spectroscopy* 74 (9): 1127–1138.

60 Liu, F., Olesen, K.B., Borregaard, A.R., and Vollertsen, J. (2019). Microplastics in urban and highway stormwater retention ponds. *Science of the Total Environment* 671: 992–1000.

61 Cowger, W., Steinmetz, Z., Gray, A. et al. (2021). Microplastic spectral classification needs an open source community: open specy to the rescue! *Analytical Chemistry* 93 (21): 7543–7548.

62 Cabernard, L., Roscher, L., Lorenz, C. et al. (2018). Comparison of Raman and Fourier transform infrared spectroscopy for the quantification of microplastics in the aquatic environment. *Environmental Science & Technology* 52 (22): 13279–13288.

63 Lenz, R., Enders, K., Stedmon, C.A. et al. (2015). A critical assessment of visual identification of marine microplastic using Raman spectroscopy for analysis improvement. *Marine Pollution Bulletin* 100 (1): 82–91.

64 Periyasamy, A.P. (2023). Environmentally friendly approach to the reduction of microplastics during domestic washing: prospects for machine vision in microplastics reduction. *Toxics* 11 (7): 575.

65 Shan, J., Zhao, J., Liu, L. et al. (2018). A novel way to rapidly monitor microplastics in soil by hyperspectral imaging technology and chemometrics. *Environmental Pollution* 238: 121–129.

7

Positive Actions Toward Reductions in Microplastics: Part I

7.1 Introduction

To reduce the amount of microplastics that originate from textiles, it is necessary to take a strategy that takes into account several different issues, such as increasing consumer awareness, fostering innovation, and modifying the established regulatory framework. Throughout the course of this chapter, the principles of Safe and Sustainable by Design (SSbD) have been thoroughly explored. The goal of SSbD is to reduce the amount of microplastics that are present in textiles by promoting sustainable practices throughout their entire lifecycle. In addition, the many methods of sustainable production, such as the utilization of sustainable additives throughout the manufacturing processes, as well as the manner in which the various types of fabric finishes and their relationship to the development and reduction of microplastics have been thoroughly addressed. Materials that are environmentally friendly, such as natural fibers or bioengineered synthetics that produce fewer microplastics, ought to be given priority by the textile industry. Furthermore, advancements in fiber treatments and weaving methods have the potential to substantially minimize the amount of fragmentation that takes place after washing. For the purpose of ensuring a concentrated effort toward reducing the amount of microplastics that originate from textiles at their place of origin, public awareness campaigns and industry coalitions have the ability to induce behavioral modifications and push for systemic change. When it comes to the total release and persistence of microplastics, as well as the potential negative effects that microplastics might have on the environment, there is without a doubt that endeavors need to be made in order to prevent emissions and additional pollution.

Textile Microplastics in the Environment: Sources, Impacts, and Solutions,
First Edition. Aravin Prince Periyasamy.

7.2 Safe and Sustainable by Design Approach

Recent legislative measures, like the European Green Deal, the European Chemicals Strategy for Sustainability, and the Zero Pollution Action Plan, aim to promote the transition to climate neutrality and a nontoxic environment [1]. To accomplish this shift, it is essential to create chemicals, materials, products, and processes that are fundamentally safer and more sustainable. This signifies both a societal necessity and a substantial economic prospect for the European manufacturing industry to restore its competitiveness. The SSbD methodology emphasizes the first phases of the supply chain to produce end chemicals, materials, and products that adhere to circular economic principles, while reducing harmful characteristics and negative impacts on human health and the environment. It incorporates essential functionality with circularity, climate neutrality, and the safety of chemicals, materials, and products throughout their entire lifespan, while concurrently fostering social responsibility and guaranteeing economic progress and innovation [1]. Safe and Sustainable by Design (SSbD) is a strategy that was established by the European Union (EU) with the intention of assisting businesses in the design of goods, processes, and systems in a manner that minimizes adverse effects on the environment and human health, while simultaneously guaranteeing that they are sustainable over the long term. The strategy entails incorporating into product design the goals of minimizing the use of hazardous chemicals, lowering emissions of greenhouse gases, and encouraging the reuse and recycling of materials within the context of a circular economy. The chemical industry describes "Safe and Sustainable-by-Design" as a process that aims to invent and bring to the market chemicals, materials, products, and technologies that are not only safe but also give value to society, the environment, and/or the economy through the uses of these goods [1]. In this chapter, the concepts of SSbD seek to mitigate microplastics in textiles by advocating sustainable practices over their whole lifespan (see Figure. 7.1) [1]. Simply, the essential strategies encompass the utilization of bio-based and biodegradable fibers, the implementation of sustainable production methods, the design of durable textiles, the incorporation of advanced filtration systems in washing machines, the promotion of textile recycling and repurposing, and the education of consumers regarding appropriate care and disposal of textiles. These solutions can markedly diminish the environmental effect of textiles and foster a more sustainable future.

7.2.1 Use of Bio-Based Fibers

Microplastics originating from textiles can be mitigated or eliminated by the adoption of sustainable design and manufacturing practices, strategies to regulate microplastic emissions during usage, and enhancements in disposal and

Figure 7.1 Safe and sustainable by design approach in textile industry for the reduction of microplastics.

end-of-life management. Transitioning textile consumption to natural and mostly bio-based and biodegradable fibers has been proposed as a solution for addressing microfiber shedding. Nonetheless, inquiries have emerged over the potential of this strategy to provide a feasible substitute for synthetic fibers, which presently constitute around 60% of textile fibers utilized. Natural fibers such as cotton, linen, and wool are favored for mitigating exposure to detrimental endocrine characteristics of plastics. The full list of natural, regenerated, and bio-based fibers that are shown in Figure. 7.2. These fibers are appropriate for several situations, including athletic activities, and may be extensively replaced with bed sheets, which we engage with for one-third of our lives. Prioritizing layers that contact the skin is essential when determining exposure to natural fibers. Natural fibers can release microfibers due to degradation, and their rapid biodegradation may lead to the release of harmful chemical additives; however, this is not universally applicable, as many of these natural fibers are chemically altered with toxic substances that further diminish their biodegradability [2, 3].

Perhaps, restricting synthetic fibers is advantageous for the environment, since the majority of natural fibers are biodegradable and diminish the quantity of

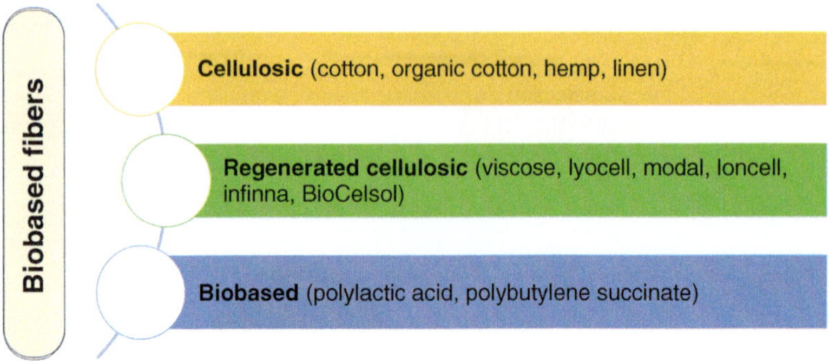

Figure 7.2 List of natural, regenerated, and bio-based fibers.

microplastics reintroduced into our seas and landfills. When selecting optimal natural fabric materials, evaluate their availability and practicality. Cotton is the most adaptable natural fiber, appropriate for warm climates and moisture absorption. Linen, derived from flax plant fibers, possesses comparable characteristics but is somewhat more durable and less pliable. Wool, particularly Merino wool, has superior insulating characteristics and yields a softer, more elastic outcome. Silk is recognized for its delicate threads and luxurious texture, making it ideal for bed linens and pillow coverings. Hemp, a textile derived from plant stems, possesses superior heat retention ability compared to cotton and linen. Kapok, a lightweight natural substance, is frequently utilized as filler for bedding duvets and pillows [4, 5].

Developing and using bio-based or recycled fibers can help minimize the environmental impact [6]. For example, fibers made from cellulose or other plant-based materials can be a sustainable alternative. Chemical recycling of cotton constitutes a crucial advancement for a sustainable textile industry, as fibers produced through this process provide biodegradability. This characteristic is beneficial for microfiber shedding, as these recycled fibers, derived from existing materials, are more likely to decompose in the environment without causing significant ecological harm [7–9]. These technologies can mitigate environmental impact, enhance resource efficiency, and facilitate a circular economy by converting trash into new fibers. As these processes advance and expand, they possess the capacity to transform textile production and consumption, facilitating a more sustainable future. Evrnu's NuCycl technology transforms cotton-rich textile waste into lyocell fiber, providing a sustainable substitute for virgin cotton and polyester. Renewcell produces dissolving pulp from discarded cotton, generating new cellulosic fibers with minimum ecological effect. Aalto University's Ioncell technology employs ionic liquids to transform cotton waste into premium fibers [10]. Lenzing's Refibra method integrates cotton waste with wood pulp to manufacture new lyocell fibers, enhancing resource efficiency and sustainability [11].

The United Nations 2030 Agenda for Sustainable Development has the objective of reducing microfibers in textiles. Frequently discussed the various alternatives to the existing synthetic polymers, including the poly lactic acid (PLA) [12], polyhydroxyalkanoate (PHA) [13], poly caprolactam (PCL) [14], polybutylene succinate (PBS), and polyhydroxybutyrate (PHB). Nonetheless, not all these alternatives are appropriate for textile applications, and the majority remain uncommercialized. An extensive study remains required on the degradation of these polymers when colored and chemically treated under various conditions. It is expected that the production of bioplastics would increase to 7.6 MMT by the year 2026, in contrast to the 1.6 MMT that were produced in 2021. The textile industry makes use of fewer than 10% of these polymers for a variety of reasons, the most important of which being the high expense of these polymers [15]. Life-cycle assessments (LCAs) have demonstrated that biopolymers are more ecologically sustainable and energy efficient than petroleum-based polymers [16, 17]. They may be discarded through several methods, including recycling, domestic and industrial composting, landfilling, and incineration. Nonetheless, it is impractical to substitute all synthetic materials with bioplastics because of the substantial production volume needed. Additional processes, like textile finishing, require enhancement to mitigate their adverse impacts on microfiber emissions and the discharge of hazardous chemicals into the environment. In the move to biodegradable fibers, the industry must evaluate two critical factors: expense and the quantity of feasible applications. Biodegradable fibers are recognized for their capacity to improve sustainability and reduce hazardous pollution in both marine and terrestrial ecosystems. The incorporation of functional additives, including dyes and finishing chemicals, has been noted to produce microfiber emissions with reduced biodegradability. The manufacture of entirely biodegradable fibers necessitates the integration of biodegradable additives, including natural colorants and functional finishes. Transitioning to plastic-free textile materials may seem daunting; nonetheless, it is essential to prioritize efficiency to save both labor and financial costs [3, 4].

7.2.2 Eco-friendly Manufacturing Processes

7.2.2.1 Use of Natural Dyes

In order to preserve the biodegradability of natural fibers, it is beneficial to use manufacturing methods that are more environmentally friendly. These methods include the utilization of natural and bio-based substances for the coloring and finishing of natural fibers. In recent years, there has been a trend in textile industry to move toward sustainable and eco-friendly processes. Perhaps, the textiles are colored with over 100,000 commercially available dyes, and annually, 280,000 tons of textile dyes are discharged into wastewater due to the dyeing and printing process, resulting in significant pollution and detrimental effects on aquatic ecosystems [18–20], also it provides considerable hazards to health and the environment, are widely used in the textile industry [21]. In addition to this, the various types of dyes, such as azo, diazo, anthraquinone, nitro, diphenylmethane, indigoid, thionindigoid, anthraquinoid, xanthene, phthaleins, and metal complex dyes [22], are utilized in the process of coloring textiles. Many of these dyes are considered as toxic and hazardous. When considering the generation of microplastics by cross-linked dyes, it is one of the series of threats to the environment. Additionally, dyes such as monochlorotriazine create a covalent bond with cellulose (like cotton), and these bonds are generally not able to be broken, which reduces the degradation properties of cellulose.

The term natural dye refers to all coloring substances obtained from natural sources, including plants, animals, and minerals. Plants are the primary source of natural dyes. Various components of the same plant, including leaves, flowers, seeds, roots, and bark, have been shown to produce distinct hues. Various mineral pigments, including gypsum, cinnabar, red and yellow ochre, malachite, ultramarine, blue, charcoal black, and gypsum, are utilized for the purpose of coloring fabrics, paintings, and murals. Microorganisms have the potential to be utilized as sources to produce bio pigments due to the fact that they provide benefits in terms of availability, stability, cost efficiency, labor, yield, and downstream processing. In addition to their use as coloring agents, many pigments have medicinal benefits, including qualities that are antiinflammatory, antioxidant, and antibacterial when combined with their coloring capabilities [23, 24]. A growing number of people are becoming interested in the process of obtaining natural pigments from discarded food at the moment. Examples of food waste that can be used for dye include avocado peels and pits, onion skins, pomegranate rinds, red cabbage, various herbs, carrot tops, beetroot, walnut hulls, and a diverse range of other foods. Using kitchen waste from restaurants and domestic waste to extract dyes has several benefits, as the utilization of natural dyes also contributes positively to the economic development of rural dyer communities [25].

7.2.2.2 Use of Bio-Based Finishing Additives

Natural dyes are regaining prominence in the textile dyeing industry; nonetheless, natural finishing needs more emphasis. The many categories of bio-based additives and their pathways are clearly shown in Figure. 7.3. Similar to natural dyes, bio-based finishing agents have some benefits and limits, which are biodegradable and may save carbon emissions. The utilization of nature lowers wastewater discharge and protects land and water from contamination. In contrast to synthetic finishes, natural finishes are sustainable and mainly renewable. Natural

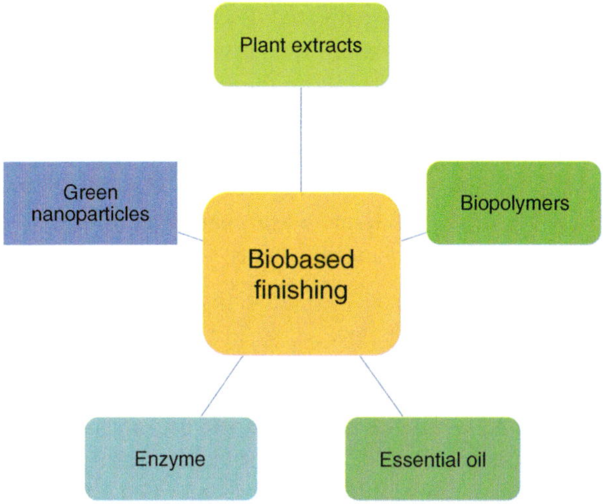

Figure 7.3 Different types of bio-based additives and their routes.

finishes exhibit biocompatibility and are nontoxic. Certain natural dyes have dual characteristics; they not only provide color to the cloth but also confer functional capabilities such as antimicrobial and UV protection [8, 26].

Textile materials must possess durability, since eco-design facilitates prolonged consumer usage, which reduces their consumption of single-use type of garments. This overarching concept has three design elements: design for disassembly, design for enduring aesthetics, and design for durability. It is essential to utilize sturdy, high-quality materials with other components, ensuring that seams are made to prevent ripping. The second concept considers the fashion sensibilities of consumers and evolving fashion trends. Designers mitigate the risk of product rejection due to perceived obsolescence by meticulously choosing colors and designs deemed timeless and broadly attractive. Extended use and repurposing are further promoted by the heightened resale value of enduring items. The assembly of a textile product influences its ease of disassembly for repair, maintenance, component reuse, and recycling. Numerous fashion businesses and shops are contemplating methods to render their products environmentally sustainable. Designers must start with more ethical or sustainable textile fibers and low-impact dyes and chemicals to manufacture more ethical or sustainable apparel.

7.2.2.3 End of Life, Recycling, and Disposal Phase

To mitigate microfiber emissions, it is essential to gather discarded clothing waste and identify alternative uses for it. Pre-owned garments may be transformed into many goods, including industrial rags, decorative furnishings, purses, backpacks, and advertising fabrics. A systematic approach for the disposal and processing of used textiles should be formulated and disseminated. Recycled PET enhances tensile strength and abrasion resistance in textiles, rendering fiber-to-fiber recycling a viable alternative. Informing the public about the significance of textile recycling is crucial. The disposal and end-of-life processing strategy involves improving the reuse and recycling of textiles, hence reducing microplastic emissions throughout production and usage. The aggregation of textile waste and its disposal can alleviate littering, inadequate waste management, and the airborne distribution of textiles from open landfills, hence reducing secondary microplastic contamination [27].

7.2.2.4 Consumer Awareness

Consumer awareness is essential for mitigating microplastic contamination, given the rising demand for synthetic fabrics, single-use plastics, and convenience-driven items. Comprehending the origins of microplastics enables customers to make more ecologically responsible choices. Informed consumer decisions may encompass selecting natural fibers, utilizing microfiber-catching laundry bags or filters, eschewing products containing microbeads, minimizing single-use plastics by choosing reusable options, and endorsing sustainable brands that emphasize eco-friendly materials and production methods. Extensive consumer awareness propels companies toward innovation and sustainability, resulting in the creation of biodegradable fabrics, closed-loop recycling, eco-labeling, and green technologies to filter microplastics throughout manufacturing and utilization. Informed consumers may influence public policy by campaigning for laws on microplastics in consumer products, endorsing bans on microbeads, and promoting enhancements to wastewater treatment infrastructure. Educational initiatives via schools, media, NGOs, and governmental bodies are crucial for disseminating awareness. These programs can educate customers about microplastic contamination, advocate for appropriate washing practices, stimulate involvement in cleaning activities, and cultivate collective action that engenders enduring behavioral change. By doing so, we may mitigate microplastic emissions from residences and contribute to a more pristine environment. For the purpose of controlling and reducing the discharge of microplastics, the most significant problem consists in the creation of robust and unified laws, as well as the penalties of nations that do not comply with these directives [12].

The hierarchy of the four Rs, which are reduction, reuse, recycling, and recovery, is the foundation around which waste management needs to build for the microplastic-free environment. The most effective method for preventing the production of microplastics is to cut down on the quantity of waste plastic that is produced. This may be accomplished by lowering the amount of synthetic textile materials that are manufactured, making use of alternative polymers from natural origins, or other items that are recycled or biodegradable, and recycling textile waste by transforming them into new products. A further alternative is recycling, which not only helps conserve resources and energy but also lessens the amount of pollution that is released into the atmosphere, which is beneficial to both society and the environment. The use of SSbD to develop long-lasting textile products with high quality and an efficient recycling process is inevitable.

Designers must prioritize the utilization of recycled fibers in fashion goods as a fundamental approach for sustainable and circular design. This transition is essential and significant as it diminishes environmental effect, curtails microplastic contamination, promotes circular design principles, addresses increasing consumer demand, and fosters industrial innovation. Recycled fibers, including recycled polyester, cotton, and nylon, necessitate reduced water, energy, and raw resources compared to virgin fibers, therefore diverting textile waste from landfills and diminishing reliance on fossil fuels. They also diminish the fabrication of new synthetic materials, hence decreasing overall microplastic synthesis over time. Designers can create garments with modular designs for easier repair and recycling and choose mono-materials or compatible blends for end-of-life recycling. Utilizing recycled fiber enables designers to conform to environmentally sensitive market trends, foster brand loyalty, and gain a competitive advantage. By using recycled materials, designers create apparel while also innovating for the future. In summary, designers significantly influence the sustainability of the fashion business, rendering recycled fibers both technically feasible and ethically imperative in the current climate-challenged environment. Current mechanical, thermomechanical, and chemical recycling technologies should be further innovative to reduce their environmental impact. Rather than removing color, it is highly sustainable to recycle fibers with dyestuffs so that there is no need for a second coloration. This also avoids dye disposal issues. Another sustainable innovation is to blend recycled fibers with varying colors to develop new products with different and vivid colors. Such innovations in postconsumer recycling can lead to significant reductions in the use of virgin fibers and enhance the circularity of textiles.

7.2.2.5 Domestic Washing Phase

Domestic laundry constitutes a major contributor to microfiber pollution. Reducing microfiber emissions from washing machines necessitates a synthesis of enhanced washing techniques and efficient collection technology [28]. Enhancing laundry conditions include decreasing microfiber discharge at the source by using cold water, utilizing mild cycles, avoiding prolonged spins, lowering washing frequency by airing garments when feasible, and opting for liquid detergent over powder [28]. These behavioral modifications can substantially decrease microfiber emissions. However, these are not sufficient techniques as the microplastic generations are quite exponential and need other technological solutions encompassing internal devices (drum-level prevention) embedded within the washing drum and external devices (post-wash capture systems) affixed to the washing machine's outlet hose [29]. These filters provide elevated fiber retention rates, accommodate a diverse array of fiber diameters, and are compatible with any machinery. Installation is necessary. Filter cartridges must be updated and disposed of correctly, and there are elevated startup expenditures. In summary, tackling microfiber pollution from household laundering necessitates a comprehensive approach, encompassing enhanced washing practices, and the adoption of both internal and exterior filtering technology. It is speculated that these methodologies can markedly diminish the ecological impact of microplastic fibers originating from residences [29].

Internal devices, such as the Cora Ball® and GUPPYFRIEND® Washing Bags, are employed to trap microfibers released during the laundering process. The Cora Ball® is a plastic sphere featuring tiny protrusions that may eliminate up to 31% of microfibers generated in the drum during laundering processes [29]. Nonetheless, its efficacy is contingent upon the fiber dimensions, with increased lengths yielding enhanced trapping efficiency. Laundry bags, like the GUPPYFRIEND® Washing Bag, serve as an alternative method for mitigating microfiber discharge (Figure 7.4). These bags are composed of PES monofilaments that inhibit the loss of microfibers at the source, safeguarding clothes during the washing process. External devices, such as PlanetCare®, are filters positioned outside the washing machine drum that extract microfibers from water flow prior to their discharge into the sewage system [30]. These filters operate on the fibers that have been discharged during the washing cycles. PlanetCare® has replaceable cartridges that may be returned to the manufacturer for comprehensive refurbishment once they are filled with microfibers. In conclusion, internal devices such as Cora Ball® and GUPPYFRIEND® effectively capture microfibers discharged during laundry; yet

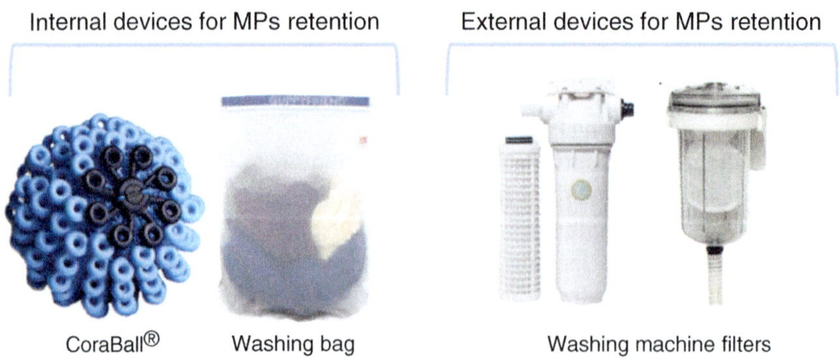

Figure 7.4 Internal and external devices for capturing microplastics during domestic laundering. (*Source:* [29] / with permission of Elsevier / CC BY-NC-ND 4.0.)

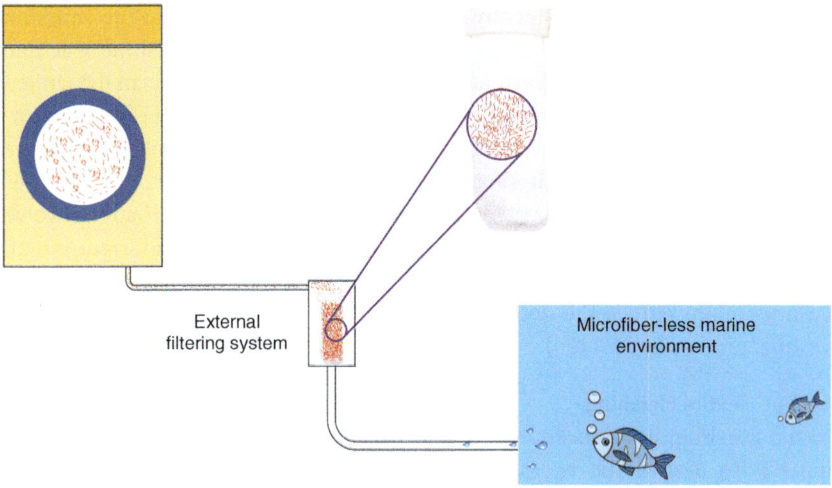

Figure 7.5 Diagram of additional filtering system for washing machines. (*Source:* Reproduced with permission of [26] / MDPI / CC BY 4.0.)

they cannot be regarded as definitive solutions because of their limited efficacy in trapping shorter fibers. Installing a detachable stainless-steel filter at the drain connection or employing an external filtration system at specific pipe junctions can serve as further protective measures. A filter with a 150–200 μm mesh may capture 87% of the total microfibers in washing effluent [26]. These procedures can mitigate the quantity of microfibers in the domestic wastewater and the filtering system is illustrated in Figure. 7.5.

Internal devices have a reduced retention rate of microfibers relative to exterior devices, perhaps because of their varied locations and processes. Internal devices passively gather microfibers discharged during laundering, whilst exterior filters aggressively compel water through their filtering systems. Design variations may impact their efficacy, since external filters often possess finer mesh sizes that may capture tiny microfibers, whereas devices such as the Cora Ball® are limited to catching only longer fibers. Comparing these devices is difficult owing to inconsistencies in research that utilize various materials and laundering settings. Both device types exhibit less efficacy in keeping smaller microfibers, raising environmental concerns due to the increased difficulty of removal and their propensity to survive in the ecosystem. The simultaneous utilization of internal and exterior devices might constitute a more efficacious approach to reducing microfiber emission during household washing; nevertheless, enhancements in performance and user-friendliness are essential for wider implementation.

France is the inaugural nation to enact legislation aimed at mitigating microfiber pollution from laundry, requiring all new washing machines to be equipped with microfiber filters by 2025. Research has shown that the emission of microfibers may be reduced by up to 80% with the use of a washing machine filter. Nonetheless, there are currently no state or federal regulations in the United States pertaining to microfiber pollution resulting from laundering [26]. California is contemplating legislation to initiate a pilot program aimed at assessing the effectiveness of microfiber filtering devices at 10 state-operated laundry facilities [31]. The most efficacious ways for mitigating microfiber contamination involve evaluating filter mesh dimensions and the ability for filtration and capture [32]. Moreover, employing laundry balls specifically engineered to collect fiber during washing might reduce the quantity of fiber in the wastewater generated by the washing process by 26% [26].

7.2.2.6 Textile Finishing

Textile finishing is a procedure employed in the production of fiber, fabric, or apparel. To bestow the necessary functional qualities to fiber or fabric, it is standard practice to submit the material to various mechanical and chemical treatments. A wash-and-wear finish is essential for cotton fabric to render it crease-free or wrinkle-free. Similarly, mercerizing, singeing, flame retardant, water repellent, waterproof, antistatic finish, and peach finish are significant treatments given to textile fabrics. In general, the various finishes have both positive and negative effects on the generation of microplastics. For example, some of the mechanical finishes on the fabric might generate more microfibers, while other finishes might reduce the generation of microfibers. As a result, this section offers a significant amount of information regarding the various finishes and the effects that they have on the generation of microplastics [26].

7.2.2.7 Mechanical Finishing

Mechanical finishes such as singeing or calendaring diminish microfiber release, since singeing effectively eliminates loose, projecting fibers from the cloth surface, resulting in a smooth look. This is achieved by employing a regulated flame to eliminate the microfibers present on the fabric's surface (see Figure. 7.6). Calendaring is the procedure in which fabric is crushed by passing it between two or more rollers under regulated parameters of time, temperature, and pressure [34], which resulting fibers are somehow bound to the fabrics structure and temporarily resist the microfibers generations (see Figure. 7.6).

Concurrently, alternative mechanical finishes such as shearing, elevating, and brushing produce a greater quantity of microfibers. Cai et al. [35] examined the PET fabric regarding the impact of mechanical finishing (i.e. shearing and brushing) on microfiber shedding. Shearing and brushing are mechanical surface treatments that render the cloth supple and voluminous. During the shearing process, the surface fibers (protruding fibers) are evenly trimmed to the specified length. The brushing procedure facilitates the extraction of fibers from the yarn

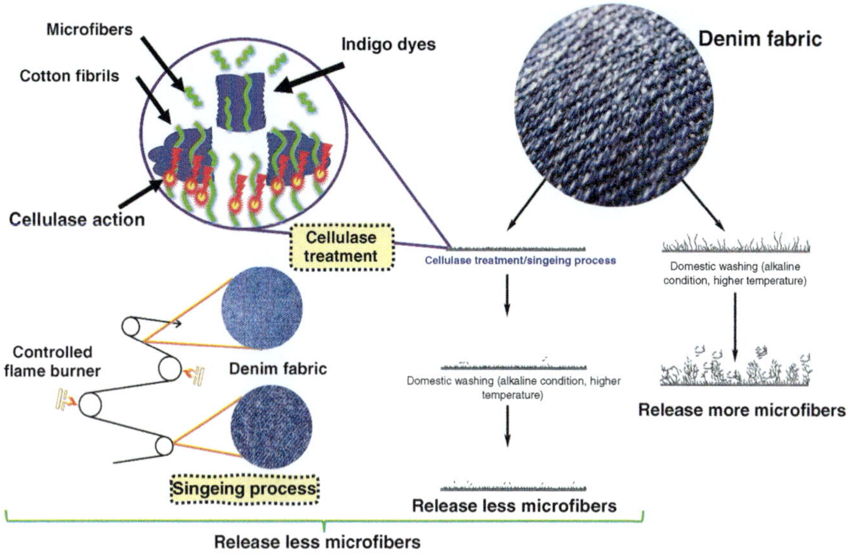

Figure 7.6 The mechanism for the microplastic generations from denim and their possible finishing techniques to reduce the microplastic generations. (*Source:* Reproduced with permission of [33] / American Chemical Society / CC BY 4.0.)

structure. Both finishes result in a notable alteration in the fabric's surface structure, hence facilitating the release of microfibers. This study suggests that increased microfiber shedding during the initial washes is attributable to the trapping of microfibers produced during the manufacturing process. In fact, for denim fabrics, various finishing procedures, including stone washing, sandblasting, scraping, whiskers, tacking, grinding, and ripping cuts, were employed to eliminate indigo stains from the surface of denim clothes. All these finishes open the yarn structure, hence increasing the quantity of microplastics and microfibers released during subsequent washing.

7.2.2.8 Chemical Finishing

7.2.2.9 Enzymatic Treatment

The problem of microfiber release can be addressed by surface changes of textiles as a preventative strategy. Biopolishing on cellulosic garments involves the elimination of fibrils or microfibrils from the surfaces of cellulosic fabrics using cellulase enzymes. Cellulase is one of the most employed enzymes in the biopolishing process within the textile industry. The end goods encounter issues due to the presence of fibrils, resulting in pilling and a faded color look. The original hue is preserved by the enzymatic elimination of fibrils, hence yielding

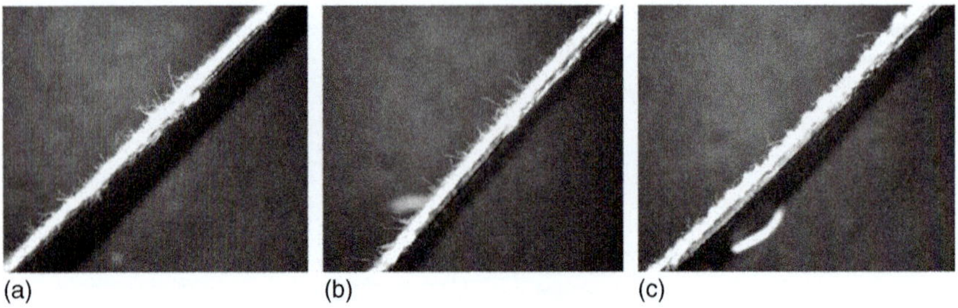

Figure 7.7 Microscopical image of protruding fibers on pristine fabric surface, (a) fabric treated with soluble cellulase, (b) and immobilized cellulase-treated fabric (c). (*Source:* [37] / with the permission of Elsevier.)

softer and cleaner products [36]. It is indicated that the immobilized cellulase has significant activity on the projecting fibers of cotton fabrics and it can be seen in Figure. 7.7 [37]. Following the enzymatic treatment, the projecting fibers on the fabric surface have diminished relative to the original fabric. Soluble enzyme treatment is more efficient than immobilized enzyme treatment because the free mobility of the enzyme facilitates a more effective interaction with the fibers, enhancing their presence on the fabric surface. Nevertheless, there is no data about the fiber protrusion resulting from wash and wear cycles in relation to the longevity of the enzymatic treatment.

By using silicones, polyurethane, and acrylic polymers as finishing and/or coatings, among other materials that are applied to textiles for the purpose of conferring functional or aesthetic features, it is possible to limit the amount of microfiber that detaches from the textiles. The surface may be simply covered by a chemical finishing that acts as a protective layer. This allows the surface to be covered. Several studies, on the other hand, have been conducted to evaluate the impact that a variety of standard finishing procedures have on the number of microfibers that are shed from textiles. To reduce the number of microfibers that are released, chemical treatment can be used to fix the protruding fibers and ensure that they are strongly linked to the surface of the cloth.

7.2.2.10 Finishing with Resins

The initial investigation on the reduction of microfibers through the use of the textile finishing process was titled "Mapping and estimation of laundry effluent microplastics release" [38]. The various synthetic fabrics were treated with a variety of resins, including acrylic resins (Polyacril 97, Polyacril 73, and Polyacril56 ECO), polyurethane resins (Purlastic 8189 and Purlastic 8139), and silicone emulsion (Polysilk CTE). These resins were applied using padding and exhaustion techniques. In addition, scientists utilized acid-based enzymes, namely Polyzyme EA 50, to reduce the amount of microfibers that were present. A test that involved pilling and washing was carried out to investigate the formation of microfibers. The acrylic resins display excellent pilling resistance out of chosen auxiliaries with all of the textiles (PET, PP, PAN, and PA), and this trend was noticed in both woven and knitted fabrics. The report indicates that acrylic resins have this resistance. The microfiber emission from the fabric that has been treated with acrylic resin, on the other hand, is significantly lower than that of other finished fabrics. In a similar vein, the softness of the fabric that was made available by silicon emulsion (Polysilk-CTE) was more than that of the pure PET fabric; nevertheless, it was observed that the pilling was more severe. In comparison to other finishes, the microfiber release from cloth that was treated with Polysilk-CTE was significantly lower. This was demonstrated by the results of the washing tests. The silicon finish on PET fabric creates a soft surface, which is the primary reason for this phenomenon. During the washing process, the friction between the fabric and the detergent is reduced, which brings about this phenomenon. On the other hand, the authors discovered that it was difficult to reproduce the findings that were almost identical. The endurance of these coatings has not been the subject of any research as of yet, even though such research would give more insight into the potential applications of these finishes in the future.

7.2.3 Pectin as Finishing Agent

Using a simple pectin-based finishing treatment, De Falco [39] was able to decrease the amount of microfibers that were released from the polyamide (PA) fabric. (i) the synthesis of PEC-glycidyl methacrylate (GMA, 97%) and (ii) the grafting of PEC-GMA on the fabric (PEC-GMA-PA) were the two techniques that were utilized in order to accomplish this goal. Because of its wide availability and its capacity to react with textiles, pectin was chosen as the material of choice for further consideration in light of the sustainability issues. A combination of emulsions is produced by first treating PEC with GMA and then applying the PEC-GMA emulsion to the PA fabric. This process is repeated until the mixture is complete. Maior et al. [40] provided a helpful explanation of the chemistry that exists between PEC and GMA. In Figure. 7.8, the typical amount of fibers that are released throughout the process of doing laundry is shown. The textiles that were pristine-PA and PEC-GMA-PA were washed for 45 min at a temperature of 40 °C using commercial detergent. The washing process was carried out in accordance with the ISO 105-C06:2010 standard. Upon analysis, it has been shown that the pristine-PA fabric produces microfibers with a length of $12 \pm 222\,\mu m$, a mean diameter of $18 \pm 3\,\mu m$, and a microfiber emission rate of $0.359\,g\,kg^{-1}$ of wash. PEC-GMA-PA, on the other hand, has the ability to neutralize microfibers with a length of $550 \pm 384\,\mu m$, a mean diameter of $16 \pm 4\,\mu m$, and a microfiber content of $0.058\,g\,kg^{-1}$ of fabric. This represents a decrease of microfibers by 90%.

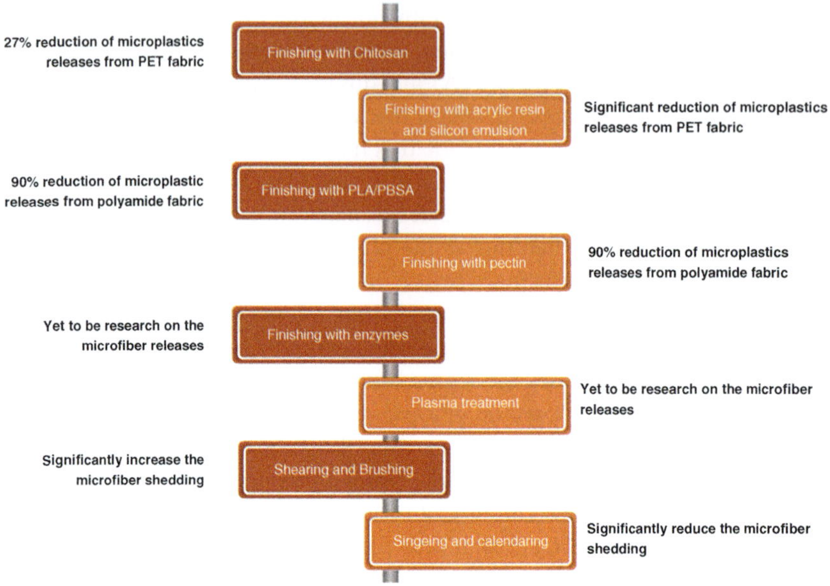

Figure 7.8 Influence of mechanical and chemical finishing on microfiber releases.

7.2.3.1 Finishing with Biodegradable Polymers

The electrofluidodynamic (EFD) coating approach was utilized in order to apply biodegradable polymers such as PLA and poly(butylene succinate-co-butylene adipate) (PBSA) to the PA fabric [41]. A significant reduction in microfiber releases, amounting to 90%, may be observed in the finished fabric when compared to the clean fabric. The uniformity of the coating achieved using the EFD coating process has the potential to efficiently restrict the release of microfibers. Some of the advantages of these approaches include the ability to generate nano-sized particles through the utilization of EFD technology for the purpose of completing the surface of PA fabric without compromising the fundamental qualities. In terms of microfiber release per gram of washed fabric, PLA and PBSA finished PA fabric have been found to have around 428 ± 92 and 456 ± 120, respectively. The pure PA fabric may release around 3966 microfibers per gram of fabric, with a standard deviation of 1425 microfibers. In addition, this research has indicated a reduction of 90% in the amount of microfibers that are lost from the PA fabric, in comparison to the research that was done before (using pectin-based finishing) [39]. The finishing process has resulted in the creation of a consistent surface coating on the fabric. This coating serves to protect the material from both mechanical and chemical reactions that may occur during the laundering process. In immaculate

PA fabric, the amount of microfibers that might be released after laundry is predicted to be 0.35 g kg^{-1}. On the other hand, it was found that completed PA fabric with PLA and PBSA released microfibers at a rate of 0.033 and 0.081 g kg^{-1}, respectively, upon the PA fabric. The finishing procedure, on the other hand, might diminish the mechanical qualities by 15–20%, according to the authors. Furthermore, the endurance of the coating can be demonstrated to withstand five washing cycles, and PLA coatings are more robust than PBSA coatings. No research has been done to determine whether or not these finishes are effective on polyester.

7.2.4 Finishing with Chitosan

Using chitosan-finished PET fabric, researchers investigated the phenomenon of microfiber shedding [38]. Among the microfibers that were released by the pristine PET fabric, there were 3047 microfibers per gram of fabric. On the other hand, the chitosan-completed PET fabric emitted 1726, 2497, and 2237 microfibers per gram of fabric at concentrations of 1%, 2%, and 3% chitosan, respectively. On the other hand, a decrease of microfibers of 27% and 18% is seen for chitosan concentrations of 3% and 2%, respectively. However, when the results are duplicated by repeated testing, there is no consistent response from the individual. In light of this, it is observed that its efficiency is low in comparison to that of other finishes, such as pectin [39] and PLA/PBSA [41]. The prospective major findings of various mechanical and chemical finishing on microfiber reduction were given in Fig. 7.8. However, enzymatic and plasma finishing still need to be investigated prior to their use.

A few years ago, the literature had just a relatively small number of study publications that were not particularly extensive. Along the same lines, shedding is decreased, demonstrating a favorable impact brought about by various finishes such as compacting and singeing. As far as polymeric chemical finishes are concerned, the findings are generally favorable. According to the findings of the majority of the research, the most effective course of action would be to either change the surface of the fiber or fabric or to coat it with a protective layer. There is a strong indication that the utilization of acrylic silicon emulsion, pectin, PLA, and PBSA-based finishes is successful. In spite of this, it is very necessary to use environmentally friendly chemicals and polymers for the process of sustainable finishing in order to prevent further pollution of the environment. Nevertheless, there are a great number of alternatives, such as enzyme therapy, plasma treatment, and other biopolymers, which have never been tested before and provide the researchers with the opportunity for future research.

7.3 Conclusion

Microplastic pollution from textiles necessitates systemic changes in the textile value chain. This chapter outlines actions to reduce these emissions, focusing on innovations from design to end-of-life management. By integrating the SSbD framework, textiles can be functional, market-ready, environmentally conscious, and health-safe. Shifting from petroleum-based fibers to bio-based alternatives enhances biodegradability, while natural dyes and plant-derived finishing agents reduce hazardous substances. These materials uphold the SSbD philosophy by aligning product design with ecological safety. End-of-life considerations such as biodegradability, recyclability, and safe disposal prevent textile products from persisting as pollutants. A multi-tiered approach rooted in SSbD, bio-based innovation, and sustainable engineering can reduce microplastic pollution from textiles. Collaboration among scientists, manufacturers, policymakers, and consumers is essential for this transition. As awareness grows and regulations tighten, the textile industry must align profitability with planetary health, making sustainable textiles imperative.

References

1 Apel, C., Kümmerer, K., Sudheshwar, A. et al. (2024). Safe-and-sustainable-by-design: state of the art approaches and lessons learned from value chain perspectives. *Current Opinion in Green and Sustainable Chemistry* 45: 100876. https://doi.org/10.1016/j.cogsc.2023.100876.

2 Seisl, S. and Hengstmann, R. (2021). Manmade cellulosic fibers (MMCF) – a historical introduction and existing solutions to a more sustainable production. In: *Sustainable Textile and Fashion Value Chains*, 3–22. Cham: Springer International Publishing.

3 Periyasamy, A.P. and Militky, J. (2020). Sustainability in regenerated textile fibers. In: *Sustainability in the Textile and Apparel Industries*, 63–95.

4 Woodings, C. (2003). Fibers, Regenerated Cellulose. In: *Kirk-Othmer Encyclopedia of Chemical Technology*. American Cancer Society ISBN 9780471238966.

5 Hearle, J.W.S. (2001). Textile fibers: a comparative overview. In: *Encyclopedia of Materials: Science and Technology (Second Edition)* (ed. K.H.J. Buschow, R.W. Cahn, M.C. Flemings, et al.), 9100–9116. Oxford: Elsevier ISBN 978-0-08-043152-9.

6 Santos, A.S., Ferreira, P.J.T., and Maloney, T. (2021). Bio-based materials for nonwovens. *Cellulose* 28: 8939–8969. https://doi.org/10.1007/s10570-021-04125-w.

7 Periyasamy, A.P. and Tehrani-Bagha, A. (2022). A review on microplastic emission from textile materials and its reduction techniques. *Polymer Degradation and Stability* 199: 109901. https://doi.org/10.1016/j.polymdegradstab.2022.109901.

8 Periyasamy, A.P. (2025). Functionalized textile microplastics: a closer look at the issues, strategy, and legislation on the microplastic reduction. *Kuwait Journal of Science* 100395. https://doi.org/10.1016/j.kjs.2025.100395.

9 Muthusamy, L.P., Periyasamy, A.P., Militký, J., and Palani, R. (2021). Adaptive neuro-fuzzy inference system to predict the release of microplastic fibers during domestic washing. *Journal of Testing and Evaluation*. https://doi.org/10.1520/JTE20210175.

10 Sixta, H., Michud, A., Hauru, L. et al. (2015). Ioncell-F: a high-strength regenerated cellulose fibre. *Nordic Pulp & Paper Research Journal* 30: 43–57. https://doi.org/10.3183/npprj-2015-30-01-p043-057.

11 REFIBRA™ Technology (2019) Right out revolutionary. https://www.sndon.com/en/new/news-008.html (accessed on 23 January 2025).

12 Ren, J. (2010). Processing of PLA. In: *Biodegradable Poly(Lactic Acid): Synthesis, Modification, Processing and Applications* (ed. J. Ren), 142–207. Berlin, Heidelberg: Springer Berlin Heidelberg ISBN 978-3-642-17596-1.

13 Degeratu, C.N., Mabilleau, G., Aguado, E. et al. (2019). Polyhydroxyalkanoate (PHBV) fibers obtained by a wet spinning method: good in vitro cytocompatibility but absence of in vivo biocompatibility when used as a bone graft. *Morphologie* 103: 94–102. https://doi.org/10.1016/j.morpho.2019.02.003.

14 Meng, Q., Hu, J., Zhu, Y. et al. (2007). Polycaprolactone-based shape memory segmented polyurethane fiber. *Journal of Applied Polymer Science* 106: 2515–2523. https://doi.org/10.1002/app.26764.

15 Truscott L. Preferred Fiber & Materials Market Report 2022. https://textileexchange.org/knowledge-center/reports/preferred-fiber-and-materials (accessed on 16 January 2023).

16 Periyasamy, A.P. and Militky, J. (2020). LCA (life cycle assessment) on recycled polyester. In: *Environmental Footprints of Recycled Polyester* (ed. S.S. Muthu), 1–30. Singapore: Springer Singapore ISBN 978-981-13-9578-9.

17 Periyasamy, A.P., Wiener, J., and Militky, J. (2017). Life-cycle assessment of denim. In: *Sustainability in Denim*, 83–110. Elsevier.

18 Sachidhanandham, A. and Periyasamy, A.P. (2020). Environmentally friendly wastewater treatment methods for the textile industry. In: *Handbook of Nanomaterials and Nanocomposites for Energy and Environmental Applications*, 1–40. Cham: Springer International Publishing.

19 Periyasamy, A.P. (2022). Natural dyeing of cellulose fibers using Syzygium Cumini fruit extracts and a bio-mordant: a step toward sustainable dyeing.

Sustainable Materials and Technologies 33: e00472. https://doi.org/10.1016/j.susmat.2022.e00472.

20 Sadannavar, M.K., Periyasamy, A., Islam, S.R. et al. (2024). Natural dyeing and antimicrobial functionalization of wool fabrics dyed with Chinese dragon fruit extract to enhance sustainable textiles. *Sustainability* 16: 6832. https://doi.org/10.3390/su16166832.

21 Periyasamy, A.P. (2024). Recent advances in the remediation of textile-dye-containing wastewater: prioritizing human health and sustainable wastewater treatment. *Sustainability* 16: 495. https://doi.org/10.3390/su16020495.

22 Singh, K. (2017). Rajani Srivastava an overview of textile dyes and their Removal techniques: Indian perspective. *Polution Research Paper* 36: 790–797.

23 Srivastava, R. and Singh, N. (2019). Importance of natural dye over synthetic dye: a critical. *International Journal of Home Science* 5: 148–150.

24 Pinheiro, L., Kohan, L., Duarte, L.O. et al. (2019). Biomordants and new alternatives to the sustainable natural fiber dyeings. *SN Applied Sciences* 1: https://doi.org/10.1007/s42452-019-1384-5.

25 Jain, A.; Kansal, S.; Sisodia, N. (2021) *International conference on sustainability in fashion design and manufacturing (SIFDM)-2021 At: department of fashion technology, PSG college of technology, Coimbatore, India*. https://www.researchgate.net/publication/360243897_Recycling_of_Kitchen_Waste_for_Extraction_of_Natural_Dyes_A_Step_towards_Sustainable_Fashion

26 Periyasamy, A.P. (2023). Environmentally friendly approach to the reduction of microplastics during domestic washing: prospects for machine vision in microplastics reduction. *Toxics* 11: 575. https://doi.org/10.3390/toxics11070575.

27 Wiedemann, S.G., Biggs, L., Nebel, B. et al. (2020). Environmental impacts associated with the production, use, and end-of-life of a woollen garment. *International Journal of Life Cycle Assessment* 25: 1486–1499. https://doi.org/10.1007/s11367-020-01766-0.

28 Periyasamy, A.P. (2021). Evaluation of microfiber release from jeans: the impact of different washing conditions. *Environmental Science and Pollution Research*. https://doi.org/10.1007/s11356-021-14761-1.

29 Luzi, B., Carnevale Miino, M., Rada, E.C. et al. (2025). Critical review of microfiber release from textiles: results, comparative challenges, mitigation strategies, and legislative perspectives. *Chemosphere* 378: 144394. https://doi.org/10.1016/j.chemosphere.2025.144394.

30 PlanetCare. A Fully Circular, Closed-Loop Solution. https://planetcare.org/pages/how-it-works?pb=0 (accessed on 1 May 2025).

31 Symon, E. Environmental Regulations For California Owned Washing Machines Under AB 1952. https://californiaglobe.com/section-2/environmental-regulations-for-california-washing-machines-under-ab-1952 (accessed on 11 December 2024).

32 Sánchez, L.D. Australia and California Join the Fight against Microfibers from Clothes. https://www.oceancleanwash.org/2021/03/australia-and-california-join-the-fight-against-microfibers-from-clothes (accessed on 14 December 2024).

33 Periyasamy, A.P. and Periyasami, S. (2023). Critical review on sustainability in denim: a step toward sustainable production and consumption of denim. *ACS Omega*. https://doi.org/10.1021/acsomega.2c06374.

34 Hossain, M.S., Islam, M.D.M., Dey, S.C., and Hasan, N. (2021). An approach to improve the pilling resistance properties of three thread polyester cotton blended fleece fabric. *Heliyon* 7: e06921. https://doi.org/10.1016/j.heliyon.2021.e06921.

35 Cai, Y., Yang, T., Mitrano, D.M. et al. (2020). Systematic study of microplastic fiber release from 12 different polyester textiles during washing. *Environmental Science & Technology* 54: 4847–4855. https://doi.org/10.1021/acs.est.9b07395.

36 Andreaus, J., Colombi, B.L., Gonçalves, J.A., and Alves dos Santos, K. (2019). Processing of cotton and man-made cellulosic fibers. In: *Advances in Textile Biotechnology*, 185–238. Elsevier.

37 Sankarraj, N. and Nallathambi, G. (2018). Enzymatic biopolishing of cotton fabric with free/immobilized cellulase. *Carbohydrate Polymers* 191: 95–102. https://doi.org/10.1016/j.carbpol.2018.02.067.

38 Piñol, L.; Rodriguez, L.; Brouta, M.; Escamilla, M.; Escudero, R. Mapping and Estimation of Laundry Effluent Microplastics Release 2014. static1.squarespace.com/static/5afae80b7c93276139def3ec/t/5b06e71c70a6ad631bc540ee/1527179088560/OCEAN+CLEAN+WASH+GUIDE.pdf

39 de Falco, F., Gentile, G., Avolio, R. et al. (2018). Pectin based finishing to mitigate the impact of microplastics released by polyamide fabrics. *Carbohydrate Polymers* 198: 175–180. https://doi.org/10.1016/j.carbpol.2018.06.062.

40 Maior, J.F.A.S., Reis, A.V., Muniz, E.C., and Cavalcanti, O.A. (2008). Reaction of pectin and glycidyl methacrylate and ulterior formation of free films by reticulation. *International Journal of Pharmaceutics* 355: 184–194. https://doi.org/10.1016/j.ijpharm.2007.12.006.

41 de Falco, F., Cocca, M., Guarino, V. et al. (2019). Novel finishing treatments of polyamide fabrics by electrofluidodynamic process to reduce microplastic release during washings. *Polymer Degradation and Stability* 165: 110–116. https://doi.org/10.1016/j.polymdegradstab. 2019.05.001.

8

Positive Actions Toward Reductions in Microplastics: Part II

8.1 Introduction

Considering that microplastic pollution is becoming an increasingly urgent problem on a worldwide scale, notably in the textile and fashion sectors, there is a growing demand for activities that are both constructive and inventive in order to limit the amount of microplastic that is produced over the whole product lifetime. Through the incorporation of sustainable consumerism, modern digital technologies, and responsible end-of-life (EoL) management techniques into fashion systems, this chapter investigates a multimodal strategy to combat the pollution caused by microplastics. One of the most important aspects of this strategy is the transition toward hyperpersonalization and digital fashion through the Metaverse. This method provides alternatives to the consumption of physical clothes, which in turn reduces the need for fast fashion and the release of microfibers from laundry and wear.

One of the most important factors in reducing the amount of microplastics that are introduced into ecosystems is the adoption of sustainable consumption habits. These patterns include purchasing less clothing of higher quality, prolonging the lifespan of products, and selecting companies that are environmentally sensitive. At the same time, the utilization of "emerging business models" like "product-as-a-service," "rental platforms," and "resale marketplaces" helps to promote the circularity of the textile industry and reduces the amount of manufacturing that is not essential. The function of extended producer responsibility (EPR) is also highlighted as a regulatory instrument that may be used to hold companies accountable for the environmental effect of their products, including the emissions of microplastics that occur throughout the disposal phase [1].

Consumers may contribute by adopting responsible laundry habits, such as washing garments at lower temperatures, utilizing microfiber filters, and using liquid detergents over powders, which produce less fiber abrasion. These are

Textile Microplastics in the Environment: Sources, Impacts, and Solutions,
First Edition. Aravin Prince Periyasamy.

all examples of activities that consumers can use. EPR programs should also be implemented by manufacturers [1]. These programs should ensure accountability for post-consumer waste and provide incentives for closed-loop recycling systems. Governments and regulatory organizations need to implement more stringent environmental requirements, which will encourage textile firms to produce designs that are favorable to the environment and textiles that can be broken down naturally. In addition, research into innovative wastewater treatment methods, such as nanofiltration and the enzymatic breakdown of microplastics, plays an important role in reducing pollution levels before it reaches the seas and rivers. Public awareness campaigns and industry alliances have the potential to induce behavioral adjustments and push for systemic change, so assuring a concerted effort toward decreasing microplastics originating from textiles at their point of origin [2–4].

The chapter underlines that one way to transform the textile ecosystem is to combine technological innovation with legislative interventions and consumer awareness. Some examples of technological innovation include virtual try-ons, AI-driven customization, and non-physical fashion assets. This integrated strategy provides a road map for decreasing microplastic pollution while simultaneously driving forward a fashion business that is more resilient and prepared for the future. It does this by connecting commercial viability with environmental sustainability.

8.2 Sustainable Consumption and Green Purchasing

Sustainable consumerism is essential, as contemporary fashion consumption practices are exhausting natural resources, exacerbating pollution, and contributing to climate change. The textile and apparel sector contributes substantially to water consumption, greenhouse gas emissions, and landfill waste. Also, it generates huge quantities of microplastics into the ocean. Unsustainable consumption further intensifies socioeconomic inequities via exploitative labor practices [3, 5]. The transition to sustainable consumption seeks to fulfill current needs while safeguarding the capacity of future generations to satisfy their own, so fostering environmental integrity, social equality, and enduring economic stability. This entails selecting resilient, sustainable apparel; purchasing less but of superior quality; endorsing ethical labor practices; reusing, recycling, or upcycling items; and prioritizing circular fashion models over rapid fashion. It entails deliberate decisions along the product lifetime, fostering an environmentally sustainable world and equitable society [3].

In this context, customers are essential in promoting sustainability through educated purchase choices or purchasing decisions. Trends like minimalism, slow fashion, and second-hand shopping advocate for less consumption and foster durability in apparel usage. Furthermore, companies are progressively embracing environmental objectives, like minimizing carbon emissions, utilizing organic or recycled products, and guaranteeing equitable labor practices. Transparency in supply chains and sustainability certifications are becoming recognized as normal industry practices. Government initiatives, such as eco-labeling, waste management legislation, and incentives for sustainable enterprises, can facilitate a transition to sustainable consumption. International accords and directives, like those from the UN and EU, establish frameworks for comprehensive industrial reform [3, 6]. Sustainable consumption is crucial for mitigating microplastics from textiles. Below are a few significant effects and approaches [7].

- *Selecting sustainable materials*: Choosing garments composed of natural fibers such as organic cotton, wool, and hemp can markedly decrease the emission of microplastics. These materials are biodegradable and do not release synthetic fibers.
- *Minimizing consumption, opting for quality*: Decreasing total consumption by purchasing fewer, superior-quality things that endure longer might diminish the frequency of laundering and disposal, thereby mitigating microplastic contamination.
- *Advocating for sustainable businesses*: Acquiring products from businesses that emphasize sustainable practices, including the utilization of eco-friendly materials and production techniques, propels the industry toward more responsible methodologies.
- *Optimal care and maintenance*: Reducing the frequency of laundering, utilizing cold water, and adopting softer wash cycles helps decrease fiber loss. Moreover, employing washing machine filters specifically engineered to catch microfibers can inhibit their entry into wastewater systems.
- *Recycling and upcycling*: Engaging in recycling initiatives and upcycling obsolete clothes may prolong the lifespan of textiles and diminish waste. This prevents synthetic fibers from entering landfills and the ecosystem.

8.2.1 Selecting Sustainable Materials

It is important to choose sustainable materials for the fashion industry, since the selection of materials is one of the most important elements that determines the sustainability of textiles, particularly in the light of the pollution caused by microplastics. Microplastics are very small plastic fibers that are predominantly shed from synthetic fabrics like polyester, nylon, acrylic, and elastane when they are washed, worn, or discarded. These microfibers are frequently too tiny to be removed by wastewater treatment systems. As a result, they eventually make their way into the seas, rivers, and even drinking water, where they pose major risks to the health of humans and marine ecosystems [8, 9]. The problem with synthetic fibers has been thoroughly discussed in the chapters that came before this one. Despite the fact that synthetic fibers offer a lot of benefits, they are not regarded to be biodegradable and are a significant contributor to the pollution caused by microplastics. During each cycle of the washing machine, hundreds of thousands of microfibers might be released into the wastewater. Furthermore, the creation of these items requires a significant amount of energy and results in a significant amount of carbon emissions. Because of this, the fashion industry is currently searching for sustainable materials to use as an alternative. To address this, one of the most important strategies is to use materials that are sustainable, natural, or have a low effect.

Organic cotton grown organically is a sustainable and environmentally friendly fiber that places an emphasis on the preservation of biodiversity, water conservation, and soil health. It is completely biodegradable, does not include any plastic, and uses just a little amount of chemicals [10]. The opposite of this is native cotton, which is either a region-specific or heritage variety and is strongly ingrained in the agricultural systems of the area. The fact that it can be cultivated organically and requires fewer chemical inputs contributes to the increasing environmental advantages it offers. Cotton crops that are native to the area typically require less irrigation and support from outside sources, making them more sustainable. Hemp and linen are biodegradable, necessitate less water and pesticide usage, and exhibit exceptional durability. Hemp, specifically, exhibits a substantial yield per acre and inherently enhances soil quality.

Viscose and lyocell are examples of regenerated varieties of cellulose fibers that are obtained from natural plant sources such as wood pulp. To differentiate them from synthetic fibers that are sourced from petroleum, these fibers are subjected to chemical processing in order to be transformed into textile fibers. Viscose, also known as rayon, is produced via a chemically complex process that involves carbon disulfide and other potentially harmful compounds. Viscose is generated from cellulose, which is found in wood pulp. Despite the fact that it does not form microplastics, it has the potential to inflict severe damage to the environment if the management of its source and wastewater treatment is negligent. A closed-loop technique that utilizes non-toxic solvents that are recycled up to 99% is utilized in the production of Lyocell, which is also created from wood pulp. Lyocell can be made from eucalyptus, bamboo, or beechwood. Additionally, it does not produce any microplastics that are long-lasting and biodegradable. In natural conditions, fibers that are lost during washing decay in the same way as viscose does. When compared to other natural or regenerated fibers, Lyocell has a tendency to have a smoother surface and a stronger structure, which may result in a lower percentage of fiber loss. The conclusion is that viscose, lyocell, organic cotton, and native cotton are all types of fibers that are sustainable and favorable to the environment, and they provide a variety of advantages [11, 12]. Innovative bio-based materials are Innovative materials like Piñatex (derived from pineapple leaves), Mylo (extracted from mushroom mycelium), and orange fiber, exemplify sustainable, biodegradable options with less environmental impact. Table 8.1 presents a detailed comparison of several textile fibers about their biodegradability and environmental effects.

Green shopping behavior defined as the deliberate choice to acquire less, superior, and frequently more sustainably produced products significantly contributes to reducing environmental consequences, especially the emission of microplastics

Table 8.1 Comparison of various fibers and their behavior of biodegradability and environmental impact.

Fiber type	Source	Biodegradability[a)]	Environmental impact
Organic cotton	Natural (plant and cellulose)	Yes	Moderate (less chemical use, but high-water use in some regions)
Cotton	Natural (plant and cellulose)	Yes	High (pesticides, fertilizers, water-intensive)
Wool	Natural (animal)	Yes	Moderate (methane emissions, but durable and low wash frequency)
Silk	Natural (animal)	Yes	Low to Moderate (resource-intensive but low-impact when managed well)
Viscose rayon	Regenerated cellulose (wood pulp)	Yes	Moderate (chemical-intensive process, needs responsible sourcing)
Lyocell	Regenerated cellulose (wood pulp)	Yes	Low (closed-loop production, sustainable sourcing)
Acrylic (PAN)	Petrochemical	No	High (non-renewable, persistent microplastics, energy-intensive)
Polyester (PET)	Petroleum-based	No	High (non-renewable, high emissions, plastic waste)
Nylon (PA)	Petroleum-based	No	High (energy-intensive, long-lasting microfibers)

a) Biodegradability defines without any functionalization's such as dyeing or finishing.

from textiles. By selecting resilient apparel crafted from natural or low-shedding fibers, customers diminish their dependence on fast fashion items, which are generally constructed from synthetic materials such as polyester or acrylic that release microplastics throughout laundering, usage, and at the conclusion of their lifecycle. Premium clothing has reduced susceptibility to wear and tears, hence necessitating less frequent laundry, a significant contributor to microfiber contamination. Moreover, their prolonged lifespan postpones disposal, alleviating the burden on waste management systems and diminishing the likelihood of synthetic fibers infiltrating the environment. Consequently, green shopping not only endorses ethical and ecological fashion practices but also acts as a proactive strategy to mitigate the cumulative discharge of microplastics throughout clothing lifetime [7].

8.3 Use of Metaverse and Hyperpersonalization

The textile industry is undergoing a substantial transition driven by Metaverse and hyperpersonalization. Metaverse, a virtual reality environment, enables users to engage with a computer-generated setting, diminishing the necessity for physical production and lessening the demand for synthetic fibers. The transition to digital fashion offers several advantages, such as diminished physical manufacturing, sustainable consumer practices, and creative design and testing methodologies [13, 14]. Hyperpersonalization, utilizing data analytics and artificial intelligence, may profoundly influence sustainability in the textile sector. By manufacturing clothing customized to specific tastes and sizes, manufacturers may mitigate overproduction and waste, hence decreasing the quantity of unsold products that contribute to microplastic contamination. Customized suggestions inspire consumers to engage in more deliberate purchasing decisions, favoring superior-quality, long-lasting products that release less microplastics over time [13, 14]. The amalgamation of the Metaverse with hyperpersonalization within the textile sector presents significant potential for mitigating microplastics. Reduced manufacturing quantities, sustainable resources, and heightened customer awareness are the advantages. The emphasis on innovation within the Metaverse promotes the advancement and utilization of sustainable materials that are less prone to releasing microplastics. In summary, the Metaverse and hyperpersonalization are catalyzing a novel epoch of sustainable fashion by diminishing the necessity for physical manufacturing, fostering conscientious consumption, and advocating for innovative materials [13, 14].

Digital fashion in the Metaverse allows individuals to convey their individuality and style without actual clothing, thereby establishing a zero-waste, zero-microplastic fashion experience. The principal effects on microplastic mitigation are as follows:

- *Absence of physical production*: Attire done by avatars in virtual environments is exclusively digital, thereby precluding the consumption of fabric and the release of microplastics.
- *Replacement for fast fashion*: Users may choose Metaverse attire for their digital presence (social media, gaming, virtual meetings), thereby diminishing the market for inexpensive, disposable physical garments, often composed of microplastic-releasing synthetic fabrics such as polyester, polyamide, and acrylic.
- *Digital showrooms and try-ons*: Virtual fashion exhibitions and 3D garment visualizations diminish physical sampling and returns, hence reducing synthetic textile waste.

Hyperpersonalization, technology leveraging real-time data, artificial intelligence, machine learning, and behavioral analytics, has emerged as a formidable influence on the fashion business. This method not only improves customer happiness but also promotes sustainability and diminishes waste, including a decrease in microplastic contamination from synthetic clothing. By incorporating client data, including body measurements, preferences, and historical purchasing history, hyperpersonalization creates clothing with exact fit and functionality, suggests materials that resonate with consumer values, and forecasts demand without resorting to mass manufacturing. Technologies such as AI-driven recommendation systems, three-dimensional (3D) body scanning, virtual fitting, and interactive product configurators facilitate this process [15]. Hyperpersonalization corresponds to:

- Tailored fit and design prior to manufacturing
- Custom manufacturing
- Enhanced demand forecasting

Advantages of microplastic reduction

- *Reduced overproduction*: By manufacturing just what is necessary, manufacturers mitigate the surplus of synthetic apparel that leads to waste and subsequent microplastic contamination.
- *Enhanced fit and reduced returns*: Returned clothing often ends up in landfills or is incinerated. Hyperpersonalization alleviates this inefficiency. By diminishing the trial-and-error aspect of apparel manufacturing and decreasing the logistics related to returns, manufacturers decrease their carbon and water footprints.
- *Material optimization*: Algorithms can provide more sustainable, low- or no-shedding fiber (i.e. natural fibers) options according to user preferences and applications.

- *Longer lifespan*: Custom-fit clothing tends to be more valued and better maintained by users, decreasing the likelihood of frequent replacement and shedding of synthetic microfibers.

The integration of the Metaverse with hyperpersonalization presents a transformative methodology. The Digital Twins of customers provide precise size matching and virtual fitting prior to garment production [14]. User-created designs on virtual fashion platforms diminish dependence on mass-produced synthetic apparel. Hybrid models in which buyers do digital attire online while purchasing fewer, superior-quality actual clothing composed of biodegradable fibers.

8.4 EoL, Disposal Phase

End-of-life textiles, encompassing both functional and waste materials, are reused into new products. This concept entails repurposing garments or fabrics to fabricate new goods, such as tote bags. Although the product or material lacks waste classification, there is no conclusive definition about which textiles are appropriate for reuse. Key stakeholders in textile reuse collecting are non-profit organizations and private enterprises that gather a specified proportion of materials for repurposing [16–18].

The fashion industry must implement sophisticated recycling solutions to address the EoL challenges associated with apparel and fabrics that are difficult or impossible to recycle. Synthetic fibers such as polyester, nylon, and acrylic are non-biodegradable and can last in the environment for centuries, which has been well discussed in Chapter 2. The EoL phase of textiles encompasses the methods and tactics employed to control waste when the materials are no longer functional. Effective disposal and recycling techniques are essential for reducing the emission of microplastics into the environment. This phase includes post-consumer actions such as reuse, recycling, landfilling, and incineration. Inadequately handled textile waste transforms into a secondary source of microplastics, exacerbating the problem beyond the usage phase. Reusing and recycling can diminish the volume of textile waste in landfills; however, both reuse and mechanical fiber recycling increase the probability of microplastics during the user phase, as the fibers have been weakened due to prior wear and tear. Moreover, mechanical fiber recycling entails shredding the fabrics into fibers suitable for repurposing in new products, which results in reducing the need for virgin materials and mitigating microplastic contamination. However, the shredding process causes more damage to the fibers which resulting in higher possibilities of microplastic mitigation [18]. Advanced chemical methods may decompose synthetic fibers into their monomers, which can subsequently be repolymerized into new fibers. This technology can efficiently decrease microplastic emissions by converting waste into recyclable products and a detailed description can be found in Chapter 11. During disposal, these fibers degrade into microplastics due to mechanical stress, ultraviolet exposure under the sunlight, and chemical deterioration of fibers under different environments, including fresh water, salt water (i.e. sea/ocean), and soil pH. They infiltrate streams and soil from inadequately managed landfill sites and are disseminated through incineration leftovers and atmospheric deposition, exacerbating airborne microplastic contamination. Incineration may decrease the volume of textile waste, although it might simultaneously emit microplastics into the atmosphere. Energy recovery from incineration needs to be regarded as a last alternative once all possibilities for reuse and recycling have been depleted [18]. Overall, landfills and incineration are unsustainable methods, as they lead to prolonged environmental contamination. Energy recovery from incineration needs to be regarded as a last alternative once all possibilities for reuse and recycling have been depleted. Encouraging garment reuse is essential, since it reduces the landfills; nonetheless, all clothing ultimately arrives at a point

Table 8.2 Different disposal pathways of textile waste and their microplastic risk and environmental impacts.

Disposal method	Microplastic risk	Environmental impacts
Landfilling	High – Fibers break down slowly and leach into soil and groundwater	Long-term pollution, microplastics enter terrestrial and aquatic ecosystems
Incineration	Moderate – Some microfibers remain in fly ash or are released into the air	Airborne microplastic pollution, toxic emissions
Open dumping/informal disposal	Very high – Direct exposure to environment accelerates fiber release	Uncontrolled pollution, soil, and water contamination
Mechanical recycling	Moderate – Fiber degradation occurs during shredding	Potential microfiber release during processing
Chemical recycling	Low – Dissolves polymers into raw materials, filtering microfibers	Higher resource use but lower fiber leakage risk
Reuse/resale	Low – Extends garment life, delays disposal	Reduces immediate waste and fiber shedding
Downcycling (e.g. insulation, rags)	Medium – Often still ends up in landfill eventually	Delays but doesn't prevent microplastic release

of obsolescence when it can no longer be used. Attaining genuine circularity in the fashion sector necessitates sophisticated recycling solutions to address the EoL phase of the extensive inventory of garments produced through conventional linear methods, as well as textiles that are challenging or unfeasible to recycle with current technologies. The fashion sector should prioritize a pragmatic approach to circularity and enhance EoL recycling options, ensuring that textile waste contributes to the manufacture of new clothing instead of increasing landfill accumulation. The use of circular economy practices in textile waste management, including chemical recycling and upcycling, can mitigate microplastic creation from textile waste disposed of in landfills or incinerators [18]. This method prolongs the lifespan of fabrics while concurrently reducing microplastic production. Table 8.2 presents several disposal pathways for textile waste, encompassing diverse recycling and disposal methods, together with associated microplastic risks and their environmental impacts.

The influence that various recycling technologies have on textile waste and the possible relationship that these technologies may have with the formation of microplastics is thoroughly researched and explored in Chapter 11.

8.5 Different Business Models and Their Impacts

The business models of the fashion and textile sector substantially impact microplastic production. Fast fashion, defined by quick manufacture and turnover of low-cost apparel, predominantly depends on synthetic materials, resulting in microfiber release during regular laundering and disposal processes. Slow fashion prioritizes quality above quantity, advocating for the creation of lasting, timeless garments crafted from natural or biodegradable fabrics. Circular business models, including clothing rental, resale, repair, and recycling, prolong the lifespan of clothes, hence diminishing the demand for new synthetic fiber manufacture and postponing their introduction into the waste stream. On-demand and made-to-order methods, facilitated by digital technology, avert overproduction and eradicate excess inventory.

The conventional linear economy of textiles and fashion paradigm of "take, make, dispose" poses significant challenges in the textile business, characterized by mass production, brief usage, and subsequent disposal of items. Elevated manufacturing levels of synthetic fabrics, including polyester and nylon, heighten the likelihood of microplastic discharge during laundering and disposal. The short lifespan of items, propelled by quick fashion fads and inexpensive, low-durability apparel, intensifies the problem. In the absence of systemic change, the linear economy sustains a cycle of overproduction, overconsumption, and rampant microplastic pollution, highlighting the pressing necessity for a shift toward more circular and sustainable models in textile production and consumption.

In contrast to the linear economy in textiles and fashion, characterized by a "make-use-dispose" model, the circular economy seeks to prolong material utilization through reuse, recycling, repair, and upcycling [19]. This approach advocates for the prolongation of garment lifecycles in textiles, hence diminishing the necessity for ongoing manufacture of synthetic fibers, which are significant contributors to microplastic pollution. Maintaining textiles in circulation – via secondhand marketplaces, clothing rental services, or textile-to-textile recycling – reduces garment disposal and minimizes synthetic waste in landfills or incinerators, where fibers may fragment and contaminate ecosystems. Circular models promote design for durability and recyclability, frequently emphasizing natural, biodegradable, or low-shedding materials that minimize microplastic danger. Closed-loop recycling systems may effectively catch and recycle synthetic fibers, preventing their release into the environment. Ultimately, implementing circular economy concepts in the fashion sector transcends mere waste reduction; it serves as an effective technique to mitigate the aggregate release of microplastics throughout the whole textile value chain [3, 19].

Subscription and rental models are developing as viable alternatives to conventional ownership in the fashion industry [20]. Consumers opt to rent clothes for temporary usage rather than purchasing garments altogether, which is especially advantageous for special events, maternity attire, or rapidly evolving fashion trends. These models prolong the lifespan of specific clothes, guaranteeing they are utilized by several consumers prior to their landfills. By diminishing the aggregate demand for new clothing production, particularly garments composed of synthetic fibers, these models can substantially reduce the quantity of microplastics emitted into the environment via manufacture, laundering, and disposal. Moreover, rental firms frequently consolidate the laundry of apparel utilizing industrial machines fitted with microfiber filters, which assist in containing and appropriately disposing of released microfibers. Furthermore, rental models often prioritize durable, high-quality fabrics capable of enduring numerous uses and cleanings, so automatically diminishing dependence on fast fashion synthetics that more quickly release microplastics. Consequently, subscription and rental systems enhance resource efficiency and diminish textile waste, while also providing a feasible approach to alleviating microplastic pollution within a circular economy.

On-demand production is an innovative business concept that entails making clothing solely upon receipt of a verified client order, rather than engaging in preemptive bulk manufacturing of inventory. On-demand manufacturing and hyperpersonalization are pioneering practices in the fashion sector designed to enhance sustainability [21]. On-demand production emphasizes the fabrication of items solely upon receipt of a client order, hence minimizing waste and excess inventory. It mitigates microplastic contamination and diminishes inventory excess. Hyperpersonalization emphasizes the development of tailored products according to individual tastes, body types, or stylistic requirements [13]. It employs data-driven insights and artificial intelligence technology to generate tailored experiences, enhance consumer pleasure, and mitigate premature disposal. Both techniques seek to minimize waste and enhance sustainability within the garment business. On-demand production significantly mitigates overproduction, a primary contributor to textile waste and microplastic contamination [13]. In conventional methods, unsold synthetic apparel frequently concludes in landfills or is burnt, thereby contributing to the emission of microfibers into the environment. By synchronizing production with genuine demand, on-demand methods mitigate the superfluous creation of synthetic fibers, hence reducing material waste and the potential for microplastic leakage over the product lifespan. Moreover, these systems frequently use digital design and 3D modeling technologies, which minimize prototype waste and enable firms to optimize supply chains with enhanced precision and sustainability. Consequently, on-demand production fosters a more resource-efficient and ecologically sustainable textile business, according to circular principles and aiming to reduce microplastic contamination from the onset.

Finally, the sustainable and **ethical fashion brands** play a vital part in the campaign against microplastic pollution. They are doing this by reconsidering the materials they use, the manufacturing processes they employ, and the total lifetime of their individual goods. These manufacturers frequently choose natural, biodegradable fibers like organic cotton, cotton, viscose rayon, lyocell, wool, and hemp as a higher priority than synthetic alternatives. As a result, the chance of microplastics being shed during washing and disposal is greatly reduced. They invest in a responsible manufacturing value chain, including water-efficient dyeing and closed-loop production systems, which decrease fiber loss and environmental pollution. In addition, they make material selections that are environmentally friendly. A great number of ethical brands also embrace openness in their supply chains, which helps to ensure that both social and environmental standards are maintained without compromise. They encourage slower consumption habits and discourage the throwaway character of fast fashion by creating clothing for durability and longevity. This is accomplished through the design of garments. Furthermore, these brands and manufacturers frequently endorse inventions such as microfiber filters, recycling programs, and garment take-back schemes, all of which contribute to the prevention of synthetic materials from entering rivers and landfills. By doing so, sustainable and ethical businesses not only help to the promotion of a more responsible fashion industry, but they also actively contribute to the reduction of the creation and spread of microplastics across the whole lifetime of the textile industry.

8.6 Extended Producer Responsibility (EPR)

Extended producer responsibility, often known as EPR, is a policy approach that holds producers liable for the whole lifespan of their products, including the management of waste after the production of the product has been consumed [1]. The use of EPR in the textile sector has the potential to be an effective method for lowering the amount of microplastic pollution, particularly that which is caused by synthetic fibers. As a result of EPR, manufacturers are financially and legally accountable for the management of textile waste, which provides them with an incentive to create clothes that are more long-lasting, recyclable, and manufactured from fabrics that shed less or are produced from biodegradable elements. This stimulates the adoption of eco-design concepts, such as the use of fewer synthetic blends, the avoidance of hazardous chemical treatments, and the development of items that shed less microfibers during usage and washing. Furthermore, EPR has the potential to encourage the infrastructure for collecting, sorting, and recycling that is required to prevent discarded synthetic textiles from degrading into microplastics in landfills or the environment. Through the process of internalizing the environmental costs of waste, EPR could transfer responsibility away from consumers and governments and onto brands and manufacturers, therefore developing a textile system that is more circular and responsible. Ultimately, EPR regulations have the potential to encourage innovation across the whole sector while simultaneously drastically reducing the amount of microplastics that are released into terrestrial and aquatic environments.

8.7 Digital Fashion and 3D Modeling to Reduce Production Waste and Microplastic Generation

There has been a revolutionary shift in the fashion and textile sector, which is represented by digital fashion and 3D modeling. These technologies provide novel answers to some of the industry's most serious environmental concerns [22–25]. These technologies make it possible for designers and businesses to build, simulate, and visualize clothes fully within virtual settings. This eliminates the need for many physical prototypes, which often need a significant quantity of fabric, water, dyes, and energy [26]. By reducing the number of physical samples that are produced, digital fashion significantly cuts down on the amount of material waste, particularly synthetic fabrics, which are a major contributor to the pollution caused by microplastics. Not only does this virtual-first strategy reduce the length of time it takes to design a product, but it also eliminates the needless creation of clothing that would never be sold. This brings about a solution to the problem of overproduction, which is a significant contributor to pollution caused by textiles.

Moreover, the use of 3D modeling tools improves the quality of fitness and design, which in turn reduces the number of returns and unsold inventory. These items frequently find their way into landfills, where the synthetic fibers break down into microplastics. Additionally, digital fashion makes it possible to conduct virtual try-ons, fashion shows, and previews of e-commerce products, therefore reducing the environmental impact that relates to conventional marketing and retail advertising techniques [25, 26]. It is possible for the fashion industry to considerably reduce the amount of microfiber shedding that takes place throughout the process of sample laundry and disposal if it reduces its dependence on synthetic prototype clothing. Furthermore, as digital fashion continues to develop and interact with Metaverse platforms like augmented reality, it opens the door to completely virtual wardrobes and fashion experiences, therefore significantly reducing the need for actual clothing altogether. Through this approach, digital fashion and 3D modeling not only contribute to the promotion of sustainable innovation but also serve as pioneering tools in the fight against microplastic pollution on a worldwide scale.

Originally, the process of developing garments was defined by numerous rounds of physical sampling, fitting, and prototyping. Each of these components creates a large amount of material waste and also consumes lots of virgin resources. In this conventional technique, the process begins with the design and production of physical samples, followed by testing those samples, making modifications, and repeating the process until the desired end result is obtained. However, digital fashion provides a game-changing alternative by substituting these physical prototypes with virtual clothing that can be produced, modified, and tested in 3D software [27]. This allows for a more intuitive and efficient process. This change removes the need for superfluous physical resources, which results in a significant reduction in the amount of trash produced and the amount of energy that is consumed during the early phases of product development. The most important advantages of digital fashion in reducing waste and pollution caused by microplastics:

- *Elimination of physical prototypes*: Traditional clothing design relies on physical samples manufactured from a variety of materials. These samples frequently end up as trash if they are not picked for manufacturing. By allowing designers to electronically test patterns, draping, and textures, digital fashion helps designers reduce the amount of fabric that is used and the amount of waste that is created during the prototyping phase.
- *Rapidity and effectiveness*: When it comes to traditional design procedures, any modifications to patterns or textiles need repeating the sample process, which

results in revisions that demand a significant number of resources. Using 3D modeling, designers can make quick modifications and fine-tune designs in a virtual environment. This helps to reduce manufacturing delays and allows for more accurate demand forecasts, which in turn reduces inventory levels and eliminates unnecessary waste.

- *Reduction of returns*: Returns are a substantial source of waste in the fashion business. These returns are frequently produced as a result of a bad fit or expectations regarding style that are not realized. By providing consumers with the ability to virtually try on garments and accurate 3D reconstructions of those garments, digital fashion enables customers to sample the fit and style of apparel before making a purchase. Since many of the returned clothes are synthetics that are prone to shedding microplastics, this can lead to a reduction in the number of returns, which in turn can greatly reduce the amount of waste connected with transportation and the disposal of returned garments.
- *Maximum efficiency*: The capacity of digital fashion to anticipate customer demand via the use of data analytics and virtual previews is one of the most significant environmental benefits of this fashion industry. Brands are able to minimize overproduction, which is a main source of unsold inventory and textile waste, by precisely understanding the preferences of their customers before committing to large-scale production. This leads to manufacturing runs that are more efficient, with fewer surplus products being sent to landfills or incinerators rather than being thrown away.

By lowering the requirement for physical samples and providing tools for more accurate demand forecasting, digital fashion makes it possible for firms to produce items in a more environmentally responsible manner. This transformation not only reduces the physical and environmental footprint of the fashion industry but it also greatly reduces the danger of microplastic contamination by lowering the dependence on synthetic fibers in clothing that are mass produced. There is considerable environmental worry over microplastic contamination, which is mostly caused by synthetic fibers like polyester, nylon, and acrylic. These fibers shed microplastics when they are washed, worn, and even via production operations like fabric cutting. The usage of synthetic textiles in the early stages of garment creation is mitigated by the implementation of digital fashion, which offers a revolutionary method to address this issue. Because of this, it makes a direct contribution to reducing the amount of microplastic that is shed throughout the various phases of the fashion value chain. One way that digital fashion can help reduce pollution caused by microplastics:

- *A smaller number of fabric samples*: Traditionally, the process of garment development involves the creation of several physical samples, which frequently involve synthetic materials. This results in a significant amount of fiber loss throughout the cutting, fitting, and washing processes. Digital prototyping allows for the creation of virtual models to take the place of this procedure, which results in a large reduction in the requirement for actual textile samples. During the early phases of product development, there is a reduction in the amount of microplastics that are released and the amount of fabric that is lost as a result.

- *Material choices*: Additionally, digital fashion gives designers the opportunity to digitally experiment with a variety of materials before committing to production, which enables them to make more informed material choices. The use of 3D simulations allows designers to evaluate how different textiles behave in terms of drape, texture, and fit without having to use physical samples. By taking this method, they can give natural or biodegradable materials more priority than synthetic materials, which are known to contribute to the shedding of microplastics. Through the reduction of dependency on fibers that release microplastics, digital fashion works to promote the utilization of materials that are more sustainable, environmentally friendly, and do not contribute to the contamination of the environment.
- *Virtual clothing*: The capacity of digital fashion to generate clothes that are totally based in virtual worlds is one of the most significant advantages of this fashion industry phenomenon. When it comes to platforms like gaming, social media avatars, and virtual events, digital clothes are worn online, which eliminates the need for actual clothing. Consequently, there is no physical waste and there is no contamination caused by microplastics. These digital ecosystems provide an innovative solution to address the expectation of customers for style and self-expression without the environmental impact of traditional textile production. This demand is growing as consumers increasingly embrace virtual fashion.

The fashion industry is evolving toward a more sustainable future because of digital fashion's considerable contribution to the reduction of the creation and release of microplastics [14]. This is accomplished by decreasing the usage of physical synthetic fabrics, optimizing material selections, and promoting totally virtual clothes. A transformation is already taking place in the fashion industry as a result of the application of digital fashion and 3D modeling. In order to lessen the amount of waste produced during manufacturing, simplify the design process, and cut down on the amount of microplastic pollution, an increasing number of companies and inventors are embracing these technologies. A pioneer in the field of fashion that is only available in digital format, The Fabricant is a company that designs and manufactures high-end clothing that is only available in virtual surroundings. The items in question are never manufactured in physical form since they are intended for usage in digital environments, such as on social networking platforms, gaming platforms, and virtual events. Due to the fact that this model generates no material waste and does not emit any microplastics, The Fabricant is a leading example of how fashion may be reinterpreted in the digital era.

Both Zara and H&M, two of the most prominent fast fashion stores, have begun integrating 3D design tools into their early product development procedures [28]. By generating virtual samples, they are able to minimize the number of physical prototypes that are required, so dramatically reducing the amount of fabric that is used and the microplastics that are shed as a result. The change not only makes the design process more efficient and cuts down on lead times but it also helps minimize overproduction by allowing for more precise forecasts and a quicker decision-making process. These examples illustrate how both new businesses and well-established brands are utilizing digital innovation to develop more sustainable business models. They also demonstrate that the fashion industry can lessen its impact on the environment, particularly with regard to the generation of microplastics, by taking advantage of virtual design and production practices.

The growing use of digital fashion and 3D modeling presents a number of obstacles, despite the fact that these technologies provide revolutionary benefits concerning the reduction of waste and microplastic contamination. Especially for smaller to medium-sized fashion firms, the initial investment that is required in advanced design software and technology, as well as the requisite training for design teams, can be prohibitively expensive. Furthermore, there is frequently reluctance to change within existing supply chains and industrial models, which are strongly anchored in physical sampling, mass production, and procedures that have been in place for a long time.

Another significant barrier is the digital divide, which restricts access to cutting-edge technology for smaller businesses or brands in countries that are still in the process of expanding. It is possible that these stakeholders may not possess the necessary infrastructure, financial resources, or digital literacy to effectively benefit from virtual fashion technologies. This might possibly increase the sustainability gap that exists throughout the fashion sector worldwide. The momentum behind digital fashion is continuously expanding despite the obstacles that have been presented. As technology continues to become more user-friendly and accessible, as well as environmental standards continue to grow more stringent, companies are becoming increasingly encouraged to employ virtual design methodologies as part of their efforts for being environmentally sustainable. In addition, the sector is being pushed to investigate novel options that have a low impact on the environment as a result of the demand from customers for transparency and environmental responsibility. In the long run, it is anticipated that digital fashion and 3D modeling will become common components of sustainable fashion practices. These techniques will play a significant part in decreasing overproduction, textile waste, and microplastic contamination.

In the textile sector, the use of digital fashion and 3D modeling has the potential to significantly reduce the amount of waste produced during production as well as the amount of microplastic contamination. Additionally, they contribute to the elimination of excessive material usage, the promotion of sustainable fiber choices, and the creation of creative consumption models such as virtual closets. This is accomplished by enabling exact, virtual construction of clothes. Adopting these technologies is not only a step toward achieving higher efficiency but it is also an essential step toward achieving environmental responsibility in the fashion industry.

8.8 Conclusion

Mitigating microplastic pollution from textiles necessitates a comprehensive transformation of the fashion ecosystem, which involves re-evaluating materials, production methods, and the consumption, valuation, and disposal of clothing. This chapter emphasizes various effective measures that can significantly reduce microplastic release throughout the textile lifecycle by focusing on technological advancements and systemic transformation.

The fashion industry is a significant contribution to the destruction of the environment as a result of excessive production, which results in an excessive amount of trash, unsold inventory, and pollution caused by microplastics. For brands, demand forecasting that is powered by artificial intelligence may assist in aligning production with customer demand, which in turn supports sustainable practices across the supply chain. Forecasting systems that are powered by AI make use of machine learning algorithms to evaluate enormous amounts of data. This enables companies to make more accurate forecasts regarding the styles, sizes, and quantities of products to create, which ultimately results in a reduction of additional inventory. Reduced overproduction, better inventory management, fewer product returns, decreased resource efficiency, and integration of sustainability are some of the benefits that may be gained from using AI to drive forecasting for the reduction of microplastics. When brands produce only what is required, they are able to minimize the number of synthetic clothes that are manufactured, hence lowering the potential for microplastic emissions from clothing that is either unsold or abandoned. A further benefit of using AI technologies is that they can more correctly forecast customer preferences and fit, hence lowering the chance of returns that result in waste or stock of a lower quality.

The promotion of sustainable consumption emphasizes buying less and durable garments, which makes the extension of product lifespans. Changes in consumer behavior, facilitated by policy and education, are essential for disrupting the fast fashion cycle that significantly contributes to microplastic emissions. Emerging technologies, including the Metaverse and digital fashion, present disruptive alternatives to conventional clothing, enabling individuals to express their identity and style while minimizing physical waste. The increasing trend of hyperpersonalization diminishes overproduction by closely aligning products with consumer needs and preferences. During the disposal phase, strategies including enhanced collection systems, textile recycling technologies, and the use of biodegradable materials contribute to minimizing the risk of garments decomposing into microplastics in landfills or aquatic ecosystems. Innovative business models such as clothing rental, resale platforms, and product-as-a-service facilitate more efficient circulation of clothing, thereby slowing resource throughput and minimizing waste. EPR policies assign accountability to manufacturers for the EoL effects of their products, promoting design strategies that mitigate microplastic pollution from the beginning. In summary, an integrated and forward-looking approach that incorporates digital transformation, regulatory support, consumer empowerment, and circular economy principles is crucial for mitigating the textile industry's impact on microplastic pollution. Embracing positive actions enables industry to transition to a model that integrates innovation with sustainability, fostering significant advancements in environmental protection and economic resilience.

8.8.1 Obstacles and Future Outlook

The implementation of these technologies requires consideration of the energy consumption associated with digital platforms, especially in blockchain-based fashion systems. Access and affordability for small-scale creators and consumers in developing regions. Digital literacy and the cultural acceptance of virtual fashion as a valid form of self-expression. Advancements in technology will increasingly integrate sustainable materials, closed-loop systems, and circular business models, thereby augmenting the microplastic reduction benefits associated with the Metaverse and hyperpersonalization. The integration of Metaverse and hyperpersonalization signifies a transformation that extends beyond digital advancements to encompass ecological solutions. The shift from physical to virtual fashion consumption, along with the customization of manufacturing to address individual needs, notably decreases the use of synthetic fabrics, resulting in a reduction of microplastic production. The proposed implementation is crucial for a sustainable and progressive fashion industry.

References

1 Ellen MacArthur (2024) We need extended producer responsibility (EPR) policy for textiles. https://www.ellenmacarthurfoundation.org/epr-policy-for-textiles (accessed on 10 January 2025).

2 European Environment Agency (2023) EU exports of used textiles in Europe's circular economy. LU: Publications Office. https://www.eea.europa.eu/en/analysis/publications/eu-exports-of-used-textiles (accessed on 10 January 2025).

3 Jahnz, A. Gospodinova, S., Stoycheva, D., and Miccoli, F. (2022) EU strategy for sustainable and circular textiles. EU Publications. https://environment.ec.europa.eu/strategy/textiles-strategy_en (accessed on 10 January 2025).

4 Köhler, A., Watson, D., Trzepacz, S., Löw, C., Liu, R., Danneck, J., Konstantas, A., Donatello, S. and Faraca, G. (2021) Circular economy perspectives in the EU textile sector. EUR 30734 EN, Publications Office of the European Union, Luxembourg, 2021 (JRC). (accessed on 13 January 2025).

5 Munhoz, D.R., Harkes, P., Beriot, N. et al. (2022). Microplastics: a review of policies and responses. *Microplastics* 2 (1): 1–26.

6 Rahaman, M.T., Pranta, A.D., Repon, M.R. et al. (2024). Green production and consumption of textiles and apparel: importance, fabrication, challenges and future prospects. *Journal of Open Innovation: Technology, Market, and Complexity* 10 (2): 100280.

7 Nayak, R., Panwar, T., and Nguyen, L.V.T. (2020). Sustainability in fashion and textiles. In: *Sustainable Technologies for Fashion and Textiles*, 3–30. Elsevier.

8 Periyasamy, A.P. and Tehrani-Bagha, A. (2022). A review on microplastic emission from textile materials and its reduction techniques. *Polymer Degradation and Stability* 199: 109901.

9 Periyasamy, A.P. (2024). Sustainable approaches in textile finishing to control microfiber releases. In: *Microfibre Pollution from Textiles*. London UK: Taylor & Francis Group.

10 TextileExchange (2014) The LCA studies of organic cotton fiber. PE International.

11 Venkatesan, H. and Periyasamy, A.P. (2019). Eco-fibers in the textile industry. In: *Handbook of Ecomaterials*, vol. 3, 1413–1433. Cham: Springer International Publishing.

12 Periyasamy, A.P. and Militky, J. (2020). Sustainability in regenerated textile fibers. In: *Sustainability in the Textile and Apparel Industries*, 63–95. Springer International Publishing.

13 Periyasami, S. and Periyasamy, A.P. (2022). Metaverse as future promising platform business model: case study on fashion value chain. *Business* 2 (4): 527–545.

14 Periyasamy, A.P. and Periyasami, S. (2023). Rise of digital fashion and metaverse: influence on sustainability. *Digital Economy and Sustainable Development* 1 (1): 16.

15 Lage, A. and Ancutiene, K. (2017). Virtual try-on technologies in the clothing industry. Part 1: investigation of distance ease between body and garment. *The Journal of The Textile Institute* 108 (10): 1787–1793.

16 Niinimäki, K. and Hassi, L. (2011). Emerging design strategies in sustainable production and consumption of textiles and clothing. *Journal of Cleaner Production* 19 (16): 1876–1883.

17 Trunk, U., Urbancic, N., and Nguyen, A. (2024) Fashion's plastic paralysis: how brands resist change and fuel microplastic pollution. Changing Markets Foundation. https://changingmarkets.org/report/fashions-plastic-paralysis/ (accessed on 10 January 2025).

18 European Environment Agency (2024) Microplastics from textiles: towards a circular economy for textiles in Europe. *European Environment Agency*. https://www.eea.europa.eu/en/analysis/publications/microplastics-from-textiles-towards-a-circular-economy-for-textiles-in-europe (accessed on 10 January 2025).

19 European Environment Agency. Textiles in Europe's circular economy. https://www.eea.europa.eu/en/analysis/publications/textiles-in-europes-circular-economy (accessed on 10 January 2025).

20 Sillanpää, A. (2025) Chosen with care & worn with love. Finnish Textile & Fashion.

21 Fashinnovation (2025) The future of fashion manufacturing: how on-demand production is reshaping the industry. Fashinnovation. https://fashinnovation.nyc/future-of-fashion-on-demand-manufacturing/#:~:text=Unlike%20traditional%20bulk%20production%2C%20on-demand%20manufacturing%20allows%20brands,the%20charge%20in%20this%20space%20is%20Tailored%20Industry (accessed on 10 January 2025).

22 Dugal, J. (2022) Digital fashion: NFTs revolutionise the virtual fashion space. https://fashionabc.org/digital-fashion-nfts-set-to-revolutionize-the-virtual-fashion-space/ (accessed on 18 January 2025).

23 Joy, A., Zhu, Y., Peña, C., and Brouard, M. (2022). Digital future of luxury brands: metaverse, digital fashion, and non-fungible tokens. *Strategic Change* 31 (3): 337–343.

24 DressX (2022) Digital fashion sustainability. DressX. https://dressx.com/pages/sustainability (accessed on 10 January 2025).

25 Noris, A., Nobile, T.H., Kalbaska, N., and Cantoni, L. (2021). Digital fashion: a systematic literature review. A perspective on marketing and communication. *Journal of Global Fashion Marketing* 12 (1): 32–46.

26 Bertola, P. and Teunissen, J. (2018). Fashion 4.0. Innovating fashion industry through digital transformation. *Research Journal of Textile and Apparel* 22 (4): 352–369.

27 Choi, K.-H. (2022). 3D dynamic fashion design development using digital technology and its potential in online platforms. *Fashion and Textiles* 9 (1): 9.

28 Xie, Y. and Yang, Q. (2023). Comparative analysis of marketing strategies of fast fashion brands ZARA and H&M. *BCP Business & Management* 39: 318–324.

9

Standardization and Regulation for the Microplastic Reduction

Mohanapriya Venkataraman[1], Jiří Militký[1], and Aravin Prince Periyasamy[2]

[1] *Department of Material Engineering, Faculty of Textile Engineering, Technical University of Liberec, Czech Republic*
[2] *Textile and Nonwoven Materials, VTT Technical Research Center of Finland, Espoo, Finland*

9.1 Introduction

The European Commission defines microplastics as "synthetic polymer particles below five millimetres that are organic, insoluble and resist degradation" [1]. According to National Oceanic and Atmospheric Administration (NOAA), plastic debris is defined as microplastics when the particles have a diameter lower than 5 mm. The sources of microplastics also change their classification – primary microplastics if released intentionally in the environment and secondary if they are released indirectly by deterioration processes [2]. Secondary microplastics are derived from the deterioration, fragmentation, or improper disposal in the environment of large pieces of plastic, such as plastic films, household garbage, atmospheric deposition, and vehicle emissions [3].

Microplastic pollution can adversely affect the environment and human health. It is not a localized problem but rather a global challenge [4, 5]. It cannot be resolved just by the local players or a single country. It requires all the countries to come together to define and redefine the rules of the game of microplastics to mitigate the huge risk it poses [5–7]. The risk is clear and present that fragments of microplastics are already found in the bloodstream and brain of humans and animals. It is necessary to develop effective methods to measure, analyze, and control the prevalence of microplastics. The world faces the challenge of streamlining and regulating the use of microplastics. Any regulation implemented needs to be equitable in nature [6, 8]. It has broadly discussed the current impact of microplastics, the means to identify, with strategies and solutions to overcome this global menace. Their work compared and contrasted the approach

Textile Microplastics in the Environment: Sources, Impacts, and Solutions,
First Edition. Aravin Prince Periyasamy.

of different countries. The overall strategies include banning plastic microbeads, reducing the usage of biodegradable plastics, and managing the lifecycle of microplastics [9]. The global phenomenon of microplastics pollution needs to be handled through a holistic strategy, including all the stakeholders. It requires a broad awareness creation across the spectrum of the supply chain. The producers and consumers need to understand the broad-based impact the microplastics have on our environment and its impact on the flora, fauna, and humans [10–12]. More investment is required in research in multiple areas. The research areas broadly include market studies on the consumption of microplastics, identification and development of sustainable alternatives, economic impact of adoption, and the development of new regulations and methods to measure. The drastic measures required to mitigate the massive risk that microplastics pose to the environment and human health need to be initiated immediately as we do not have the cushion of time available to us [3, 10, 13].

9.2 Global Standards for Microplastics and Their Quantifications

The ubiquitous presence of microplastics has evolved into a serious issue because of their far-reaching pollution and possible threats to both humans and ecological systems. While research on microplastics has grown, there has been limited progress by governmental bodies in confronting this newly emerging group of pollutants. Effective dialogue between scientists and bureaucrats is pivotal for transforming the research into actionable insights for administration. In California, legislative demands have led to the creation of techniques and approaches for discovering and evaluating the dangers linked to microplastics in drinking water and aquatic habitats [14]. Yet, more research is required to lessen ambiguities and tackle logistical roadblocks to the regulation of microplastics. A fundamental issue in microplastics research is the discordance between particle types utilized in toxicity experiments and those identified in natural settings, potentially leading to an underestimation of the risks. As a remedy, recent research has established frameworks to accommodate these discordances and used detailed monitoring data to evaluate environmental risks [15]. These evaluations reveal that risks associated with microplastics are substantial and cannot be ignored. Regarding drinking water, the potential risks to human health from microplastics have been gauged through a specialized workshop, which established toxicity limits and offered advice on conveying potential risks to the public. Nonetheless, uncertainties still exist in evaluating human risks and additional research is needed to fine-tune hazard evaluations. To enhance the precision and trustworthiness of risk evaluations, it's essential to gather superior monitoring data and employ a variety of analytical approaches. Open-source tools and databases can support data exchange and cooperation between researchers. Furthermore, employing tiered monitoring strategies and decentralized scientific methods can optimize resource use and encourage transparent cooperation. In conclusion, the journey toward regulatory emergence of microplastics demands ongoing research, data sharing, and collaboration among scientists, regulators, and policymakers. By tackling the challenges and uncertainties associated with microplastics research, we can effectively manage the risks to protect ecosystems and human health [14].

Almuhtaram et al. [16] discuss the need for standardized methods for sampling microplastics in water matrices. The potential health hazard that microplastics pose to human health needs to be quantified accurately. There is a lack of standardized methods for collection of microplastics in water. The authors have reviewed the best practices for collection, including using adequate water volume, minimizing contamination, using positive controls, and incorporating appropriate digestion and sample processing protocols. Recent studies [16] suggest that sampling >500 L of untreated water and > 1000 L of treated drinking water is necessary to generate representative data. In-line filtration methods, such as using cartridge filters, have advantages in processing large water volumes on-site and eliminating the need to ship samples to a lab for particle separation. Enclosed "in-line" filtration also reduces the need for field blanks and allows for the analysis of laboratory blanks to quantify contamination. The ASTM method for microplastic sampling in drinking water matrices has limitations, such as susceptibility to contamination and limited total volume that can be filtered. The article suggests adopting an appropriate and standardized microplastic sampling approach that includes the use of enclosed in-line filters. Standardization of microplastic processing and analysis methods is also necessary [16]. The article highlights the different techniques used for extracting microplastics from sampling equipment and emphasizes the need for standardized processing protocols. The State of California is making efforts to address the standardization of microplastic sampling and analysis methods. Overall, standardized methods for sampling, processing, and analyzing microplastics in water matrices are crucial for accurate assessment of their occurrence and potential health risks.

The United States Microbead-Free Waters Act, signed into law in December 2015, aims to eliminate microplastic pollution caused by plastic microbeads in selected cosmetic products [17]. However, the Act has been criticized for its limited scope and for treating all plastics the same, without considering innovative bioplastics with limited environmental impact. To address these issues, the authors propose a new standard called "Ecocyclable" that includes requirements related to toxicity, bioaccumulation, and degradation/assimilation into the natural carbon cycle. The standard aims to inform future regulation and legislation to reduce pollution while encouraging the development of sustainable technologies. The authors suggest that the lack of an appropriate standard for environmentally safe microplastics led to the imperfect language in the Act [17]. They propose a relative standard that distinguishes between environmentally benign materials and those that are persistent, bioaccumulative, and/or toxic. The standard includes representative environmental conditions and tests for degradability, bioaccumulation, and toxicity. The authors recommend the implementation of this standard to guide policy-making efforts and promote the development of environmentally safe materials [17] (Table 9.1).

In a perspective article by Denise M. Mitrano and Wendel Wohlleben, the authors argue that regulations on microplastics need to be more precise in order to encourage both innovation and environmental safety. They suggest that

Table 9.1 Comparison of four methods of collection of microplastic samples.

Method		Example	Approx. vol. range (L)	Filter surface area	Need for field blanks	Pressurized filtration
In-lab	Sample collection using bottle		1–20	Not applicable	Yes	N/A
On-site	ASTM sieve filtration		NR	324.3 cm^2 (203 mm diameter sieves)	Yes	No
	Low capacity in-line (filter holder)		100–3000	17.3 cm^2 (47 mm diameter sieves)	No	Yes
	High capacity in-line (filter holder)		500–10,000	543.4 cm^2 (24.765 cm height, 6.985 cm diameter cylindrical filter)	No	Yes

Source: Almuhtaram and Andrews [16] / American Chemical Society.

regulations should consider factors such as the harm caused by microplastics, the availability of replacement materials, and the quality, cost, and hazards of alternative materials. The authors propose that policymakers carefully evaluate when and where plastic restrictions are most necessary, prioritizing cases that can stimulate the development of sustainable materials and promote environmental health and safety. The authors also discuss the challenges of assessing the risks and impacts of microplastics, as well as the role of public perception and consumer behavior in shaping regulations. They emphasize the importance of a cost-benefit approach and the need for precise and measurable criteria for biodegradability in order to drive real change in plastic usage and handling. The article concludes by highlighting the potential of regulations to encourage industry and consumers to develop and adopt more sustainable alternatives to plastic. The authors stress the need for regulations to be based on scientific evidence and risk assessment, rather than solely relying on public perception. They call for a comprehensive and coordinated approach to address plastic pollution, including reducing plastic use, improving waste management, and promoting innovation in recycling and alternative materials. Overall, the authors argue that regulations on microplastics should be more targeted and enforceable, taking into account various factors and incentivizing sustainable alternatives. They emphasize the need for further research and development to better understand the impacts of microplastics and inform effective regulations [18].

A new standard method for determining the size, distribution, shape, and concentration of microplastic particles and fibers in water samples with high to low suspended solids has been developed by ASTM Committee D19 on Water. The method utilizes a dynamic image particle size and shape analyzer, which measures size, shape, and concentration by transmitting light through a cell containing particles in a liquid medium. The particles create shadows that are captured by a camera, and the images are used to analyze the particles. The method is intended to be used in conjunction with other identification methods for microplastics, such as spectroscopy and pyrolysis-gas chromatography/mass spectrometry. The proposed method has been validated using reference materials of known particle size and concentration. The results showed acceptable precision and bias within the scope of the method. The next steps involve analyzing actual sample matrices after sampling and preparation. The publication also includes information on the apparatus used, sampling and sample digestion procedures, and data analysis. The method is applicable to various water sources, including wastewater, rivers, ambient waters, and drinking water. The publication concludes with a list of abbreviations and references [19, 20].

The report discusses the regulation and standardization of microplastic analytics. It highlights the involvement of various organizations, such as the International Organization for Standardization (ISO) and the Comité Européen de Normalisation (CEN), in developing standards for microplastic analysis. ISO consists of 167 members and has 810 technical committees and subcommittees, while CEN includes 34 members and has more than 317 technical committees. The Vienna Agreement ensures that ISO and CEN coordinate their standardization work to avoid duplication. The report also provides a list of ISO technical committees involved in various industries, including plastics, water quality, and environmental management. The report mentions two ISO standards related to microplastic analysis: CEN ISO/TR 21960:2020 and ISO/FDIS 24187:2023. CEN ISO/TR 21960:2020 provides principles for plastic and microplastic analysis in the environment, while ISO/FDIS 24187:2023 establishes an identification system for suitable analytical tools. The report also discusses the ISO/TC 147 – SC2/JWG 1 technical committee, which focuses on the standardization of methods for the characterization and quantification of plastics, including microplastics, in water. It mentions a series of standards being developed for water quality analysis of microplastics. Furthermore, the report highlights the involvement of other technical committees, such as ISO/TC 38 Textiles, ISO/TC 61 Plastics, ISO/TC 147 Water quality, and CEN/TC 444 Environmental characterization, in developing standards for microplastic analysis in their respective industries. The report concludes by emphasizing the need for international standards to ensure comparability of data and the importance of avoiding contradictions in the standards. It also invites participation in the Joint Working Group (JWG) for further collaboration. It is a global concern, especially in marine ecosystems. This article by da Costa examines the effectiveness of various regulatory instruments and initiatives in controlling plastic pollution. At the international level, agreements like UNCLOS and the Basel Convention aim to prevent marine pollution and regulate plastic waste disposal. Regional initiatives, such as the Regional Seas Conventions, also contribute to marine biodiversity conservation. Many countries have implemented their own regulations, including bans on single-use plastics, charges for plastic bags, and restrictions on microplastics. For example, Kenya, Rwanda, and India have completely banned plastic bags, while others have introduced taxes or charges to discourage their use. Local initiatives, such as city and provincial bans on plastic bags, have also been effective in reducing plastic consumption. However, challenges remain in enforcing these bans and addressing waste management infrastructure issues. Social media has played a significant role in raising awareness about plastic pollution, with hashtags like #cleanup and #trashtag encouraging people to clean up litter and share their efforts online. To effectively control plastic pollution, a combination of regulatory instruments, public awareness campaigns, and improved waste management infrastructure is necessary. Collaboration and coordination between governments, industry, and the public are crucial for achieving lasting change. Ongoing research and proposed regulations are also focused on addressing the risks of microplastics and implementing reduction strategies to mitigate plastic pollution [21, 22].

9.2.1 Standardization for the Quantification

Microplastics are an emerging contaminant in the oceans, causing physical and toxicological risk to organisms. Larger plastic particles degrade into microplastics. Textiles release fiber fragments and plastics through wear and washing, textile construction, finishing processes, washing methods, age, and fiber type. In February 2023, the ISO launched ISO 4484-1, a new test method to help the textile industry select fabrics that reduce shedding or material loss during use and laundering. ISO 4481-1 Textile and textile products – Microplastics from textile sources – Part 1: Determination of material loss from fabric during washing is a method to determine the material loss from fabrics under standardized laundering conditions. It is a modification of ISO 106-C06. The test specimen is subjected to an accelerated laundering process under appropriate conditions of temperature, time, and mechanical action. The residual wash liquid is vacuum filtered, then the material loss is collected, and its mass is reported. ISO 4484-1 can be used to assess both synthetic and natural material loss [23].

Aquafil company announced the first standardized, global measurement system for microplastics in the textile sector: ISO 4484-2:2023. Over five years of work in collaboration with CNR of Biella STIIMA and the textile commission of UNI CT 046 has led to this accomplishment, which provides the global textile community with a method to qualitatively and quantitatively measure the impact of microplastics. Aquafil hopes this international standard gives designers and stylists an opportunity to consciously create their products from an eco-design perspective. A standard methodology that would enable the textile industry to precisely and unambiguously measure the release of microplastics, systemically correlate the effects to determine what causes the release, and construct products in a way that would limit their microplastic footprint. Together, these groups were able to establish a methodology to measure the microplastic footprint of a process and/or product in order to consequently develop technological solutions that can reduce it.

ISO 4484-2:2023 is the first official standardized global method for the qualitative and quantitative determination of microplastics (including fibrous plastics) in solid, liquid, or gaseous matrices coming from the textile sector. This new ISO standard enables the textile industry to quantify microplastic pollution for the first time, serving as an initial pathway to solving the overall microplastic issue. Aquafil hopes to encourage those in the textile industry to reduce their microplastic footprint through eco-design: selecting fibers and following production processes that will minimize the release of microplastics based on our findings. ISO 4484-2 will also serve as a powerful legislative tool, guiding legislative powers to enact regulations and limitations within the textile value chain [24–28].

AATCC TM212-2021 is a standardized test technique used to measure the amount of fiber fragments released during the process of home laundry [29].

The test method, referred to as "microfibers" or "microplastics," was created to measure the weight of fiber fragments generated during a high-speed washing process. It resolves ambiguity and establishes a standardized and consistent testing procedure for the worldwide industry. The standard encompasses not only synthetic fibers but also enables the assessment of the propensity of shedding for natural fibers and fiber mixes. The standard was formulated through the cooperative efforts of the AATCC Committee RA100 Global Sustainability, The Microfibre Consortium, and the Cross Industry Agreement of European Signatories. The test technique was formulated and endorsed through consensus by a heterogeneous panel of global specialists from various industries [7]. AATCC TM212 is the inaugural international standard developed to measure the discharge of fiber fragments. It serves as a valuable and widely adopted resource for the textile industry to enhance its comprehension of this matter [29].

9.3 Policies Toward the Reduction of Microplastics

The United Nations Environment Assembly (UNEA) was established in 2013 as the successor to the United Nations Environment Programme (UNEP) Governing Council. The UNEP Governing Council has been promoting international cooperation and driving environmental policies since 1972 [30, 31]. With its global membership and commitment to including civil society and stakeholders, UNEA plays a crucial role in shaping the global environmental agenda and promoting international environmental governance. In 2014, UNEA-1 addressed sustainable development goals and the post-2015 development agenda, which included sustainable consumption and production [32]. UNEA-4 is set to take place in 2019, during which member states will decide the future direction, timing, and outcome of the Expert Group's efforts on marine plastic litter and microplastics. The resolutions on marine litter have highlighted plastic as a major contributor to marine pollution. They have also emphasized the importance of addressing land-based sources of pollution, in line with Sustainable Development Goal (SDG) 14 and its target 14.1 to reduce marine pollution from land-based activities by 2025 [33–35]. The UN SDGs address the global problem of marine plastics through Goal 14, which focuses on the conservation and sustainable use of oceans, seas, and marine resources for sustainable development. By 2025, Target 14.1 seeks to reduce marine pollution, specifically from land-based activities such as marine debris and nutrient pollution. Coastal eutrophication and floating plastic debris density will be used as indicators for measurement [34].

9.3.1 EU Regulations

The European Commission (EC) enacted regulations in September 2023 to impose limitations on the presence of microplastics in various products such as cosmetics, detergents, medications, and medical equipment [36]. The provisional agreement to eliminate microplastics and other micropollutants from urban wastewater was agreed by the EC and European Parliament in January 2024. According to this agreement, pharmaceutical and cosmetic manufacturers would assume responsibility for a minimum of 80% of the expenses associated with the elimination of pollutants throughout the sewage treatment procedure. European Union (EU) member states will be obligated to eliminate a wide range of micropollutants from urban wastewater prior to its release into the environment by 2045. Producers of cosmetics, chemicals, and medicines in the EU might experience substantial financial consequences as a result of this interim accord. Companies must bear the expenses associated with collecting and validating data regarding the impact of their products on wastewater, potential financial consequences resulting from product reformulation, and regulatory examination of new product formulations [36].

Given the proven effects of microplastics, it's imperative to engage all relevant parties in constructive conversations and actions to alleviate the threat they pose to humanity. One initiative addressing this issue is the Operation Clean Sweep® program (OCS) [37]. This proposal is designed to diminish environmental plastic pellet loss, aiming for a 54–74% decline compared to the initial amount, which equates to a 6% reduction in total inadvertent microplastic emissions. In line with the Commission's goal for a 30% overall reduction of environmental microplastic release, this initiative aids in preserving ecosystems and biodiversity, diminishing potential health risks, and benefiting local economic activities. This also has the potential to enhance the understanding of pellet losses within the pellet supply chain. Measures tailored to small and medium enterprises (SMEs) will help moderate any potential impacts on their operations [37].

With microplastics' potential human and environmental health impact, certain key areas have started the action to decrease microplastic use. For instance, the Netherlands was the first to impose a ban on microbeads in cosmetic products in 2014, followed by a joint statement from Austria, Luxembourg, Belgium, and Sweden requesting an EU-wide microplastics ban later in the year. The United States soon enacted the Microbead-Free Waters Act of 2015, which prohibited manufacturing "rinse-off" cosmetic products containing plastic microbeads from 2017, and their retail sales from 2018. The EU's actions align with the global trend of recognizing microplastic pollution as an increasingly pervasive environmental concern, highlighting the need for coordinated regulatory efforts [1]. In line with its comprehensive zero-pollution action plan, and in response to the acknowledged long-term harmful effects of microplastics on ecosystems, the EU has recently proposed several initiatives to lower microplastics pollution from various sources: intentional use in products, plastic waste and litter, and accidental releases (e.g. plastic pellet loss, tire wear and tear, or release from clothes). These regulatory measures will affect the use of microplastics in cosmetics and medicines, among other items. The proposals follow global regulatory developments concerning microplastics, as well as bans on other potentially harmful substances in drugs and cosmetics. Collectively, these proposals suggest a growing regulatory focus on making the industry financially accountable for potentially harmful elements in their products and packaging. The proposals adhere to the widely accepted "polluter pays" principle, where polluters are financially responsible for managing the pollution they create to prevent harm to human health or the environment. This alert explores the EU's new initiatives on microplastics, their potential industry impact, and relevant regulatory trends globally.

The EC defines microplastics as "synthetic polymer particles smaller than five millimeters, organic, insoluble and resistant to degradation." Microplastics are frequently added to products like the granular infill material used on artificial sports surfaces, cosmetics, medicine, and detergents; they're also used in packaging. Microplastics can journey unaltered through waterways and into the ocean. Even though microplastics are everywhere, ongoing research is still assessing their impact on human health, animal health, and ecosystems. Nonetheless, studies have found that microplastics can affect the human body by triggering the release of endocrine disruptors and can also carry other toxic substances like heavy metals and organic pollutants, which can negatively affect the human body.

For the first time, the European Commission is putting forth measures aimed at curtailing microplastic pollution resulting from unintentional plastic pellet emissions. At present, due to mishandling across the supply chain, an estimated 52,000–184,000 tons of pellets make their way into the environment annually. The proposal seeks to ensure that all entities engaged in pellet handling within the EU implement necessary precautionary measures, with a target reduction of up to 74% in pellet emissions. The introduction of unified EU-wide measures also facilitates equal conditions for business operators. Unintentional plastic pellet discharge represents one of the largest sources of microplastic pollution. Therefore, the Commission is proposing that operators prioritize their actions in the following order: avoiding pellet spills through prevention, containing any spills that occur to prevent environmental contamination, and finally, performing clean-up operations following a spill or loss event [38]. The proposal includes:

- *Best handling practices for operators*: Depending on the size of the installation or transport activity, operators will have to abide by certain best handling practices. These have already been implemented by frontrunners.
- *Mandatory certification and self-declarations*: To assist national competent authorities in verifying compliance, larger operators should obtain a certificate issued by an independent third party, while smaller companies should make self-declarations of their conformity.

- *A harmonized methodology to estimate losses*: To help operators monitor their losses and tackle some of the remaining data gaps, a harmonized methodology will be developed by standardization bodies. It should also increase accountability by increasing awareness of the impacts of different practices on the environment and human health.
- *Lighter requirements for SMEs*: As the pellet supply chain has a large share of SMEs, lighter requirements will apply, especially to micro and small operators.
- The Commission's proposal for a Regulation on preventing pellet losses will now be discussed by the European Parliament and the Council. All economic operators, both EU and non-EU, will need to comply with the requirements set out in this regulation within 18 months of its entry into force.

The intentional use of microplastics in various products like sports fields, cleaning agents, diapers, and beauty products, to mention a few, poses a significant and hazardous contribution to plastic pollution. An estimated 42,000 tons of these detrimental microplastics end up scattered globally every year. They amass in our oceans and mountains, infiltrate animals, food, our drinking water, and ultimately, our bloodstream and organs. After years of deliberations at the European Commission asked the European Chemicals Agency (ECHA), and negotiations between the European Commission and member states of the EU, the restriction was ultimately approved on 25 September. It will officially take effect after 20 days. The Rethink Plastic Alliance, which has been supportive and closely followed the restriction process from its inception, expresses relief at the final adoption of the text, and is dedicated to ensuring its swift and effective enforcement. This prohibition could considerably aid in the removal of superfluous plastics from our environment, but it necessitates the collaborative commitment of all involved parties, such as national governments, local authorities, civil society, and the industrial sector, for proper enforcement. We strongly encourage companies that incorporate microplastics in their products, particularly those exempt from the restriction or granted extended transition periods, to promptly begin the usage of safe, microplastic-free alternatives [37, 39].

9.3.2 EU Microplastics Proposal

The EU's zero-pollution action plan set a target to reduce microplastics pollution by 30% by 2030. It is a collection of proposals aimed at reducing human exposure to microplastics. On 25 September 2023, the European Commission adopted measures that restrict microplastics intentionally added to products such as cosmetics, detergents, medicines, and medical devices. The first measures, including a ban on loose glitter and microbeads, went into effect in October 2023 [36]. Most recently, on 29 January 2024, the European Council and European Parliament reached a provisional political agreement on a proposal to remove microplastics and other micropollutants out of urban wastewater. Under this proposal, at least 80% of the costs needed to remove pollutants in the sewage treatment process would be covered by pharmaceutical and cosmetic producers, in line with the "polluter pays" principle. By 2045, the provisional agreement would require EU member states to remove a broad spectrum of micropollutants from urban wastewater before releasing it into the environment. This provisional political agreement would have a significant financial impact on producers of cosmetics, chemicals, and medicines in the EU. Under the proposal:

- Companies would have to cover the costs of gathering and verifying data on how their products impact wastewater;
- Potential cost implications arising from product reformulation to meet the proposed environmental risk minimization requirements; and
- New product formulations are subject to regulatory scrutiny by medicines regulatory authorities to ensure compatibility with the new requirements without adversely affecting the safety, quality, and efficacy of the reformulated products.

As a next step, the provisional agreement will be submitted to the EU members' representatives with the European Council and to the European Parliament's environment committee. Then, if formally adopted by both institutions, the directive will be published in the European Union's Official Journal [1].

9.3.3 EU Green Deal

In 2017, the ECHA proposed measures to address the pollution caused by microplastics used in products. Within two years, the ECHA made an ambitious proposal to ban all intentionally added microplastics, unless they meet criteria for specific derogations, such as being biodegradable. However, some products, like artificial sports pitches and cosmetics, were granted excessively long transition periods within the proposal, despite the fact that sustainable and effective alternatives are already readily available on the market. Furthermore, additional legislative measures on other microplastic sources are still needed to fully meet the 30% reduction target for microplastic pollution by 2030 [38]. The charter of the European Green Deal is that pollution harms our health and our environment and is identified as the largest environmental cause of multiple mental and physical diseases and of premature deaths, especially among children, people with certain medical conditions and the elderly. In addition to affecting people's health, pollution is one of the main reasons for the loss of biodiversity. It reduces the ability of ecosystems to provide services such as carbon sequestration and decontamination. The public health, environmental, moral, and socio-economic case for the EU to lead the global fight against pollution is today stronger than ever.

On 12 May 2021, the European Commission adopted the EU Action Plan: "Towards a Zero Pollution for Air, Water and Soil" – a key deliverable of the European Green Deal. More information about the implementation of the Action Plan is made available through the Zero Pollution Stakeholder Platform. The objectives of the deal were:

- The zero pollution vision for 2050 is for air, water, and soil pollution to be reduced to levels no longer considered harmful to health and natural ecosystems, that respect the boundaries with which our planet can cope, thereby creating a toxic-free environment. This is translated into key 2030 targets to speed up reducing pollution at source.

  These targets include:

- Improving air quality to reduce the number of premature deaths caused by air pollution by 55%;
- Improving water quality by reducing waste, plastic litter at sea (by 50%), and microplastics released into the environment (by 30%);
- Improving soil quality by reducing nutrient losses and chemical pesticides' use by 50%;
- Reducing the EU ecosystems where air pollution threatens biodiversity by 25%;
- Reducing the share of people chronically disturbed by transport noise by 30%; and
- Significantly reducing waste generation and by 50% residual municipal waste.

The Action Plan aims to strengthen the EU green, digital, and economic leadership, while creating a healthier, socially fairer Europe and planet. It provides a compass to mainstream pollution prevention in all relevant EU policies, to step up implementation of the relevant EU legislation, and to identify possible gaps. The European Green Deal announced headline actions on zero pollution [5].

- *Chemical strategy for sustainability*: To better protect citizens and the environment against hazardous chemicals.
- *Zero pollution action plan for water, air, and soil*: To better prevent, remedy, monitor, and report on pollution.
- *Revising measures to address pollution from large industrial installations*: To ensure they are consistent with climate, energy, and circular economy policies [40].

Figure 9.1 Route map for the European Green Deal. (*Source:* [41] / MDPI / CC BY 4.0.)

Climate change and environmental degradation are an existential threat to Europe and the world. To overcome these challenges, the European Green Deal will transform the EU into a modern, resource-efficient, and competitive economy, ensuring:

- No net emissions of greenhouse gases by 2050
- Economic growth decoupled from resource use
- No person and no place left behind

One third of the €1.8 trillion investments from the Next Generation EU Recovery Plan, and the EU's seven-year budget will finance the European Green Deal. The European Commission has adopted a set of proposals to make the EU's climate, energy, transport, and taxation policies fit for reducing net greenhouse gas emissions by at least 55% by 2030 (Fig. 9.1), compared to 1990 levels. More information on Delivering the European Green Deal.

9.3.4 United States of America

The Break Free from Plastic Pollution Act of 2023, which has been introduced in the US Senate and seeks to tackle the escalating issue of plastic pollution by offering nationwide guidance in mitigating the creation of squandering plastic and overhauling waste and recycling infrastructures. The objective of the Act is to achieve a 25% reduction in plastic waste both in terms of weight and plastic component by 1 January 2032, followed by a 40% reduction by 1 January 2040, and a 50% reduction by 1 January 2050 [42].

9.3.5 Australia

The Australian Government intends to engage in collaborative international efforts to address the issue of marine litter and microplastic contamination by means of a newly established global accord. Collaborate with the textile and white goods industries to implement microfiber filters on new household and commercial washing machines by 1 July 2030, following an industry-led approach [43].

9.3.6 South Korea

In 2023, South Korea introduced legislation to address the growing problem of microplastic pollution caused by increased plastic production. The legislation, under the environment and labor committee, aims to create a comprehensive framework for assessing, regulating, and promoting technology to reduce and eliminate microplastics. The legislation prohibits the commercialization, production, and importing of goods that contain primary microplastics that exceed safety regulations and pose immediate contact or exposure hazards [44]. The Plastic Resource Circulation Act (PRCA) has been implemented by the Japanese Government with the aim of fostering the recycling of plastics [45]. The PRCA advised making the laundry net (i.e. for the washing machine) with the objective of mitigating the release of plastic chemical fiber particles [5]. The aperture of the net is 0.05 mm, enabling it to effectively capture smaller fiber fragments [46].

The production of microplastics derived from textiles and their careless disposal have experienced a significant and rapid rise, presenting grave risks to the environment, aquatic life, food safety, human health, and livelihood. Several policies were established to solve the issue, each with differing degrees of defects that need to be addressed in the future, as outlined in Fig. 9.2.

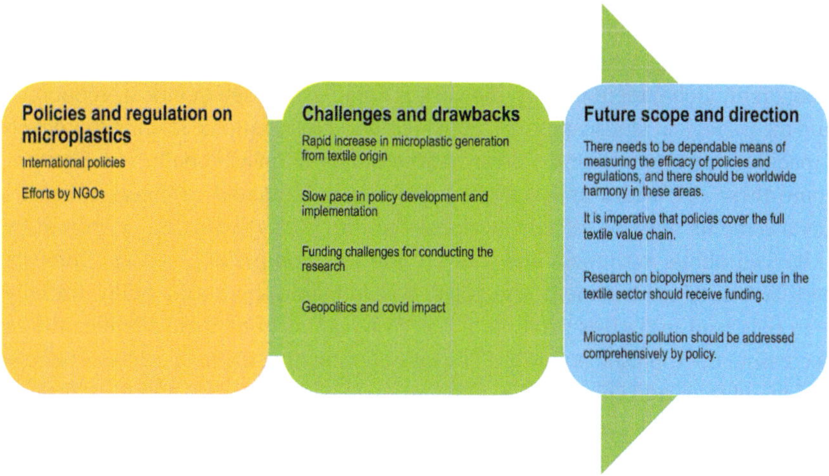

Figure 9.2 Summary of microplastic policies and regulations, challenges, and future directions. (*Source:* Periyasamy [5] / MDPI / CC BY 4.0.)

9.4 Conclusion

In conclusion, the burgeoning issue of microplastic pollution represents a deeply concerning environmental crisis that cannot be overlooked. These tiny plastic particles, barely visible to the naked eye, have insidiously permeated our terrestrial and aquatic environments, becoming an integral facet of the global ecosystem. The consequences of their ubiquitous presence are potentially wide-ranging and devastating - both for our planet's diverse bio-networks and for human health. Despite their small size, microplastics carry an environmental impact that is vast and disconcerting. They infiltrate soil and water bodies, gradually accumulating in these environments due to their durability and resistance to natural degradation processes. As they become embedded in these ecosystems, they are ingested by a variety of organisms, thereby entering food chains and bioaccumulating across trophic levels. The potential health implications of this for species higher up the food chain, including humans, may be significant, although the exact extent and nature of these implications are areas that need more comprehensive research. Furthermore, the properties that make microplastics so environmentally problematic – their durability, small size, and varied composition – also make them difficult to detect, quantify, and remove. These attributes present an array of challenges for scientists and environmental managers and highlight the pressing need for innovative technological solutions, methodologies, and strategies to better understand and manage this pervasive pollutant.

On the global stage, regulations governing microplastics are still in their formative stages. Certain regions like California and certain countries in the EU have taken a lead role in establishing legislative mandates and regulatory guidelines, setting a positive precedent. Yet, the task of harmonizing these into a coherent, global regulatory framework is complex. The geopolitical, economic, and technological landscapes vary considerably across nations, leading to uneven capacities and commitments toward addressing microplastic pollution. To overcome these disparities and achieve a globally coordinated response, scientific advancements must be effectively translated into policy measures, supported by enhanced public awareness and more responsible industrial practices. Comprehensive waste management strategies, research into environmentally friendly materials, and initiatives promoting recycling can all contribute significantly to decreasing microplastic pollution. The task of mitigating the microplastics issue is undeniably vast and complex, involving numerous stakeholders and multifaceted strategies. However, the urgency of the problem calls for swift, data-driven action, built upon the foundation of global collaboration and mutual understanding. Policymakers, scientists, industry leaders, and the public need to work in sync to navigate the challenges that lie ahead.

This chapter underscores the importance of addressing not only the knowledge and insights gleaned from current research on microplastics but also the uncertainties that remain. Only through such a holistic approach can we hope to formulate a robust, effective response to the global problem of microplastic pollution. The goal must be to protect our ecosystems and human health while striving toward a more sustainable relationship with our planet.

References

1 Tsang, L., and Kvedar, J. (2024). Macro regulation of microplastics: new EU initiatives and their global business impact. Ropes & Gray LLP.

2 Sadri, S.S. and Thompson, R.C. (2014). On the quantity and composition of floating plastic debris entering and leaving the Tamar estuary, Southwest England. *Marine Pollution Bulletin* 81 (1): 55–60.

3 Hale, R.C., Seeley, M.E., La Guardia, M.J. et al. (2020). A global perspective on microplastics. *Journal of Geophysical Research, Oceans* 125 (1): e2018JC014719.

4 Periyasamy, A.P. (2023). Microfiber emissions from functionalized textiles: potential threat for human health and environmental risks. *Toxics* 11 (5): 406.

5 Periyasamy, A.P. (2025). Functionalized textile microplastics: a closer look at the issues, strategy, and legislation on the microplastic reduction. *Kuwait Journal of Science* 52 (3): 100395.

6 Periyasamy, A.P. and Tehrani-Bagha, A. (2022). A review on microplastic emission from textile materials and its reduction techniques. *Polymer Degradation and Stability* 199: 109901.

7 Periyasamy, A.P. (2023). Environmentally friendly approach to the reduction of microplastics during domestic washing: prospects for machine vision in microplastics reduction. *Toxics* 11 (7): 575.

8 Periyasamy, A.P. (2024). Sustainable approaches in textile finishing to control microfiber releases. In: *Microfibre Pollution from Textiles.* London UK: Taylor & Francis Group.

9 Jain, R., Gaur, A., Suravajhala, R. et al. (2023). Microplastic pollution: understanding microbial degradation and strategies for pollutant reduction. *Science of the Total Environment* 905: 167098.

10 Qi, R., Jones, D.L., Li, Z. et al. (2020). Behavior of microplastics and plastic film residues in the soil environment: a critical review. *Science of the Total Environment* 703: 134722.

11 Pironti, C., Ricciardi, M., Motta, O. et al. (2021). Microplastics in the environment: intake through the food web, human exposure and toxicological effects. *Toxics* 9 (9): 224.

12 Rocha-Santos, T., Costa, M.F., and Mouneyrac, C. (ed.) (2022). *Handbook of Microplastics in the Environment*. Cham: Springer International Publishing.

13 Hidalgo-Ruz, V., Gutow, L., Thompson, R.C., and Thiel, M. (2012). Microplastics in the marine environment: a review of the methods used for identification and quantification. *Environmental Science & Technology* 46 (6): 3060–3075.

14 Coffin, S. (2023). The emergence of microplastics: charting the path from research to regulations. *Environmental Science: Advances* 2 (3): 356–367.

15 Vos, P., Meelis, E., and Ter Keurs, W.J. (2000). A framework for the design of ecological monitoring programs as a tool for environmental and nature management. *Environmental Monitoring and Assessment* 61 (3): 317–344.

16 Almuhtaram, H. and Andrews, R.C. (2022). Sampling microplastics in water matrices: a need for standardization. *ACS ES&T Water* 2 (8): 1276–1278.

17 McDevitt, J.P., Criddle, C.S., Morse, M. et al. (2017). Addressing the issue of microplastics in the wake of the microbead-free waters act – a new standard can facilitate improved policy. *Environmental Science & Technology* 51 (12): 6611–6617.

18 Mitrano, D.M. and Wohlleben, W. (2020). Microplastic regulation should be more precise to incentivize both innovation and environmental safety. *Nature Communications* 11 (1): 5324.

19 Lipps, W.C., Mizuno, Y., and Maeda, H. (2022). New standard determination of microplastics particle and fiber size, distribution, shape and concentration in waters with high to low suspended solids using a dynamic image particle size and shape analyzer.

20 Tiffin, L., Hazlehurst, A., Sumner, M., and Taylor, M. (2022). Reliable quantification of microplastic release from the domestic laundry of textile fabrics. *The Journal of The Textile Institute* 113 (4): 558–566.

21 da Costa, J.P., Mouneyrac, C., Costa, M. et al. (2020). The role of legislation, regulatory initiatives and guidelines on the control of plastic pollution. *Frontiers in Environmental Science* 8: 104.

22 Usman, S., Abdull Razis, A.F., Shaari, K. et al. (2022). The burden of microplastics pollution and contending policies and regulations. *International Journal of Environmental Research and Public Health* 19 (11): 6773.

23 SGS Société Générale de Surveillance SA (2023). ISO launches new test method to determine material loss from fabrics during washing. SGS Société Générale de Surveillance SA.

24 Ivleva, N.P. (2021). Chemical analysis of microplastics and nanoplastics: challenges, advanced methods, and perspectives. *Chemical Reviews* 121 (19): 11886–11936.

25 Sharma, P., Sharma, P., and Abhishek, K. (2024). Sampling, separation, and characterization methodology for quantification of microplastic from the environment. *Journal of Hazardous Materials Advances* 14: 100416.

26 Abdolahpur Monikh, F., Darbha, G.K., Vijver, M.G., and Peijnenburg, W.J.G.M. (2022). An overview of methodologies for tracing and quantifying microplastics in environmental samples. In: *Advances in Toxicology and Risk Assessment of Nanomaterials and Emerging Contaminants*, 21–46. Singapore: Springer Singapore.

27 Cui, T., Shi, W., Wang, H., and Lihui, A.N. (2022). Standardizing microplastics used for establishing recovery efficiency when assessing microplastics in environmental samples. *Science of the Total Environment* 827: 154323.

28 Aquafil. (2023). New international standard for measuring microplastics. *Aquafil.*

29 Wyman, D. (2021). *AATCC Method Quantifies Fiber Shedding*. AATCC.

30 United Nations. (2013). Change of the designation of the Governing Council of the United Nations Environment Programme: resolution/adopted by the General Assembly.

31 United Nations. (2012). Report of the Governing Council of the United Nations Environment Programme on its 12th special session and on the implementation of section IV.

32 United Nations. (2022). Report of the United Nations Environment Assembly of the United Nations Environment Programme, resumed 5th session.

33 United Nations. (2019). Report of the United Nations Environment Assembly of the United Nations Environment Programme, 4th session.

34 Walker, T.R. (2021). (Micro)plastics and the UN sustainable development goals. *Current Opinion in Green and Sustainable Chemistry* 30: 100497.

35 Deng, L., Cai, L., Sun, F. et al. (2020). Public attitudes towards microplastics: perceptions, behaviors and policy implications. *Resources, Conservation and Recycling* 163: 105096.

36 European Commission. (2024). Microplastics. https://environment.ec.europa.eu/topics/plastics/microplastics_en

37 European Commission. (2023). Preventing Plastic Pellet Losses to Reduce Microplastic Pollution. Explanatory Memorandum to COM (2023)645. PDC Informatie Architectuur.

38 Oprea, A. (2023). EC Proposes Measures to Reduce Microplastic Pollution from Plastic Pellets. Green Forum.

39 Rethink Plastic. (2023). European Commision finally restricts intentional use of microplastics in first concrete victory for ecosystems and human health. *Rethink Plastic alliance.*

40 Directorate-General for Environment Zero Pollution Action Plan. Zero pollution action plan.

41 Directorate-General for Communication The European Green Deal.

42 Merkley, J. (2023). Break free from plastic pollution act of 2023.

43 Australian Government. (2023). Plastics in our oceans and waterways.

44 Korean Ministry of Environment. (2023). South Korea – proposed bill on reduction and management of microplastics.

45 European Commission. (2022). The new law to reduce single-use plastics. https://environment.ec.europa.eu/topics/plastics/single-use-plastics_en

46 Ministry of the Environment, Japan. (2023). Good practices for reducing microplastics. https://www.env.go.jp/content/900453449.pdf.

10

Elimination of Microplastics from Marine Environments

10.1 Introduction

Small plastic particles that are less than 5 mm in size are known as microplastics, and they represent a substantial and persistent pollution in marine environments. They originate from the breakdown of bigger plastic goods or the direct discharge of microbeads and fibers from personal care products, synthetic clothes, and industrial trash. Both processes are directly responsible for their formation. Their presence in marine and freshwater habitats poses a significant risk to the health of humans as well as to the biodiversity and ecology of such environments. Pollution caused by microplastics has propagated from aquatic ecosystems to other parts of the environment, resulting in negative consequences. As the level of pollution throughout the world continues to rise, it is more important than ever to do research on the removal and reduction of microplastics. Microplastics can be consumed by a variety of animal species as well as by people, with the biological impact mostly focused on the physical damage and the leaking of chemical components. The problem arises from the fact that wastewater treatment facilities (WWTPs) do not focus their attention on the removal of microparticles. However, because of the vast amounts of wastewater that were treated, it was possible that some kinds of microplastics were kept. The removal of all microplastics from the influents that are carried to wastewater treatment plants is not a given. The methods that are now utilized for the treatment of wastewater are not intended to retain minute elements. This is because these methods were first developed for the purpose of treating various types of contaminants, including biological agents, heavy metals, and organic and inorganic substances. This chapter looks at several different methods used to remove microplastics from the marine environment. The chapter has a specific emphasis on the physical, chemical, and biological methods of reduction (Fig. 10.1), with a particular emphasis on the various adsorption, enzyme hydrolysis, and microbial breakdown methods.

Textile Microplastics in the Environment: Sources, Impacts, and Solutions,
First Edition. Aravin Prince Periyasamy.

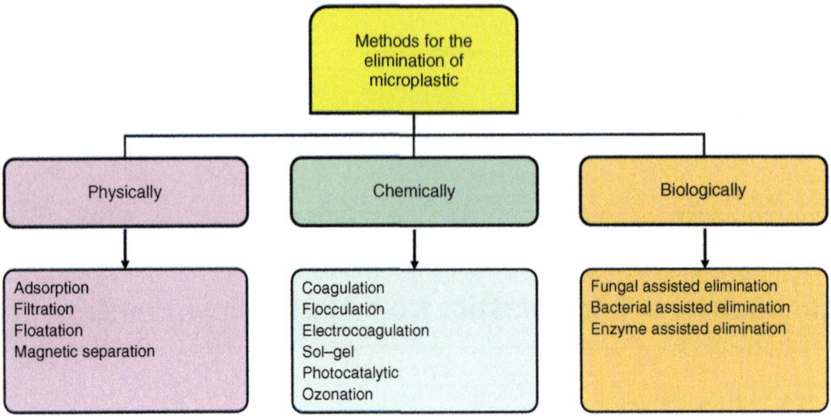

Figure 10.1 Methods for the elimination of microplastics.

10.2 Physical Method

Several distinct physical approaches and the impact that they have on the removal of microplastics are covered in this section. For example, adsorption is a process that is used to remove dissolved contaminants from solid surfaces and liquids. Furthermore, it has been expanded to include the ability to catch microplastics that are floating or suspended, particularly those that have hydrophobic or charged surfaces. According to the surface chemistry, porosity, hydrophobicity, and functional groups of the adsorbents, the success of this approach is dependent on these characteristics [1].

In addition to being biodegradable and inexpensive, biochar is a carbon-rich substance that may be generated by the pyrolysis of biomass. It can also be enhanced with magnetic nanoparticles to facilitate its recovery. Carbon nanotubes (CNTs) are nanostructures that are tubular in shape and possess an outstanding surface area and adsorption capability. These nanostructures have the potential to be useful for the capture of very tiny particles. Sponges and polymeric foams are examples of porous polymeric sponges that have been developed with hydrophobic or amphiphilic coatings. These sponges are perfect for floating plastics in marine surface layers. A magnetic adsorbent is a composite material that combines adsorptive materials with magnetic nanoparticles. It is designed to capture microplastics by adsorption, which is then followed by magnetic separation. In addition to providing surface interaction sites, fibrous adsorbents also provide physical entrapment [1, 2]. In the process of filtration, microplastics are physically separated from one another on the basis of size exclusion, variations in density, or affinity to the filter material. Sand filters could capture particles through the processes of interception, sedimentation, and adhesion. However, their effectiveness is diminished when it comes to microplastics that are smaller than 100 μm. Magnetic filtration makes use of magnetically sensitive materials that attract or bind microplastics, whereas biofilters contain biological media that encourage the development of microorganisms. Traditionally utilized in the mineral processing industry, froth flotation is a surface-based separation technology that has recently been adopted for the removal of particles. In this process, air bubbles are injected into water, and hydrophobic microplastic particles adhere to the bubbles, causing them to rise to the surface as froth. Bubble size and distribution, collection agents, pH and water chemistry, and the capability to be merged with adsorption for better efficiency are some of the most important elements [1].

10.2.1 Adsorption of Microplastics in Marine Environments

Adsorption is the process by which particles attach to the surface of another substance, with both natural and artificial materials serving as adsorbents to collect microplastics in marine environments. Efficient removal methods using adsorption processes include chemically produced sponge materials [3], graphene materials [4], and biochar materials, which may efficiently eliminate microplastics and nanoplastics from aquatic ecosystems. Figure 10.2 describes the different types of adsorbents for the removal of microplastics. The kinetics of adsorption and desorption may differ based on the kind of microplastics, including polyethylene (PE), polypropylene (PP), polystyrene (PS), and polyvinyl

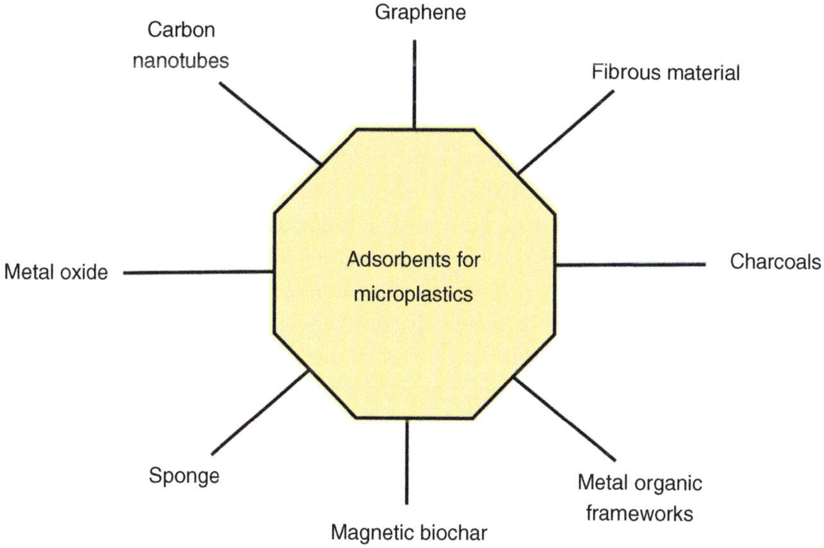

Figure 10.2 Different types of adsorbents for the removal of microplastics.

chloride (PVC) [5]. The adsorption of polyfluoroalkyl substances (PFAS) on microplastics is governed by electrostatic and hydrophobic interactions, with electrostatic interactions prevailing for short-chain PFAS and hydrophobic interactions for long-chain PFAS. The ambient presence of organic matter may alter the adsorption of both long-chain and short-chain PFAS on microplastics. This may happen via the interaction of organic matter-adsorbed microplastic surfaces with PFAS via electrostatic or hydrophobic forces. Hydrophobic organic matter has been shown to improve per- and polyfluoroalkyl substances (PFAS) retention in granular activated carbon (GAC) adsorption, but hydrophilic organic matter does not substantially influence PFAS adsorption on either GAC or anion exchange [6]. Numerous studies indicate that the presence of organic matter may diminish the sorption of PFAS on activated carbon fiber, maybe due to competition for active sites or pore obstruction. An elevation in organic matter content to 500 mg L^{-1} resulted in little sorption, indicating competitive sorption between organic matter and PFOA, as well as pore obstruction of activated carbon fiber [7]. A separate investigation indicated a 40% decrease in adsorption equilibrium for the elimination of perfluorinated carboxylates (PFCAs) when organic matter was present [8]. The adsorption processes of pollutants, including PFAS, on microplastics seem to be influenced by the presence of organic matter. The fundamental processes of contaminant adsorption on microplastics are not well comprehended, necessitating future research to alleviate contaminant adsorption on microplastics and eliminate microplastics from water.

The use of sponge materials has the potential to effectively remove a wide variety of toxins from wastewater. At the moment, several studies have utilized natural organic compounds as raw materials in order to manufacture sponge materials for the purpose of adsorbing microplastics in aquatic environment. The chemical cross-linking of plant proteins (amino acids) resulted in the synthesis of a green sponge material that is both natural and biodegradable, and it has exceptional mechanical properties, which are shown as a promising material for the removal of microplastics. Because the amino acids include a large number of active side chains, the protein sponge is able to demonstrate an extraordinary adsorption capacity for microplastics. A removal efficacy of up to 81.2% has been achieved by Wang et al. [9], demonstrating a positive outcome. In another study [3], chitin and graphene oxide have been utilized to fabricate sponges, demonstrating satisfactory performance and reusability. Every material exhibited above 70% removal efficiency for pure polystyrene and its derivatives.

Aerogels are extensively porous structures characterized by a substantial specific surface area and elevated porosity. Polydopamine-modified magnetic chitosan material has been created to exhibit excellent adsorption capabilities for PET, PE, and PS. The mechanism is governed by electrostatic attraction and physical adhesion between aerogels and magnetic nanoparticles.

A quaternary ammonium salt-modified cellulose nanofiber aerogel exhibits a high adsorption capacity for polystyrene microplastics; nevertheless, its stability and reusability require enhancement. Cellulose fibers exhibit significant adsorption ability for nanoparticles due to their elevated specific surface area. Polyethyleneimine-coated cellulose fibers exhibit enhanced adsorption capabilities for polymethyl methacrylate (PMMA), PVAc, and PVC nanoparticles from contaminated water samples. Polyethyleneimine-coated cellulose fibers facilitate the introduction of amino groups into the surface of cellulose fibers, enhancing electrostatic interactions [10].

Microplastics may be efficiently eliminated from aquatic solutions by adsorbing them via surface charge interactions with organic materials such as GAC and untreated coffee grounds [11, 12]. The adsorption processes of microplastics by chemically produced sponge materials depend on hydrogen bonds and electrostatic interactions between the microplastics and the sponge materials. Microplastics in aquatic ecosystems possess a negative charge, facilitating their interaction with positively charged molecules [3]. The adsorption techniques for microplastics utilizing micromotors are predicated on phoretic interactions, shoveling, non-contact shoveling, and adsorptive bubble separation. The synergistic adsorption processes of surface and bubble separations are essential in the adsorption of microplastics [13]. Biochar, originating from plant biomass, is environmentally sustainable and potentially biodegradable. Biochar, produced from inexpensive biomass, is more possibly produced from agrowaste, which is considered as an underutilization, and is extensively utilized for the adsorption of environmental pollutants owing to its physicochemical properties, including a substantial specific surface area, abundant functional groups, elevated porosity, and environmental sustainability. The adsorption capability of biochar is contingent upon its nanostructure and surface chemistry [14]. High-temperature biochar has an enhanced surface area and pore density, a reduction in carbonyl functional groups, and improved efficacy in the removal of polysaccharides. Surface functionalization also affects the adsorption efficacy of biochar. For instance, the corncob biochar, oxidation under the high-temperature pyrolysis enhances oxygen-containing functional groups and specific surface area, which promote hydrogen bonding and substantially aid in the adsorption of polystyrene microplastics [15]. A biochar adsorption-hydrothermal method has been established to effectively decompose polystyrene nanoparticles on biochar, yielding low molecular weight oligomers. The degrading process maintains biochar characteristics and demonstrates excellent reusability. This adsorption-degradation approach is crucial for the valorization of plastic waste in the production of essential chemicals [15]. Innovative advancements in the application of biochar have emerged to address the challenges associated with its post-adsorption separation, leading to the development of magnetic biochar endowed with

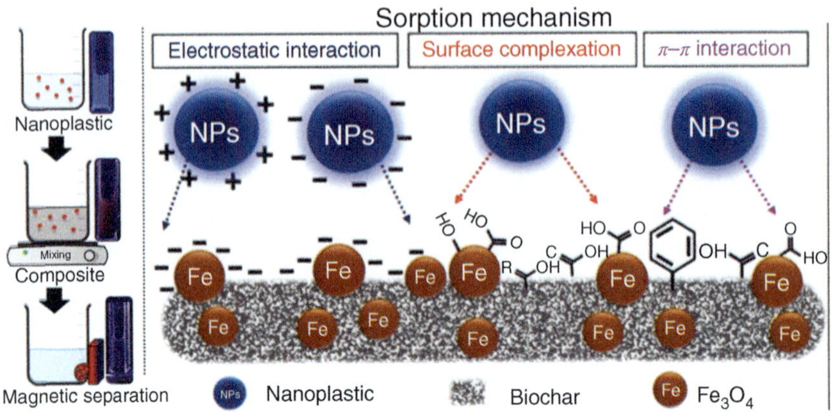

Figure 10.3 The mechanism of adsorption process of nanoplastics on Fe-modified biochar. (*Source:* Singh et al. [16] / with permission of Elsevier.)

magnetic properties and metal functionality. Either nano- or microplastics-loaded Fe-modified biochar, for instance, can be efficiently extracted from reaction solutions via magnetic separation techniques [16]. The presence of Fe species, predominantly Fe_3O_4, significantly enhances the adsorption performance by providing active sites for nanoparticle surface complexation. This Fe-modified biochar demonstrates rapid elimination of all nanoparticles from water within a mere 10 min and maintains commendable reusability across four cycles (Fig. 10.3).

Various carbon materials, including activated carbon, graphene oxide, algae, and seaweed, tea waste, coffee waste, agrowaste, and magnetic CNTs, have been used for the adsorption of microplastics and nanoplastics. These materials differ in their adsorption methods; for example, graphene oxide or reduced graphene oxide adsorbs polystyrene microplastics through strong π–π interactions. Untreated coffee waste, which demonstrates significant adsorption capacity within 40 min. This provides evidence of reutilization of biowastes due to the cost-effectiveness of adsorbents [12]. Algae and seaweed serve as natural filters, particularly hydrophobic varieties, by adsorbing microplastics. Synthetic adsorbents, such as polystyrene, polyethylene, and polypropylene, have an affinity for microplastics and have been engineered for selective extraction. Magnetic nanoparticles, altered with certain functional groups, can attract and eliminate microplastics from marine habitats by specifically targeting and segregating them using magnetic fields [17].

The efficacy of microplastics removal is affected by several parameters, including pH, temperature, types of adsorbents, dissolved organic matter, and ions. Generally, pH and temperature are the two crucial factors, since they affect the surface charge of microplastics and the adsorbent. Temperature influences the diffusion rate of the adsorbate and the equilibrium capacity, with elevated temperatures resulting in increased adsorption. Various adsorbent materials exhibit distinct electrostatic and hydrogen bonding interactions with microplastics, hence modifying their adsorption capabilities. Dissolved organic matter alters the interaction between adsorbents and microplastics, whereas ions affect the electrostatic attraction between microplastics and the adsorbent. The adsorption removal of microplastics has benefits like elevated adsorption capacity, little energy usage, and reusability. Nonetheless, adsorbents must be eluted post-usage, presenting a potential danger of reintroduction into the environment. Adsorption is predominantly successful for suspended microplastics, although they may encounter difficulties in addressing microplastics that have settled in sediment or are firmly lodged in aquatic creatures. The efficacy of adsorbents can be affected by variables like as salinity, temperature, and the dimensions and morphology of microplastics. Opportunities involve investigating the integration of natural adsorbents with modern materials like as biochar and magnetic nanoparticles to enhance the efficiency and sustainability of microplastic removal. When it comes to the adsorption of contaminants onto microplastics in marine environments, salinity is one of the parameters that carries influence. Additionally, the surface roughness may have an effect on the sorption capacity; nevertheless, the adsorption process takes place at the molecular level, and at this level, the porosity of the adsorbent plays a significant role.

In addition to these adsorbents, there are many types of adsorbents, such as natural adsorbents for microplastics, which include clay minerals like bentonite and kaolinite, charcoal, algae, and seaweed. Clay particles have a significant surface area and the ability to exchange cations, which enables them to effectively trap microplastics in ecosystems that contain both freshwater and saltwater.

10.2.2 Filtration

The removal of suspended particles from fluids is accomplished by the use of filtration, which is a physical separation method that is extensively utilized. This process involves passing the fluids through a porous medium or filter. When it comes to the issue of microplastic contamination, filtration is an extremely important component in the process of absorbing and eliminating microplastics from wastewater, surface water, stormwater runoff, and even drinking water systems. Filtration systems need to be carefully designed or changed in order to successfully target a wide variety of particles, particularly those that are on the nanometer to micrometer scale. This is because microplastics vary in size, shape, density, and have different surface chemistry.

It has been demonstrated that the utilization of ozonation in conjunction with GAC filtration results in a substantial reduction in the quantity of microplastics present in the GAC filtering phase, offering removal efficiencies above 55% [18]. In spite of this, further study is required to completely comprehend the manner in which GAC filtration eliminates microplastics. Ultrafiltration has the potential to enhance the removal of microplastics in wastewater treatment plants (WWTPs), but it is not completely efficient since microplastics are released on a regular basis. To further understand the parameters that influence the effectiveness of microplastics removal, additional study is required. Additionally, membrane bioreactors and other advancements in treatment technology have the potential to enhance the efficacy of WWTP for microplastics removal. However, no study has been able to attain a removal efficiency of 100% for microplastics. When developing technologies for the last stage, it is important to take into account the complexity of the wastewater, the population and activities in the surrounding area, as well as the cost-effectiveness of modern technology in order to improve the effluent quality of wastewater [19]. Sand filtration is a common method for treating wastewater; nevertheless, it is not regarded to be the most efficient way for removing microplastics from the water. Microplastics of a greater size are found to be maintained in waste material, and it has been demonstrated via research that some microplastics can become entrapped in sand granules or attach to the surface of sand granules [20].

In WWTPs, disc filters are a popular tool for removing contaminants and particles. They work well to lower the quantity of microplastics in wastewater effluent. According to research, disc filters have an 80–98% removal rate for microplastics, with comparable mesh sizes reaching up to 89% [21]. To improve particle recovery, Talvitie et al. (2017) employed 24 pilot-size disc filters with an iron coagulant and a cationic polymer [21]. Water quality, filter design, and pore size all affect how effective disc filters are. The efficiency of disc filters in eliminating microplastics can also be increased by using additional treatment methods like flocculation, coagulation, and advanced oxidation [22].

Biofilters are extensively employed to eliminate pollutants from stormwater and wastewater effluent, comprising media such as GAC and anthracite sand. Biofilters are now utilized to remove microplastics, effectively removing polyamide, polyethylene, polypropylene, and polystyrene, with retention rates exceeding 97% in charcoal and bark filters [23]. Biofilters composed of several plant species, including *Armeria maritima*, *Hippophae rhamnoides*, *Juncus effusus*, and *Festuca rubra*, demonstrate efficacy in the removal of microplastics exceeding 10 μm in size [23]. Biochar, a solid carbon-rich substance produced from biomass, has use in water and wastewater treatment, soil quality improvement, carbon sequestration, and energy storage systems. Its extensive pores facilitate the adsorption of microplastics via physisorption. Experiments using activated carbons have shown significant success in the removal of microplastics, whereas non-activated biochar is employed for the extraction of bigger particles [24]. Biochar may be produced from several materials, including maize bark, pine, hardwood, and spruce. Research indicates that biochar filters may efficiently eliminate microplastics under diverse circumstances. Biochar-based adsorbents derived from spruce and pine bark achieve 100% efficacy in removing microplastics; nevertheless, this method does not substantially diminish microplastic particles at the micrometer scale [24]. Additional organic resources, including pinewood and sugarcane, have been utilized for biochar production. Research indicates that biochar produced from jujube waste may eliminate more than 99% of nylon and ethylene from water. Biochar derived from rice husk demonstrated a microplastic removal effectiveness of 99% in aquatic environments [25]. In summary, biofilters and biochar provide effective techniques for the elimination of microplastics from rainwater and wastewater effluent. Additional investigation is required to comprehensively comprehend their processes and efficacy.

The hydrophobicity-based method for oil film extraction effectively removes microplastic particles from water. Researchers have attained recovery rates as high as 96% with canola oil extraction [26]. Different types of edible oil have been used, and for instance, castor oil shows the better removal efficiency on the microplastics, which accounts for more than 90% [27], perhaps it shows the promising oil film extraction. A technique utilizing vegetable oil was devised for the extraction of microplastics from water samples, employing high-speed agitation to generate bubbles that adhere to oil films, then removed with reagent alcohol. This approach necessitates an additional stage for elevated biogenic loads and demands supplementary cleaning techniques. The amount of oil and the salinity of water are the primary parameters influencing the efficacy of oil film techniques [27].

10.2.3 Froth Flotation Method

An effective mineral processing method called froth flotation adhesively bubbles to minerals only on demand. With materials whose hydrophobic outer surface floats as froth aggregations and their hydrophilic counterparts transported as underflow, surface wettability is the fundamental component of froth flotation. More likely to become buried with a bubble cluster and have a hydrophobic low-energy face are microplastics, including Polystyrene, PET, PVC, PLA, and PBS [28]. Small microplastic particles from wastewater can be effectively removed by froth floatation including ultrafine bubbles. Combining floatation and coagulation will increase microplastic removal efficiency. An advanced nanobubble flotation technique raises attachment probability and promotes bubble-particle collision. Natural cationic polymer chitosan (CTS) was chosen to increase polystyrene separation by air flotation, so attaining removal efficiency up to 95% [29]. Electrostatic attraction, polymer types, hydrophobicity, bridging adsorption, aeration volume, treatment time, pressure, and pH all influence froth floatation. Large concentrations of heavy metals, drugs, plasticizers, and other persistent organic pollutants can be found among microplastics. Foreign functional groups introduced on microplastic surfaces could influence their capacity for hydrophilization and floatability. Enhanced flotation potential is displayed by small microplastics. Because of its strong magnetic qualities and capacity to bind with microplastics, magnetic separation is a very efficient approach for eliminating microplastics from wastewater. This approach is under control by electrical friction, hydrogen bond formation, and complexation and employs magnetic seeds including CNTs and iron nanoparticles. For small-sized microplastics, its efficiency is 93%; for polystyrene removal with modified Maifanite by rotating magnetic field, it ranges up to 98% [30]. While other pollutants reduce efficacy and selectivity, the form and size of microplastics also affect the separation process. Better adsorption capacity, lower sludge waste, and enhanced separation resulting from long-range magnetic force are just a few of the benefits magnetic separation provides. More study is required, though, to improve the efficiency of magnetic removal [31].

10.3 Chemical Method

The use of a number of different chemicals in a series of reactions that are designed to improve water purification by removing the microplastics from the waterbodies is known as chemical treatment. Microplastics that are present in water are eradicated by the use of chemical treatment, which includes the incorporation of chemicals that are specifically targeted. Chemical treatments are frequently necessary in situations when microplastics cannot be removed from wastewater through the use of physical and biological methods [31]. The harmful microplastics are converted into molecules that are less poisonous using these chemical methods.

10.3.1 Coagulation and Flocculation

Since coagulation is one of the most important technologies for the removal of microplastics in waste water treatment plants. In general, the flocculation/coagulation process is designed to separate pre-existing colloidal particles by neutralizing their charge, generating floccules, and removing them by sedimentation or filtering. This is the fundamental purpose of the technique to remove microplastics from the waterbodies. Among the coagulants that are utilized most frequently are ferric chloride, ferric sulfate, and aluminum sulfate [32]. When it comes to eliminating polystyrene, $FeCl_3$ demonstrated a removal effectiveness of 77%. At a rate of up to 92%, 84%, and 96%, respectively, the removal of low-density polyethylene, high-density polyethylene, and polypropylene

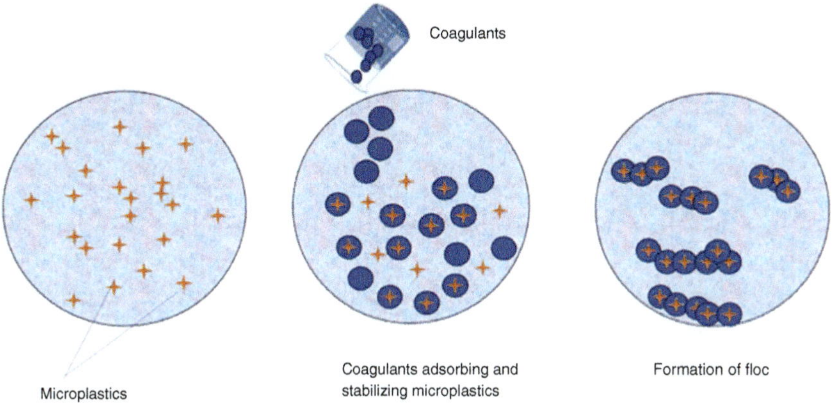

Figure 10.4 Mechanism of coagulation method for the removal of microplastics. (Source: Dayal et al. [34] / with permission of Elsevier / CC BY-NC 4.0.)

from synthetic stormwater may be accomplished by the utilization of alum in conjunction with polyacrylamides [33]. For the removal of microplastics, it is believed that a combination of coagulation and flocculation techniques is the most successful way. The coagulation process is a widely utilized technique for wastewater treatment; nonetheless, it has operational disadvantages, including the aggravation of environmental problems due to sludge accumulation [34]. Future studies must determine the appropriate coagulants and flocculants, as well as the optimal conditions for the removal of microplastics. Notwithstanding comprehensive investigations on microplastics elimination, research deficiencies persist. Future research should concentrate on standardizing detection methodologies, enhancing pilot-scale testing, investigating the properties of microplastics and environmental variables, examining synergistic effects, formulating rapid, safe, and effective coagulants, and advancing integrated coagulation technologies to address the escalating microplastic pollution. Apart from that, it is possible that conventional coagulation and flocculation only remove a small percentage of pristine microplastics (less than 10%) [34]. One possible explanation for this low clearance rate is that the coagulant and the surface of the clean plastic do not interact with one another. As a result of changes in their surface roughness, weathered microparticles demonstrated a greater removal efficiency through coagulation and flocculation than fresh particles did under the same conditions [34]. According to some reports, the coagulation of aluminum is more effective than the coagulation of iron. The mechanism of the coagulation technique for microplastic removal is well illustrated in Fig. 10.4.

10.3.2 Electrocoagulation

Electrocoagulation is an economical and eco-friendly technique for eliminating microplastics from wastewater effluents. It employs electrical current to eliminate microplastic fibers, providing great removal efficiency and minimal investment expenses. Electrodes (Fe^{3+} or Al^{3+}) are generated during the electrochemical process, which, upon contact with pollution particles, this mechanism is well explained in Fig. 10.5. Conventional electrode electrocoagulation encounters challenges such as electrode passivation and metal sludge contamination. Electrocoagulation is a potential method for water treatment because of its non-toxic and innocuous characteristics. It may also be utilized to retrieve sludge for the of useful TiO_2 nanoparticles [35]. Ti anodes demonstrate decreased overpotentials for chlorine and oxygen evolution processes, hence decreasing energy consumption throughout the process. Electrocoagulation generates oxidizing agents such as hydroxyl radicals and active chlorine, demonstrating enhanced corrosion resistance relative to iron and aluminum electrodes. In another work, the Al-Al electrode electrocoagulation technique effectively removes microplastics of

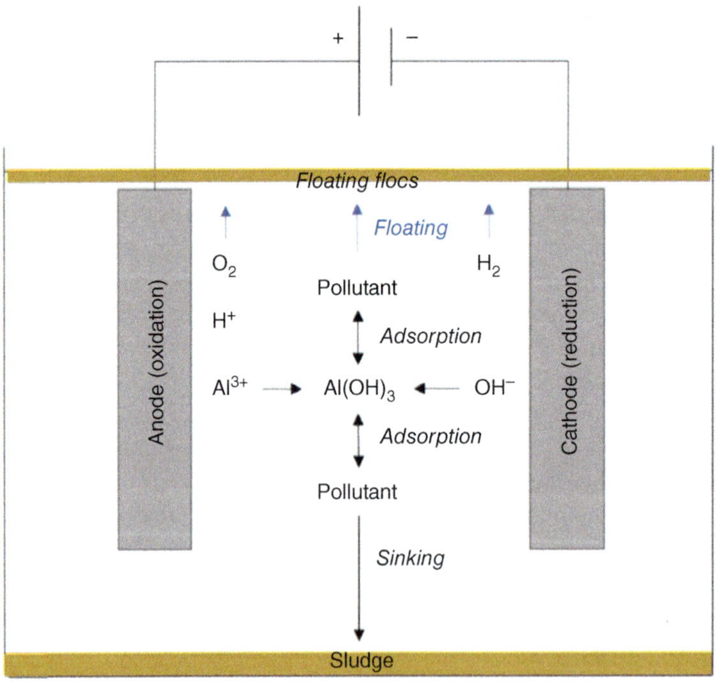

Figure 10.5 Reactions involved in electrocoagulation. (*Source:* Tang and Hadibarata [37] / with permission of Elsevier / CC BY-NC-ND 4.0.)

different size ranges, with removal efficiencies above 90%. Electrocoagulation may be effectively utilized across diverse wastewater characteristics, yielding less sludge waste and water with diminished dissolved solids. Researchers are exploring this economical and eco-friendly approach as a potential alternative to traditional coagulation methods [36].

10.3.3 Electrochemical Oxidation

By converting organic pollutants like microplastics, and pigments into harmless byproducts like carbon dioxide and water vapor, electrochemical oxidation is a cost-effective method for treating wastewater [38]. This method removes the need for biological agents, which is a need for the treatment of wastewater. Indirect cathode and anode oxidation is the method that is utilized in this environmentally beneficial and cost-effective technology. There are several different types of electrodes that have been shown to be effective in the removal of microplastics. In order to reduce the dimensions of PVC and to induce surface fissures in microplastics, composite anodes have been utilized. It has been established that electrodes made of TiO_2 and stannous oxide that have been doped with rare earth elements have a 77% efficiency rate in the electro-oxidative removal of microplastics from water [39]. Boron-doped diamond (BDD) electrodes have demonstrated a remarkable effectiveness of 65% in the removal of polystyrene from wastewater. There are a number of factors that can affect the efficiency of electrochemical oxidation. These factors include surface area, material composition, current intensity, type, electrolyte content, and the length of time that the degradation reaction is carried out [38].

10.3.4 Ozonation

Ozonation is a tertiary treatment that is utilized in WWTPs with the purpose of degrading the microplastics that contain oxygen compounds [40]. Through the modification of water characteristics, the facilitation of plasticizer migration into water, and the induction of structural changes in microplastics, it is able to efficiently eliminate them. A study conducted by Kye and colleagues examined the surface modification of microplastics both before and after being subjected to ozone treatment [41]. The findings of the study indicated that the pristine microplastics displayed smooth surfaces, which served as reference materials for comparison. A gradual rise in surface abrasion and the formation of cracks was noticed as the ozone dosage rose from 0 to 45 mg L^{-1} (Fig. 10.6). This could be seen as the ozone dose increased. Based on these morphological changes, it appears that exposure to ozone layers causes the surface layers of microplastics to get oxidized. The oxidation of the surface layers of the microplastics, as compared

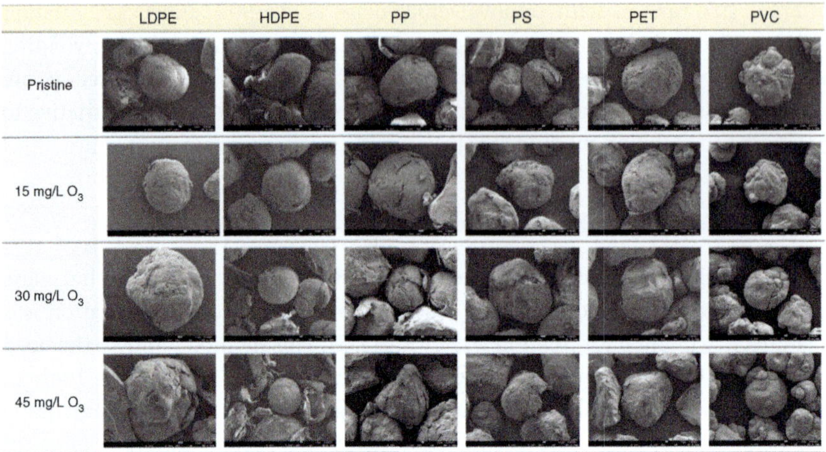

Figure 10.6 Microscopical images of microplastics exposed to varying ozone doses. (*Source:* Kye et al. [41] / Elsevier / CC BY-NC-ND 4.0.)

to their original condition, is shown by the formation of abrasion and cracks on the surface of the microplastics, which occurred as the dose of ozone increased.

Overall, ozonation treatment causes surface abrasion, fractures, and the creation of oxygen-containing functional groups, which ultimately results in an increase in the carbonyl index from microplastics to microplastics. Because of this alteration, their contact with contaminants is enhanced, which raises worries about the way they will behave in the environment over the next years. To get a better understanding of these findings and to devise methods to reduce the amount of microplastic contamination in aquatic environments, continuous monitoring and further study are required, respectively. The eradication of microplastics, on the other hand, may result in high operational expenses. The most significant disadvantage is the possibility of producing intermediate chemicals, which, as a result of the generation of reactive oxygen species, may be hazardous to both human health and the environment [40, 41].

10.3.5 Sol–Gel Technique

The sol–gel technique is a novel treatment technology that utilizes induced polymer aggregates in wastewater to create huge aggregate particles. This method is economical and has robust chemical stability, rendering it appropriate for the elimination of contaminants from wastewater [31]. The sol–gel method is cost-effective and unaffected by variables such as pH, temperature, or the type, size, and concentration of pollutants among influencing factors. Microplastic flocculants are generated by fusion, facilitating the removal of wastewater via sand trap-like segregation techniques. Organosilanes are affixed to microplastic particles to create an insertion complex, which encapsulates them in large agglomerates, facilitating their removal from water. Investigations are underway to determine whether specific organosilanes can serve as an additional measure alongside the chemical-induced agglomeration. For instance, the inorganic–organic hybrid silica gels as an effective and straightforward approach for the elimination of microplastics from wastewater [37]. The suggested substance serves dual purposes as a filter and a binding agent for the elimination of microplastics. The elimination processes consist of two steps: the creation of an insertion element and a capture unit, which can interact with the material to be included via functional groups. A robust, highly cross-linked inorganic–organic macromolecule is synthesized by sequentially hydrolyzing and condensing the precursors during the sol–gel process. This approach has demonstrated efficacy for microplastics measuring up to 1 mm and exhibits the greatest potential for the removal of three- to five-carbon chain atoms. In another technique for extracting microplastics from water utilizing organosilanes instead of traditional flocculants [42]. The chemical properties and surface chemistry of microplastics substantially affect the removal process and physical interactions with organosilanes. The efficiency of organosilanes can be enhanced by utilizing greater quantities of these compounds. Additional research on the amalgamation of various organosilanes is essential for the efficient elimination of polar and non-polar polymer mixtures.

10.3.6 Photocatalytic Degradation

Photocatalytic degradation is a technique utilizing semiconductor materials such as titanium dioxide or zinc oxide to decompose organic contaminants, including microplastics, in the presence of light. This approach has become significant as it provides an option for converting microplastics into less hazardous compounds, hence reducing the introduction of persistent organic pollutants and persistent, bioaccumulative, and toxic chemicals into the trophic chain [43]. Advanced oxidation mechanisms, including photocatalysis, ozonation, and Fenton-like methods, have considerable efficacy in microplastic breakdown. Sunlight, a natural energy source, gradually destroys plastic trash; however, photocatalytic degradation can markedly expedite this process owing to the photocatalyst's capacity to promote photochemical reactions. The iron-zinc oxide nanocomposite is an important semiconductor material utilized in the photocatalytic degradation of polyethylene. Research indicates that the mass losses during the photocatalytic degradation of polyethylene terephthalate varied from 9% to 16% [43, 44]. Polyamide microplastics can be nearly entirely decomposed following 10 days of irradiation in an aqueous $FeCl_3$ solution [44]. The rate of photocatalytic degradation is influenced by the solution's pH, termination rate, microplastic particle size, duration, and the efficiency of free radical generation required for hydrogen atom extraction from polymer chains. Photocatalytic degradation is an effective technique for the treatment of harmful organic contaminants, such as microplastics, in wastewater.

10.4 Biology Method

A key problem that affects aquatic systems and wastewater is the presence of microplastic contamination, and biological techniques have the potential to assist ameliorate this issue. In order to contribute to the breakdown of microplastics that are present in fresh water and wastewater, biological methods that include a wide variety of plant and animal species can be utilized. Biological methods for the removal of microplastics are now the subject of continuous research, and a variety of biological species are being exploited in the course of these investigations. Microbial degradation is the process by which the different polymers such as PET, PE, and PVC are broken down under saltwater circumstances [45, 46]. Microorganisms, such as bacteria, fungi, and algae, are responsible for this process. This process results in the production of enzymes, which further contributes to the breakdown of the plastics into smaller molecules that can be broken down by biological processes. Microplastics may also be dissolved by microbial communities in freshwater environments; however, the amount of time it takes for these communities to do so is slower since there are fewer microbes that are especially adapted to degrade plastic and a smaller diversity of bacteria. Surface attachment, extracellular enzyme production, and the formation of biofilms are all mechanisms that contribute to the destruction of microorganisms. These mechanisms enable bacteria to break down plastic polymers into smaller molecules and create an environment that is favorable to deterioration. This section explores various biological methods to remove microplastics from wastewater.

10.4.1 Fungal-Based Degradation of Microplastic

Fungi, which are creatures that can be either obligatory parasites, opportunistic parasites, or saprotrophs, exhibit extraordinary plasticity and can flourish in a variety of terrestrial and aquatic habitats and environmental situations. Fungi can also be classified as saprotrophs. In addition to being able to tolerate potentially harmful chemicals and metals, they are also capable of producing enzymes from outside the cell. These enzymes include organic biosurfactants such as hydrophobins, which are able to break down complicated polymers into simpler monomers. In addition, these fungi provide electrons and carbons to microorganisms that are accountable for the decomposition and mineralization of complicated contaminants. *Aspergillus and Cladosporium* are two examples of fungi that are capable of decomposing PE microplastics [47]. These fungi rely on polyethylene as their primary carbon source. As a result of their ability to reduce the hydrophobicity of plastic particles, these fungi facilitate the creation of chemical interactions. The largest number of these fungi, which have a broad dispersion and efficient reproduction, has a substantial impact on the growth and degradation of microplastics. This is because these fungi have their own unique characteristics. *Aspergillus and Fusarium* species have been discovered to be capable of biodegrading low-density polyethylene [48]. While *Aspergillus flavus* showed high-density polyethylene degrading qualities, *Aspergillus tubingensis* demonstrated polyethylene degradation capabilities that were both effective and efficient [48]. The emergence of omics and molecular technologies, such as polymerase chain reaction, in vitro transcription, and high-throughput sequencing, has been of considerable assistance to the research that has been conducted on fungal groups that are capable of breaking down microplastics. Several distinct fungal isolates have shown that they are capable of degrading a wide variety of polymers, the majority of which are originally derived from a variety of settings. For instance, the plastic waste disposal sites have demonstrated the ability to degrade LDPE and PU, and *Zalcrion maritimum* from marine sediments has the ability to degrade PE pellets in as little as 28 days [45, 49].

10.4.2 Bacterial-Based Degradation of Microplastic

Bacteria are a varied class of microorganisms that may be found in soil, water, and the environment. Most people are aware of their capacity to digest pollutants such as microplastics or other similar substances [50]. For the purpose of evaluating the breakdown of microplastics, studies have concentrated on the utilization of pure bacterial cultures in tightly regulated laboratory conditions [50]. The vast majority of bacterial cultures may be discovered in collections of cultures or acquired by enrichment culture from sludge, sediment, and wastewater. Researchers have the ability to investigate metabolic processes and employ pure lines to evaluate the consequences of these processes, in contrast to ecological variables that influence the breakdown of microplastics. By using this technology, it is possible to monitor the entire process of microplastics through the changes that are caused by active bacteria [51]. Significant progress has been achieved in identifying the molecular processes that are responsible for partial bacterial degradation as a result of research conducted on the bacterial breakdown of microplastics. Bacteria such as *Acinetobacter, Bacillus, Pseudomonas*, and *Klebsiella* have been steadily isolated from plastic waste, sludge, and wastewater by the use of enrichment and culture techniques. The process of plastic breakdown involves the identification of a vast variety of genes and enzymes, which are constantly being discovered

[52, 53]. In another work, the strain of *Ideonella sakaiensis* was capable of totally degrading polyethylene terephthalate (PET) in a span of six weeks [54]. This strain was intended for use with polyester polymers. This bacterium is responsible for the production of hydrolytic enzymes known as polyesterase [55]. These enzymes catalyze the breakdown of PET into monomers, which in turn facilitates further catabolism [55].

It is possible for the micro-bacterial population that has a high cell surface hydrophobicity to more effectively cling to the surface of PE and develop biofilm on its surface when it is found in nature. In the natural world, several studies have demonstrated that bacteria often work together in communities. This means that a wide range of microorganisms are involved in the degrading process, which results in a higher level of efficiency when compared to the performance of a single bacterium [56]. High-density polyethylene (HDPE) and low-density polyethylene (LDPE) are the two types of polymers that are utilized the most frequently in the field of polyethylene (PE) manufacture [56]. LDPE, in particular, has emerged as one of the most significant plastic grades for contemporary polymers. With regard to the decomposition of PE, fungi were found to be more successful than bacteria, according to the findings of a study between bacteria and fungi that were isolated from the same trash. The calculations of weight loss and carbon mineralization of LDPE film are used to assess whether or not the utilization of LDPE as a carbon source for the development of bacteria is successfully accomplished [56]. *Bacillus cereus* and *Bacillus gottheilii*, two types of Bacillus bacterial strains that were isolated from mangrove sediments, were used in the study by Auta et al. [57] to explore the removal of various microplastics including PE, PS, PP, and PET.

10.4.3 Biodegradation with the Assistance of Enzymes

Cutinases, polyesterases, laccases, and peroxidases are some of the enzymes that are utilized in the process of microplastic breakdown [45, 46, 55]. In the process of enzyme hydrolysis, certain enzymes are utilized to break down polymers into monomers that are smaller and easier to work with. As enzymes that can break down plastics are discovered and tailored to target the chemical linkages in plastic polymers, the enzymatic degradation of microplastics is a fast-increasing area of research. Polyesterases are enzymes that break down polyester-based polymers such as PET, converting them into simpler chemicals that can be further destroyed or mineralized by other organisms when they are found in the environment [55]. By stimulating the oxidation of carbon—carbon bonds, laccases and peroxidases, which are found in fungi and bacteria, are able to degrade plastic polymers such as polyethylene and polypropylene. Additionally, present in fungi and bacteria, cutinases are responsible for the degradation of polyesters by severing the ester linkages. Enzyme hydrolysis is a challenging and focused technique because of the tremendous complexity and diversity of microplastics found in marine habitats. This presents a number of challenges. There are a lot of polymers that are resistant to being broken down by enzymes, and enzymes frequently need particular circumstances in order to function at their best. The manufacture and deployment of enzymes may be expensive, particularly when scaled up for large-scale environmental cleaning activities. The creation of genetically modified microorganisms that are capable of generating extremely effective enzymes for the degradation of plastic, the optimization of enzymes for usage in a variety of conditions, and the utilization of enzymes in conjunction with physical approaches to boost the overall efficiency of microplastic removal are all opportunities [55].

The biocatalysts, including hydroxylases, laccases, peroxidases, and reductases that are responsible for the degradation of PE have been discovered in actinomycetal, bacterial, and fungal sources [58]. Among the enzymes that are involved in the breakdown of PE by *Phanerochaete chrysosporium* and *Trametes versicolor*, manganese peroxidase is the most important enzyme. It is also noteworthy that alkane hydroxylase plays a role in the breakdown of PE by *Pseudomonas* sp. E4. Alkane hydroxylase, rubredoxin, and rubredoxin reductase are included within an enzyme system that is found in *Pseudomonas aeruginosa* [59]. This system is responsible for the breakdown of low molecular weight PE. Also exhibiting considerable PE degradation are laccases derived from actinomycetes and fungi such as Aspergillus flavus and Pleurotus ostreatus [60]. These laccases may be responsible for the breakdown of PE by the oxidation of the PE hydrocarbon backbone.

Joo et al. [61] suggest a mechanism for the breakdown of PET (see Fig. 10.7), which starts with the binding of polyesterase from the bacteria to the surface of the PET material by using the hydrophobic surface of the PET material. First, there is nick creation, and then there is terminal digesting and these are the two stages of the process. During the phase of nick generation, a scissile ester bond is formed between each of the four MHET moieties that are brought together to form each subsite. Due to the cleavage of a single ester bond, a nick is formed in the PET, which leads to the production of two PET chains with distinct endpoints. These chains are the TPA-terminal from subsite I and the HE-terminal from subsite II. It is suggested that a comprehensive method for the degradation of polyethylene terephthalate (PET) into bis(2-hydroxyethyl) terephthalate (BHET), monohydroxyethyl terephthalate (MHET), terephthalate (TPA), and ethylene glycol (EG). *I. sakaiensis* is responsible for the secretion of polyesterase, which first attaches to the surface of the PET material by utilizing its hydrophobic, flat surface that features a substrate binding cleft [61]. After that, it goes through additional degradation by way of a nick production stage and a final digesting process, which ultimately results in the accumulation of four monomers including BHET, MHET, TPA, and EG. BHET is capable of being further broken down into MHET and EG [61].

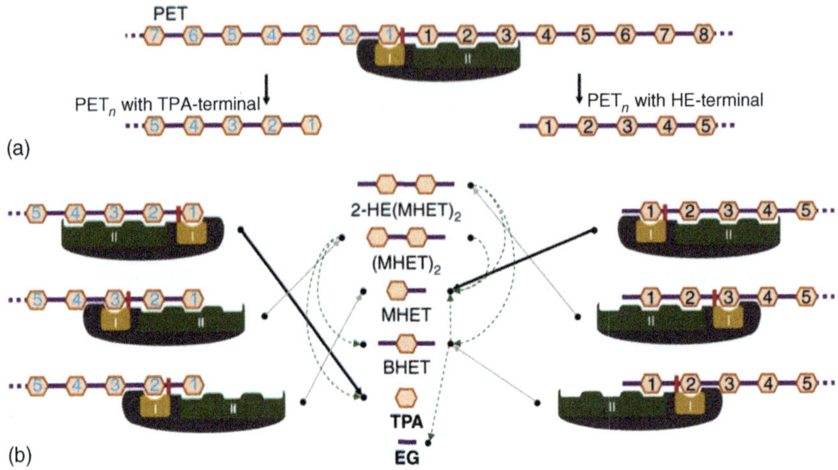

Figure 10.7 PET breakdown process by enzyme, nick generation (a) and terminal digestion step (b). (*Source:* Reproduced with permission of Joo et al. [61] / Springer Nature / CC BY 4.0.)

There is a significant amount of polyurethane (PU) fibers that are utilized in the textile industry. It has been demonstrated that a variety of enzymes, such as cutinases, esterases, lipases, laccases, peroxidases, proteases, and ureases, derived from bacterial and fungal sources, are capable of degrading PU [62]. In general, the process of enzymatic degradation of polymers by hydrolysis is a two-step procedure: first, the enzyme forms a link with the polymer by means of a hydrophobic domain, and then it catalyzes a hydrolytic breakdown [63]. The identification and improvement of these enzymes derived from bacteria and fungi has been the primary focus of recent research [64]. The urethane bond has been subjected to microbial assault, according to the reports that have been received. The hydrolysis of the ester bonds in the process of PUs biodegradation has been hypothesized as the mechanism of PU biodegradation. The breakdown products of PUs are formed from the ester bonds that are hydrolyzed and broken [63]. Ester-type polyurethanes are susceptible to degradation by soil microbes, but polyether-type polyurethanes are more resistant to the attack of microorganisms. Research has also been conducted to investigate the efficacy of commercial enzymes in the degradation of various types of polyurethane. This research has taken into consideration aspects such as the structural composition and complexity of the material undergoing degradation [64]. In a consortium of microorganisms, a number of enzymes that are involved in the breakdown of PU were discovered through the utilization of proximity ligation-based metagenomic analysis. According to Gaytán et al. research from 2020, the enzymes that were found to be responsible for the metabolism of numerous PU intermediates included a variety of dioxygenases, decarboxylases, dehydrogenases, transferases, ligases, hydrolases, isomerases, and peroxidases [65]. On the other hand, it has been observed that the majority of the PU-active enzymes that have been investigated up until this point have been active on ester-linked polyurethane ethers, less probable on polyurethane ethers [66]. On the other hand, the carbonyl group that is present in the ester-linked PU makes them more susceptible to hydrolysis brought on by enzymatic activities. This is in contrast to the ether-linked PU, which is more resistant to hydrolysis. In conclusion, the many enzymes found in the PU samples exhibited a variety of different degradation processes. A synergistic impact was observed when lipase and peroxidase were combined, since this combination proved to be very efficient. In a subsequent study, it is recommended to investigate additional enzyme combinations. The investigation of commercial enzymes is of utmost importance since these enzymes provide scalable methods for the breakdown of PU. In the future, sustainable treatment of PU waste may be possible because of developments in enzyme technology and genetic engineering.

Over the course of several years, research has been conducted on the microbial degradation of polystyrene (PS) by a wide variety of bacteria and fungi [67]. However, the primary enzymes that are involved in the depolymerization of the polymers have not been discovered with absolute certainty [68]. On the other hand, it has been demonstrated that PS can be broken down by an extracellular esterase that is produced by *Lentinus tigrinus* [68]. In addition, it has been observed that some polymerases, which were obtained from *Bacillus* and *Pseudomonas* species, are also accountable for the breakdown of PS [69]. Styrene monooxygenase, styrene oxide isomerase, phenylacetaldehyde dehydrogenase, and phenylacetyl coenzyme. The depolymerization of the polymer into styrene, the oxidation of styrene into phenylacetate, and the integration of phenylacetate into the Krebs cycle are all processes that are thought to be responsible for the involvement of these enzymes in a sequence of reactions. The side-chain breakage of PS polymers is the starting point for Zhang et al. [67] suggested process of microbial degradation of PS, which is illustrated in Fig. 10.8. Despite the fact that monooxygenases and aromatic ring hydroxylases are possible candidates for breaking the aromatic ring of PS, the specific degradation process and enzymes that are involved have not yet been uncovered.

There have been discoveries made about enzymes that are capable of digesting polyamides that originate from bacterial and fungal sources [70]. Among these enzymes is a manganese peroxidase that was isolated from a white-rot fungus which shows much better results with respect to their degradation. Also, this enzyme was demonstrated to remove the surface of Nylon 6 and to promote the creation of deep horizontal grooves in the polymer. Peroxidases are a category of enzymes that include heme and are responsible for catalyzing the oxidation of a

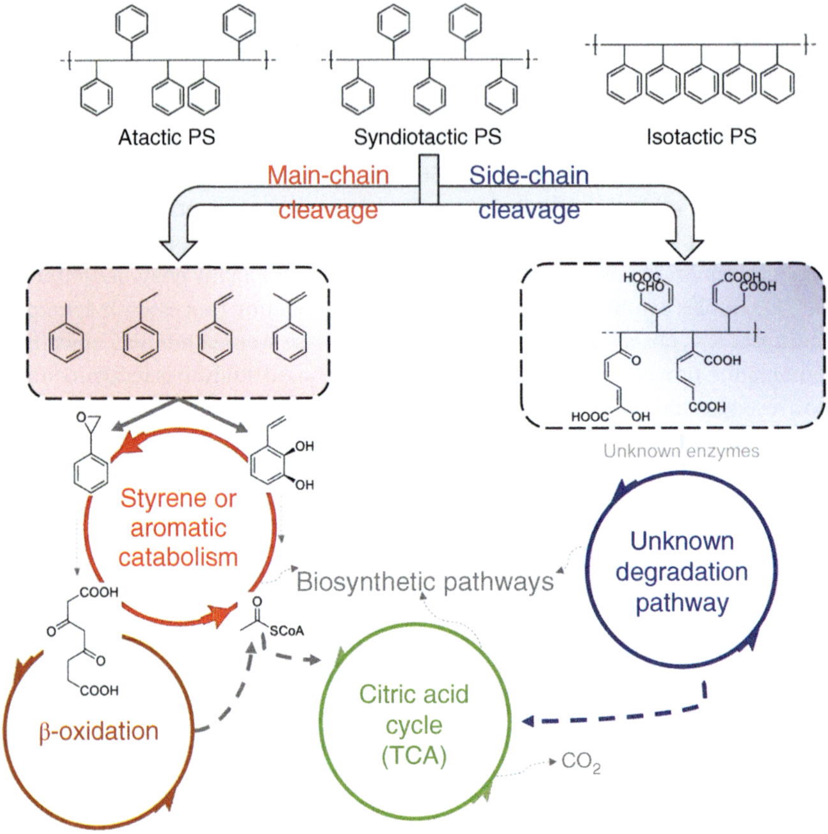

Figure 10.8 Biodegradation of polystyrene. (*Source:* Danso et al. [67] / with permission of Elsevier / CC BY 4.0.)

wide variety of organic and inorganic substrates. Hydrogen peroxide is used as the electron acceptor in this process.

10.4.4 Biodegradation of Microplastics: Impact of Different Factors

There are a number of factors that can influence the microbial biodegradation of plastic polymers. These factors include the properties of the polymer, ambient conditions, and polymer chemistry. These elements make future microbial activity easier to achieve by increasing the surface area, making the substance more hydrophilic, and lowering the molecular weight. The rate of microbial destruction decreases with increasing molecular weight, which is one of the features of polymers. This is because polymers need to be carried across the cellular membrane to fulfill their metabolic functions. Monomers, dimers, and oligomers are examples of smaller units of polymers that are more easily decomposed and mineralized than larger molecules. The morphology of the polymer, which includes branching, crystallinity, and physical shape, is another factor that has a substantial impact on the pace of degradation. Polymers made of plastic that have a greater number of side chains and enhanced branching are less absorbed by microbes, which means they are less susceptible to deterioration [71].

A significant impact on the rate of microbial breakdown of the corresponding polymer is exerted by the melting temperature of the polymer. There is a relationship between the change in enthalpy of melting (ΔH) and the change in entropy of melting (ΔS) and the melting temperature. It is possible that the glass transition temperature (T_g) of synthetic polymers will make it easier for microorganisms to attack them. An increase in the material's hydrophilicity leads to an increase in the degree of early microbial colonization of the plastic polymers. On the other hand, an increase in the material's hydrophobicity leads to an inhibition of the activity of extracellular enzymes. As a result of their greater wettability, hydrophilic surfaces have higher surface energies and offer lower contact angles with water. This, in turn, encourages the adhesion of microorganisms to the surface of the polymer, which in turn speeds up the pace of degradation. In the presence of environmental conditions, such as exposure to ultraviolet light due to the sunlight, it has been shown that the contact angle with water decreases while the hydrophilicity of the polymers were increased, for instance polyamide like Nylon 6, which is moderately hydrophilic, as compared to olefin fibers like PP and esters like PET, which is very hydrophobic, had the least amount of biodegradative potential. To a significant extent, the biodegradation process carried out by bacteria is dependent on the initial individual or synergistic activity of several ambient factors toward the polymers [71]. In a dry environment, humid air, landfill, compost, or a marine environment, the pace of microbial action will vary. This is because the properties of each habitat are drastically different from one another. There are a number of factors that have been discovered to boost bond scission, structural homogeneity, and the synthesis of novel functional groups. These factors include light, heat, moisture, pH, and biological activity [71].

Melting in the environment will lead to an increase in the solubility and hydrolysis rate of plastic polymers, which will result in an increase in the number of chain scissions and increased sites of microbial activity on the polymer chains, which will lead to an improvement in biodegradation. High temperatures have also been reported to cause thermal degradation of plastic, which results in disorderliness in the polymeric structure and causes chemical interactions between the components. This phenomenon has been seen. The molecular weights of the polymers are altered as a result of thermal degradation, which also diminishes ductility and embrittlement, initial color, and cracking, among other effects. The process of biodegradation of polymers is significantly influenced by the chemical reagents and additives that are utilized. As a result of their influence on functional groups and hydrophilicity/hydrophobicity, they can either activate, inhibit, or catalyze the process. During the reprocessing or deterioration of plastic, certain additives, such as dibutyl tin dilaurate, can diminish the microplastic's recalcitrance. Other substances, such as starch and palmitic acid, contribute to the process of biodegradation by supplying bacteria with sources of nutrients. Hydrochloric acid, hydrogen peroxide, sulfuric acid, and nitric acid are examples of oxidative agents that can oxidize polymer surfaces, which in turn increases the rate of biodegradation [71].

10.5 Conclusion

Microplastics provide a substantial ecological and public health risk in maritime ecosystems. A complete elimination process is required to eradicate these particles. Physical methods like filtration, magnetic separation, sedimentation, and froth flotation are useful for large-scale operations, but must be combined with other approaches. Coagulation, electrocoagulation, and ozonation are chemical treatments that may agglomerate or destroy microplastics; nevertheless, they need careful control and consideration of secondary pollutants. Biological techniques, such as fungi, bacteria, and enzymes, provide a low-impact and ecologically beneficial alternative. Early research suggests that synthetic polymers may degrade over time in both freshwater and saline environments. Advances in biotechnology, synthetic biology, and environmental management hold the promise of a cleaner, more sustainable future where microplastic pollution can be effectively mitigated and ecosystems restored. The successful eradication of microplastics will need synergistic use of various approaches, adapted to local ecological circumstances and pollution levels. Future success will be determined by the integration of physical capture technology with enhanced chemical and biological therapies, as well as monitoring systems, legislative frameworks, and public awareness.

References

1 Yang, J.W., Park, C., and Jho, E.H. (2024). Removal technologies of microplastics in soil and water environments: review on sources, ecotoxicity, and removal technologies. *Applied Biological Chemistry* 67 (1): 106.

2 Singh, S., Kalyanasundaram, M., and Diwan, V. (2021). Removal of microplastics from wastewater: available techniques and way forward. *Water Science and Technology* 84 (12): 3689–3704.

3 Sun, C., Wang, Z., Chen, L., and Li, F. (2020). Fabrication of robust and compressive chitin and graphene oxide sponges for removal of microplastics with different functional groups. *Chemical Engineering Journal* 393: 124796.

4 Yuan, F., Yue, L., Zhao, H., and Wu, H. (2020). Study on the adsorption of polystyrene microplastics by three-dimensional reduced graphene oxide. *Water Science and Technology* 81 (10): 2163–2175.

5 Alimi, O.S., Farner Budarz, J., Hernandez, L.M., and Tufenkji, N. (2018). Microplastics and nanoplastics in aquatic environments: aggregation, deposition, and enhanced contaminant transport. *Environmental Science & Technology* 52 (4): 1704–1724.

6 Gagliano, E., Sgroi, M., Falciglia, P.P. et al. (2020). Removal of poly- and perfluoroalkyl substances (PFAS) from water by adsorption: role of PFAS chain length, effect of organic matter and challenges in adsorbent regeneration. *Water Research* 171: 115381.

7 Lin, A.Y.-C., Panchangam, S.C., Chang, C.-Y. et al. (2012). Removal of perfluorooctanoic acid and perfluorooctane sulfonate via ozonation under alkaline condition. *Journal of Hazardous Materials* 243: 272–277.

8 Du, Z., Deng, S., Chen, Y. et al. (2015). Removal of perfluorinated carboxylates from washing wastewater of perfluorooctanesulfonyl fluoride using activated carbons and resins. *Journal of Hazardous Materials* 286: 136–143.

9 Wang, Z., Sun, C., Li, F., and Chen, L. (2021). Fatigue resistance, re-usable and biodegradable sponge materials from plant protein with rapid water adsorption capacity for microplastics removal. *Chemical Engineering Journal* 415: 129006.

10 Zhuang, J., Rong, N., Wang, X. et al. (2022). Adsorption of small size microplastics based on cellulose nanofiber aerogel modified by quaternary ammonium salt in water. *Separation and Purification Technology* 293: 121133.

11 Ramirez Arenas, L., Ramseier Gentile, S., Zimmermann, S., and Stoll, S. (2021). Nanoplastics adsorption and removal efficiency by granular activated carbon used in drinking water treatment process. *Science of the Total Environment* 791: 148175.

12 Yen, P.-L., Hsu, C.-H., Huang, M.-L., and Liao, V.H.-C. (2022). Removal of nano-sized polystyrene plastic from aqueous solutions using untreated coffee grounds. *Chemosphere* 286: 131863.

13 Wang, L. and Villa, K. (2021). Self-propelled micro/nanomotors for removal of insoluble water contaminants: microplastics and oil spills. *Environmental Science. Nano* 8 (12): 3440–3451.

14 Panahi, H.K.S., Dehhaghi, M., Ok, Y.S. et al. (2020). A comprehensive review of engineered biochar: production, characteristics, and environmental applications. *Journal of Cleaner Production* 270: 122462.

15 Zhu, N., Yan, Q., He, Y. et al. (2022). Insights into the removal of polystyrene nanoplastics using the contaminated corncob-derived mesoporous biochar from mining area. *Journal of Hazardous Materials* 433: 128756.

16 Singh, N., Khandelwal, N., Ganie, Z.A. et al. (2021). Eco-friendly magnetic biochar: an effective trap for nanoplastics of varying surface functionality and size in the aqueous environment. *Chemical Engineering Journal* 418: 129405.

17 Grbic, J., Nguyen, B., Guo, E. et al. (2019). Magnetic extraction of microplastics from environmental samples. *Environmental Science & Technology Letters* 6 (2): 68–72.

18 Wang, Z., Lin, T., and Chen, W. (2020). Occurrence and removal of microplastics in an advanced drinking water treatment plant (ADWTP). *Science of the Total Environment* 700: 134520.

19 Tadsuwan, K. and Babel, S. (2022). Microplastic abundance and removal via an ultrafiltration system coupled to a conventional municipal wastewater treatment plant in Thailand. *Journal of Environmental Chemical Engineering* 10 (2): 107142.

20 Kaetzl, K., Lübken, M., Nettmann, E. et al. (2020). Slow sand filtration of raw wastewater using biochar as an alternative filtration media. *Scientific Reports* 10 (1): 1229.

21 Talvitie, J., Mikola, A., Koistinen, A., and Setälä, O. (2017). Solutions to microplastic pollution – removal of microplastics from wastewater effluent with advanced wastewater treatment technologies. *Water Research* 123: 401–407.

22 Komorowska-Kaufman, M. and Marciniak, W. (2024). Removal of microplastic particles during municipal wastewater treatment: a current review. *Desalination and Water Treatment* 317: 100006.

23 Johansson, G., Fedje, K.K., Modin, O. et al. (2024). Removal and release of microplastics and other environmental pollutants during the start-up of bioretention filters treating stormwater. *Journal of Hazardous Materials* 468: 133532.

24 Siipola, V., Pflugmacher, S., Romar, H. et al. (2020). Low-cost biochar adsorbents for water purification including microplastics removal. *Applied Sciences* 10 (3): 788.

25 Wu, J., Yang, C., Zhao, H. et al. (2022). Efficient removal of microplastics from aqueous solution by a novel magnetic biochar: performance, mechanism, and reusability. *Environmental Science and Pollution Research* 30 (10): 26914–26928.

26 Zhang, Y., Jiang, H., Bian, K. et al. (2021). A critical review of control and removal strategies for microplastics from aquatic environments. *Journal of Environmental Chemical Engineering* 9 (4): 105463.

27 Mani, T., Frehland, S., Kalberer, A., and Burkhardt-Holm, P. (2019). Using castor oil to separate microplastics from four different environmental matrices. *Analytical Methods* 11 (13): 1788–1794.

28 Zhang, Y., Jiang, H., Bian, K. et al. (2021). Is froth flotation a potential scheme for microplastics removal? Analysis on flotation kinetics and surface characteristics. *Science of the Total Environment* 792: 148345.

29 Kharraz, J.A., Jia, M., Farid, M.U. et al. (2024). Determination of microplastic pollution in marine ecosystems and its effective removal using an advanced nanobubble flotation technique. *Journal of Water Process Engineering* 57: 104637.

30 Shi, C., Wu, H., Wang, W. et al. (2024). Microplastic removal from water using modified maifanite with rotating magnetic field affected. *Journal of Cleaner Production* 434: 140111.

31 Amparán, M.A.A., Palacios, A., Flores, G.M., and Olivera, P.M.C. (2025). Review and future outlook for the removal of microplastics by physical, biological and chemical methods in water bodies and wastewaters. *Environmental Monitoring and Assessment* 197 (4): 429.

32 Akinnawo, S.O., Ayadi, P.O., and Oluwalope, M.T. (2023). Chemical coagulation and biological techniques for wastewater treatment. *Ovidius University Annals of Chemistry* 34 (1): 14–21.

33 Monira, S., Bhuiyan, M.A., Haque, N., and Pramanik, B.K. (2021). Assess the performance of chemical coagulation process for microplastics removal from stormwater. *Process Safety and Environmental Protection* 155: 11–16.

34 Dayal, L., Yadav, K., Dey, U. et al. (2024). Recent advancement in microplastic removal process from wastewater – a critical review. *Journal of Hazardous Materials Advances* 16: 100460.

35 Zhao, Y.X., Gao, B.Y., Shon, H.K. et al. (2011). Effect of shear force, solution pH and breakage period on characteristics of flocs formed by titanium tetrachloride (TiCl4) and Polyaluminum chloride (PACl) with surface water treatment. *Journal of Hazardous Materials* 187 (1–3): 495–501.

36 Subair, A., Priya, K.L., Chellappan, S. et al. (2024). Evaluating the performance of electrocoagulation system in the removal of polystyrene microplastics from water. *Environmental Research* 243: 117887.

37 Tang, K.H.D. and Hadibarata, T. (2021). Microplastics removal through water treatment plants: its feasibility, efficiency, future prospects and enhancement by proper waste management. *Environmental Challenges* 5: 100264.

38 Du, H., Xie, Y., and Wang, J. (2021). Microplastic degradation methods and corresponding degradation mechanism: research status and future perspectives. *Journal of Hazardous Materials* 418: 126377.

39 Zheng, W., Wang, B., Liu, Z. et al. (2024). La-doped Ti/Sb-SnO_2 electrode enhanced removal of microplastics by advanced electrocatalysis oxidation process (AEOP) strategy. *Desalination and Water Treatment* 320: 100622.

40 Sangkham, S., Pitakpong, A., and Kumar, R. (2025). Advanced approaches to microplastic removal in landfill leachate: advanced oxidation processes

(AOPs), biodegradation, and membrane filtration. *Case Studies in Chemical and Environmental Engineering* 11: 101056.

41 Kye, H., Yoon, Y., and Hwang, T.-M. (2024). Changes in physical and chemical properties of microplastics by ozonation. *Process Safety and Environmental Protection* 192: 1062–1072.

42 Herbort, A.F., Sturm, M.T., Fiedler, S. et al. (2018). Alkoxy-silyl induced agglomeration: a new approach for the sustainable removal of microplastic from aquatic systems. *Journal of Polymers and the Environment* 26 (11): 4258–4270.

43 Liu, P., Qian, L., Wang, H. et al. (2019). New insights into the aging behavior of microplastics accelerated by advanced oxidation processes. *Environmental Science & Technology* 53 (7): 3579–3588.

44 Zhong, Y., Zhuo, L., and Lu, W. (2024). Analysis of photocatalytic degradation of polyamide microplastics in metal salt solution by high resolution mass spectrometry. *Journal of Environmental Sciences* 139: 473–482.

45 Temporiti, M.E.E., Nicola, L., Nielsen, E., and Tosi, S. (2022). Fungal enzymes involved in plastics biodegradation. *Microorganisms* 10 (6): 1180.

46 De Jesus, R. and Alkendi, R. (2023). A minireview on the bioremediative potential of microbial enzymes as solution to emerging microplastic pollution. *Frontiers in Microbiology* 13: 1066133.

47 Sowmya, H.V., Ramalingappa, Krishnappa, M., and Thippeswamy, B. (2015). Degradation of polyethylene by Penicillium simplicissimum isolated from local dumpsite of Shivamogga district. *Environment, Development and Sustainability* 17 (4): 731–745.

48 Sangeetha Devi, R., Rajesh Kannan, V., Nivas, D. et al. (2015). Biodegradation of HDPE by aspergillus spp. from marine ecosystem of Gulf of Mannar, India. *Marine Pollution Bulletin* 96 (1–2): 32–40.

49 Paço, A., Duarte, K., da Costa, J.P. et al. (2017). Biodegradation of polyethylene microplastics by the marine fungus Zalerion maritimum. *Science of the Total Environment* 586: 10–15.

50 Cai, Z., Li, M., Zhu, Z. et al. (2023). Biological degradation of plastics and microplastics: a recent perspective on associated mechanisms and influencing factors. *Microorganisms* 11 (7): 1661.

51 Gao, W., Xu, M., Zhao, W. et al. (2024). Microbial degradation of (micro)plastics: mechanisms, enhancements, and future directions. *Fermentation* 10 (9): 441.

52 Park, S.Y. and Kim, C.G. (2019). Biodegradation of micro-polyethylene particles by bacterial colonization of a mixed microbial consortium isolated from a landfill site. *Chemosphere* 222: 527–533.

53 Hou, L., Xi, J., Liu, J. et al. (2022). Biodegradability of polyethylene mulching film by two pseudomonas bacteria and their potential degradation mechanism. *Chemosphere* 286: 131758.

54 Yoshida, S., Hiraga, K., Takehana, T. et al. (2016). A bacterium that degrades and assimilates poly(ethylene terephthalate). *Science (1979)* 351 (6278): 1196–1199.

55 Guo, R.-T., Li, X., Yang, Y. et al. (2024). Natural and engineered enzymes for polyester degradation: a review. *Environmental Chemistry Letters* 22 (3): 1275–1296.

56 Eck, M., Schwab, S.T., Nelson, T.F. et al. (2023). Biodegradable high-density polyethylene-like material. *Angewandte Chemie International Edition* 62 (6): e202213438.

57 Auta, H.S., Emenike, C.U., and Fauziah, S.H. (2017). Screening of bacillus strains isolated from mangrove ecosystems in peninsular Malaysia for microplastic degradation. *Environmental Pollution* 231: 1552–1559.

58 Iiyoshi, Y., Tsutsumi, Y., and Nishida, T. (1998). Polyethylene degradation by lignin-degrading fungi and manganese peroxidase. *Journal of Wood Science* 44 (3): 222–229.

59 Gyung Yoon, M., Jeong Jeon, H., and Nam Kim, M. (2012). Biodegradation of polyethylene by a soil bacterium and AlkB cloned recombinant cell. *Journal of Bioremediation & Biodegradation* 3 (04): 1–8.

60 Zhang, J., Gao, D., Li, Q. et al. (2020). Biodegradation of polyethylene microplastic particles by the fungus aspergillus flavus from the guts of wax moth galleria mellonella. *Science of the Total Environment* 704: 135931.

61 Joo, S., Cho, I.J., Seo, H. et al. (2018). Structural insight into molecular mechanism of poly(ethylene terephthalate) degradation. *Nature Communications* 9 (1): 382.

62 Magnin, A., Pollet, E., Phalip, V., and Avérous, L. (2020). Evaluation of biological degradation of polyurethanes. *Biotechnology Advances* 39: 107457.

63 Nakajima-Kambe, T., Shigeno-Akutsu, Y., Nomura, N. et al. (1999). Microbial degradation of polyurethane, polyester polyurethanes and polyether polyurethanes. *Applied Microbiology and Biotechnology* 51 (2): 134–140.

64 Katnic, S.P. and Gupta, R.K. (2025). Assessment of the effectiveness of commercial enzymes for polyurethane biodegradation. *Journal of Polymers and the Environment* 33 (4): 1689–1704.

65 Gaytán, I., Sánchez-Reyes, A., Burelo, M. et al. (2020). Degradation of recalcitrant polyurethane and xenobiotic additives by a selected landfill microbial community and its biodegradative potential revealed by proximity ligation-based metagenomic analysis. *Frontiers in Microbiology* 10: 2986.

66 Danso, D., Chow, J., and Streit, W.R. (2019). Plastics: environmental and biotechnological perspectives on microbial degradation. *Applied and Environmental Microbiology* 85 (19): e01095-19.

67 Zhang, Y., Pedersen, J.N., Eser, B.E., and Guo, Z. (2022). Biodegradation of polyethylene and polystyrene: from microbial deterioration to enzyme discovery. *Biotechnology Advances* 60: 107991.

68 Tahir, L., Ishtiaq Ali, M., Zia, M. et al. (2013). Production and characterization of esterase in lantinus tigrinus for degradation of polystyrene. *Polish Journal of Microbiology* 62 (1): 101–108.

69 Mohan, A.J., Sekhar, V.C., Bhaskar, T., and Nampoothiri, K.M. (2016). Microbial assisted high impact polystyrene (HIPS) degradation. *Bioresource Technology* 213: 204–207.

70 Yamano, N., Kawasaki, N., Ida, S., and Nakayama, A. (2019). Biodegradation of polyamide 4 in seawater. *Polymer Degradation and Stability* 166: 230–236.

71 Amobonye, A., Bhagwat, P., Singh, S., and Pillai, S. (2021). Plastic biodegradation: frontline microbes and their enzymes. *Science of the Total Environment* 759: 143536.

11

Textile Recycling and Microplastics

11.1 Introduction

There have been several factors that have led to the rapid expansion of the global textile industry. Some of these factors include growing populations, the rise of fast fashion, and evolving consumer behavior. The number of used textiles that are discarded each year is in the millions of tons, with a substantial portion of this waste being disposed of in landfills or combustion facilities. The conventional methods of waste disposal led to the waste of valuable resources, the emission of greenhouse gases, the leaching of chemicals, and the pollution that is produced by microplastics. In recent years, the recycling of textiles has emerged as a significant method for mitigating the adverse impacts of environmental repercussions, protecting raw resources, and fostering the development of a circular economy. Recycling textiles entails collecting and reusing fibers, yarns, or textiles from pre- or post-consumer textile waste. The transition from a linear "take-make-dispose" paradigm to a circular, regenerative system is facilitated by this process, which can take on a variety of forms, including mechanical, chemical, and thermal processes.

This chapter discusses the various approaches in recycling textiles and the influence those approaches have on the production of microplastics. In it, the difficulties associated with recycling are discussed, notably those associated with the complexity of mixed materials and economic viability. The recycling process is made more difficult by the fact that textiles are created from a wide variety of fibers, including natural (such as cotton and wool) and synthetic (such as polyester and nylon) fibers, as well as mixes. The attempt to recycle might be hampered by contamination problems, such as the presence of dyes and chemicals. The present infrastructure frequently lacks the ability to manage huge amounts of textile waste, which makes efficient collection and sorting of textile waste a serious problem that requires continued attention. Because the market for recycled fibers is still in its early stages of development, economic feasibility is another issue

Textile Microplastics in the Environment: Sources, Impacts, and Solutions,
First Edition. Aravin Prince Periyasamy.

that must be overcome. The cost of recycling textiles may be rather expensive. It is necessary to encourage innovation as well as investments in technology and infrastructure in order to guarantee the economic sustainability of textile recycling. Fiber mechanical recycling, chemical recycling, and thermos-mechanical recycling are some of the several types of recycling methods that may be used for textile processing. The shredding process and subsequent usage of recovered fibers can both result in the production of microplastics, which can be a consequence of fiber mechanical recycling of materials. Even though chemical recycling helps lower the size of synthetic fibers, it still has the potential to contribute to the pollution caused by microplastics if it is not controlled appropriately. In the event that melted materials are not controlled and handled in an appropriate manner, thermal recycling has the potential to contribute to emissions of microplastics.

11.2 Recycling of Textiles

The worldwide textile and fashion business is under heightened examination over its environmental and social consequences, especially with textile waste. The industry generates about 100 million tons of garments each year, with a substantial fraction ultimately discarded in landfills or incinerated after a few years of use. The typical consumer in developed nations disposes of around 15–40 kg of textiles annually [1–3], often after just a few uses. This disposable society, driven by fast fashion and consumerism, burdens landfill capabilities and emits detrimental toxins into the environment [2]. Textile waste poses significant challenges due to its composition of both natural and synthetic fibers, as well as dyes, coatings, and finishes [4, 5]. Natural fibers, like cotton and wool, are subject to biodegradation over time, but synthetic fibers such as polyester, nylon, and acrylic last for centuries, breaking down into microplastics that contaminate soil and aquatic ecosystems [6]. The ecological impact of textile manufacturing is significant, representing up to 10% of worldwide carbon emissions. The recycling of textiles may significantly diminish the need for new raw materials, thereby reducing energy use, greenhouse gas emissions, and water usage. Recycling polyester may save up to 75% of the energy necessary for its production from raw petrochemicals. It diverts garbage from landfills and mitigates the total ecological impact of apparel [6, 7]. Natural resources, including water, land, and fossil fuels, are becoming more limited. Textile recycling adheres to the tenets of a circular economy, whereby materials are used for extended durations to maximize their worth prior to recovery and regeneration. Textile recycling facilitates the repurposing of fibers, textiles, and clothes, therefore completing the loop and reducing reliance on new resources. This also facilitates the emergence of novel business models such as clothing rental, resale, and product-as-a-service, all of which encourage extended garment lifespans and more gradual trend cycles [7]. The advent of chemical recycling technology offers investment prospects for inventors and entrepreneurs, harnessing the potential of previously non-recyclable textile fabrics like polyester and nylon significantly contribute to microplastic contamination, particularly during laundering and disposal [8, 9]. Properly managed textile recycling may alleviate this problem by prolonging the lifespan of clothes, decreasing the manufacture of new synthetic materials, and integrating filtering devices during processing [10]. Regulations like the European Union's textile strategy for sustainable and circular textiles mandate that all textile goods introduced to the market must be durable, repairable, and recyclable by 2030 [10]. With ongoing technological advancements, the efficiency, purity, and economic feasibility of textile recycling will be enhanced, making it more accessible and appealing to industry stakeholders. The need for recycling is now a practical and immediate concern; a unified and strategic effort is imperative. Investment in technology, infrastructure, education, and policy will dictate our ability to effectively expand textile recycling and attain a sustainable fashion future.

11.2.1 Fiber Mechanical Recycling of Textiles: Challenges and Opportunities

Cutting, shredding, carding, and re-spinning fibers into new yarns and fabrics are all examples of several physical processing procedures that are included in the mechanical recycling of textiles, which is a technology that has gained widespread acceptance and is able to produce at an industrial scale. Because of their consistent polymer structure and mechanical robustness, mono or multi-materials are favored for the use of this approach. However, despite the fact that mechanical recycling helps redirect trash away from landfills and incineration, it also has substantial limits in terms of the retention of material quality, the control of contamination, and the impact on the environment, most notably the accidental release of microplastics. The fibers are subjected to significant mechanical strains during the process of mechanical recycling. The fibers become shorter with each cycle of shredding, cutting, and re-spinning, which ultimately results in a gradual decline in the structural and mechanical properties of the fibers. This is a decrease in the tensile strength of the fibers as well as their flexibility. The low quality of the yarn, which limits its applicability to lower-value items (such as insulation materials, mattress stuffing, and non-woven textiles) rather than high-end clothes [7, 11]. In terms of the economy, a decrease in the quality of the material has a negative impact on both the market demand for recycled fibers and their economic viability. The nature of the problem is that the majority of contemporary clothing is made up of fiber mixes (such as polyester-cotton, polyamide-wool, and polyester-elastane) that are intended to improve performance attributes such as stretchability, moisture management, or durability. When it comes to shredding, fiber mixes are physically separate from one another. The quality of the recycled product is diminished when the qualities of the respective materials are incompatible with one another (melting point, elasticity, degradation behavior). Mixed fibers provide inferior mechanical performance, which makes it more difficult to re-spin them into yarn and frequently requires them to be downcycled into products with a lesser value. The nature of the problem is that mechanical recycling processes, in particular shredding, cutting, and melting, induce the breakdown of synthetic fibers, which in turn results in the release of microplastics and nanoplastics. Microplastic particles have the potential to escape into the atmosphere, water systems, and soil if they are not successfully trapped. Microplastics are persistent pollutants that accumulate in aquatic environments. They have the ability to infiltrate food chains and pose a threat to the health of both marine animals and humans that are exposed to them. The increased presence of airborne microplastics within recycling plants may potentially have ramifications for the health of construction workers. The nature of the problem is that textiles that are destined for recycling streams are frequently polluted with dirt, oils, dyes, chemical coatings, and mixed fibers. Contaminants have the potential to impede the functioning of processing machinery, increase the amount of maintenance that is required, and decrease the efficiency with which operations are carried out. When recycled fibers have a lower purity, it has an effect on their ability to absorb dye, their mechanical strength, and their overall market worth. Additional washing and cleaning processes are necessary, which results in an increase in the amount of water used, the amount of energy consumed, and the risks of microplastic leakage. The nature of the problem is that the existing recycling infrastructure, particularly in economies of the West, is largely designed for recycling PET bottles and not for recycling complicated textile streams. A lack of specialist technology for dealing with difficulties that are unique to the textile industry (such as mixed fibers and pollutants). The processing capacity is low in comparison to the sheer volume of textile waste that is produced each year. There is a requirement for additional investments in recycling technologies that have been exclusively developed for the textile industry. These technologies include fiber identification, fine shredding that does not cause undue damage, and mixed fiber handling. It is possible for mechanical recycling methods to unintentionally become sources of contamination with microplastics.

The proper sorting and identification of fibers is essential to the successful recycling of textiles, particularly those that contain or shed microplastics. But there are substantial technology gaps, including the optical sorting techniques, such as near-infrared (NIR) spectroscopy, which have difficulty dealing with composite materials, fabrics that have been highly dyed, and dark-colored textiles [12]. Current technologies are unable to differentiate between fiber blends, coated textiles, or materials that are loaded with microplastics in a reliable manner. Processing is slowed down, and prices are increased when there is a high level of contamination and a diverse range of fibers. The absence of worldwide standards for sorting quality has an impact on the consistency and scalability of the results of recycling. In addition, the development of sophisticated artificial intelligence-based optical sorting systems that are able to differentiate between many fibers. The incorporation of chemical markers or tracers while textiles are being manufactured in order to facilitate simpler sorting at the end of their useful lives [13].

Furthermore, the repetitive mechanical recycling of textiles results in a considerable degradation of the fiber quality, which in turn leads to the manufacture of materials that are weaker and less environmentally friendly [14]. The fibers undergo a reduction in their mechanical strength and aesthetic characteristics as a result of each cycle of mechanical treatment, which includes shredding, tearing, carding, and re-spinning [15]. This causes the fibers to become shorter, rougher, and more brittle to a greater extent. This deterioration restricts the usage of recycled fibers to applications with lesser value, such as padding, insulation, or industrial textiles. Instead of reintroducing them into high-performance garment markets, recycled fibers are used for these applications. In addition, the mechanical recycling that is involved, particularly the shredding and re-melting operations, has the potential to cause the delicate synthetic fibers to get detached. Microplastics have the potential to be released into the environment if suitable containment measures are not applied [15]. These methods include sophisticated filtration systems and closed-loop water management. If these measures are not implemented, microplastics may enter water bodies, soils, and even the air. The environmental advantages of recycling are undermined by this accidental contamination caused by microplastics, which poses substantial dangers to both the environment and human health [16].

It is still unclear, and there is no conclusive study that has been conducted about the recycled textiles that produce a greater level of microplastics in comparison to the virgin materials. This is due to the fact that some of the research has shown inconsistent findings [16–18]. On the other hand, it is important to note that the researchers found that the mechanically recycled fibers that were used in the knitted textiles produced around 2.5 times more microplastics than the virgin fibers that were used in the same fabric and structure [19]. A significant number of microscopical pictures are presented in Fig. 11.1, which gives information on the various stages of fiber mechanical recycling.

11.2.2 Thermo-Mechanical Recycling

Thermo-mechanical recycling is a hybrid method that integrates mechanical and heat processing to reclaim materials, especially synthetic fibers such as

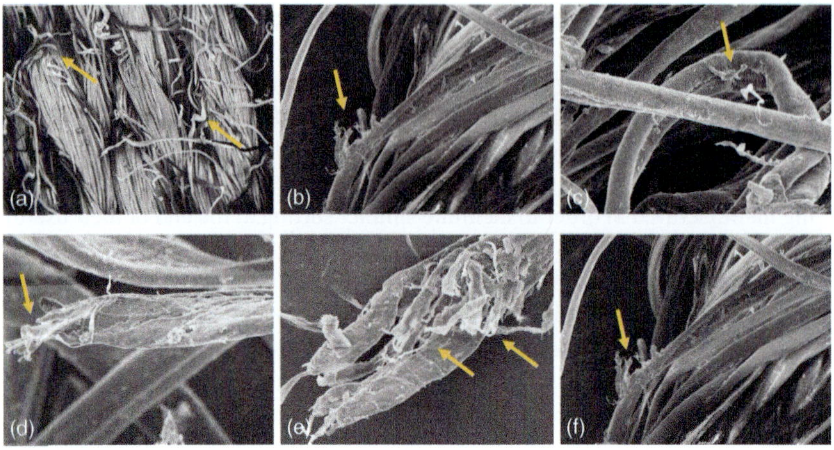

Figure 11.1 SEM images of the fiber damages in different stages of fiber mechanical recycling; appearance of textile waste (a), during the cutting (b), shredding (c and d), after spinning (e, d, and f).

polyester (PET) [20]. This procedure entails the liquefaction of textile fibers or fabrics, followed by re-extrusion into granules or pellets, and further reprocessing into novel products such as yarns, films, or non-woven materials. Recycled PET is utilized in fibers for apparel, packaging films, non-woven geotextiles, and automotive components. Recyclable polyester fabrics are subjected to temperatures adequate to liquefy the polymer (~250–280 °C). The molten polymer is extruded into pellets, subsequently spun into fresh yarns or molded into items. Thermo-mechanical recycling encounters considerable technological and environmental obstacles. Requirements for material purity, such as the necessity for 100% pure materials, restrict the feedstock accessible for recycling and heighten the demand for effective sorting systems.

Thermal degradation of polymers, including polymer chain sclerosis, can result in material implications such as diminished tensile strength, reduced fiber durability, and worse dyeability and processability in later applications. The persistence and volatilization of microplastics pose significant obstacles in thermo-mechanical recycling. Partial melting may not eradicate all microfibers contained inside the material, while volatilization and equipment contamination can exacerbate environmental pollution. Elevated energy consumption and carbon emissions present problems, rendering thermo-mechanical recycling more carbon-intensive compared to alternative processes. Enhancing thermo-mechanical recycling may necessitate the advancement of sophisticated compatibilizers, thermal separation technologies, low-temperature processing, microplastic management, treatment of process water and air, integration of renewable energy, and process innovation. Through the innovation of technologies that facilitate blended material recycling, enhance microplastic containment, and maximize energy efficiency, thermo-mechanical recycling has the potential to transform into a more sustainable and circular approach for global textile waste management [20].

11.2.3 Chemical Recycling of Textiles

Chemical recycling provides a revolutionary remedy for the textile sector by decomposing polymers into their monomeric constituents, facilitating the retrieval and restoration of superior resources. In contrast to mechanical recycling, which frequently deteriorates fibers and restricts the functionality of the recovered product, chemical recycling rejuvenates materials at a molecular level, providing the possibility of producing fibers that match the quality of their virgin equivalents. The technique is intricate, differing markedly based on the fiber type and encountering several technological, economic, and environmental challenges [21–23].

The chemical recycling process varies depending on whether the cloth is composed of cellulose or synthetic materials. The procedure for cellulose-based fibers such as cotton, viscose, and lyocell generally includes dissolving in designated solvents, succeeded by regeneration. Solvents like N-methylmorpholine N-oxide (NMMO), utilized in the Lyocell process, and sodium hydroxide (NaOH), employed in conventional viscose processing, are used to dissolve cellulose fibers [24]. Subsequent to dissolution, the material is reconstituted by a wet spinning technique employing an appropriate solidification solvent, therefore producing new fibers from the recovered cellulose. Synthetic fibers including polyester (PET), nylon, and polyurethane need distinct chemical treatments. Methods include hydrolysis, glycolysis, methanolysis, and solvent-based dissolution are employed to decompose polymer chains into monomers or oligomers [25]. Glycolysis of polyester entails utilizing glycol to decompose PET into bis-hydroxyethyl terephthalate (BHET), which may then be re-polymerized into fresh PET.

This diversity in chemical recycling, customized for fiber type, is especially beneficial for managing the intricate mixtures common in contemporary fabrics. Garments frequently have blends of cotton and polyester or nylon and elastane, materials that are challenging to disaggregate and recycle only by mechanical processes [26, 27]. Chemical recycling enables the recovery of materials from previously unrecyclable composites, therefore substantially diminishing textile waste. Notwithstanding its promise, chemical recycling encounters considerable obstacles, mostly owing to the intricacy of textile compositions. Contemporary apparel seldom consists of pure fibers. Rather, they frequently comprise colors, coatings, elastane fibers, waterproof treatments, and flame retardants. These additions disrupt the chemical recycling process, necessitating costly purification procedures to extract the necessary monomers free from contaminants [28, 29]. In cotton-polyester mixes, the cotton component must be selectively dissolved without impacting the polyester, or vice versa. The inclusion of elastane or substantial dye loading might render chemical reactions unpredictable, resulting in diminished purity of recovered components and requiring further procedures for cleaning and purification of the outputs. This introduces complexity, duration, and expense to the recycling process [28].

A notable issue in chemical recycling is the emission of microplastics during the initial phases. Prior to depolymerization, textiles often undergo shredding or rigorous washing to prepare them for chemical treatment. These mechanical operations can detach microfibers, releasing them into the wastewater flow. If inadequately handled, this microplastic contamination might undermine the environmental advantages of recycling. Therefore, chemical recycling plants must possess sophisticated wastewater treatment systems that can capture microplastics and avert their discharge into natural ecosystems. Additionally, several chemical recycling methods need elevated temperatures, specific solvents, and energy-demanding purification procedures. The restrictions may lead to a significant carbon footprint, potentially negating certain environmental benefits of recycling, unless the facilities utilize renewable energy sources and implement efficient process designs.

Chemical recycling is presently more costly than the production of fresh fibers. This economic reality constitutes a significant obstacle to wider adoption. The elevated expenses are ascribed to the intricacy of the chemical processes, the requirement for costly solvents and catalysts, and the energy demands of the procedures. The infrastructure for chemical recycling is nascent. Only a limited number of large-scale chemical recycling plants are present globally [30, 31], with other operations still in the pilot or demonstration phase. Certain initiatives have encountered insolvency, underscoring the substantial financial risks involved. Absent significant investment in operational scale and process optimization, chemical recycling will have challenges in competing with virgin fiber production. Notwithstanding these obstacles, the prospective advantages of chemical recycling are substantial. A significant benefit is its capacity to process intricate textile mixes that cannot be mechanically recycled. Chemical recycling can effectively handle cotton-polyester and nylon-elastane blends, facilitating the recovery of valuable elements from clothing that would otherwise be sent to landfill or burning. When completed correctly, chemical recycling can yield monomers of exceptional purity, which can then be re-polymerized into fibers that rival virgin materials in strength, durability, and aesthetics. This skill is essential for generating recycled textiles that satisfy the stringent performance criteria required by sectors such as fashion, automotive, and construction.

Moreover, enhancing chemical recycling might redirect substantial quantities of textile waste from conventional disposal practices. Annually, millions of tons of textile waste are produced, with the predominant portion disposed of in landfills or incinerated. Chemical recycling may significantly contribute to establishing a circular textile economy by completing the loop on textile manufacturing, therefore substantially diminishing the industry's environmental impact. Chemical recycling is in complete harmony with the tenets of a circular economy, which aims to minimize waste, optimize resource efficiency, and decrease reliance on new raw materials. By closing the material loop, chemical recycling not only mitigates the demand for fossil fuel–derived feedstocks but also cuts down the greenhouse gas emissions associated with raw material extraction and fiber production. Prominent businesses and environmental programs are progressively acknowledging the significance of chemical recycling in attaining zero-waste objectives and carbon neutrality aspirations. With suitable regulatory backing, technological advancement, and investment, chemical recycling may emerge as a fundamental element of sustainable textile production in the forthcoming decades.

Chemical recycling offers a promising pathway for sustainable textile waste management, especially given the complexity of modern textiles. Despite encountering significant technological, economic, and environmental obstacles, its capacity to extract high-quality components from intricate mixtures establishes it as a pivotal technology for the future. Investments in operational expansion, enhancement of process efficiency, and establishment of comprehensive environmental protections, especially for microplastic management, will be essential. Chemical recycling has the potential to revolutionize the textile business, becoming it more circular, robust, and less reliant on scarce natural resources in the long term. Table 11.1 delineates the several steps of chemical recycling and their effects on the release of microplastics. The comparative analysis of fiber mechanical and chemical recycling emphasizes process type, suitability, quality of outcomes, energy consumption, and economic feasibility, as presented in Table 11.2.

11.2.4 Challenges in Textile Recycling

The recycling process can be made significantly more difficult by the fact that modern fabrics are sometimes composed of a complicated mixture of elements. Certain mixes of natural and synthetic fibers, such as cotton-polyester, wool-polyamide, and polyester-elastane, are examples of the types of blends that may be manufactured. The application of chemicals finishes with a wide variety of fabrics in order to bestow features such as waterproofing, flame retardancy, or anti-bacterial effects adds another layer of complexity to the situation.

Table 11.1 Different stages of chemical recycling and their impacts on release of microplastics.

Stage of chemical recycling	Risk of microplastic release	Mitigation strategy
Collection/sorting	Medium	Improved pre-sorting, dry cleaning
Shredding/cutting	High	Enclosed systems, suction and filtration units
Washing	High	Advanced filtration, microfiber capture systems
Melting (thermo-mechanical)	Moderate to high	Melt filtering, vapor extraction systems
Depolymerization (chemical)	Low to moderate	Process optimization, controlled environment
Pelletization/extrusion	Medium	Closed systems, inline microplastic capture

Table 11.2 Comparative analysis of fiber mechanical and chemical recycling.

Feature	Mechanical recycling	Chemical recycling
Process type	Physical reshaping and re-melting	Molecular breakdown and repolymerization
Suitability	Mono-materials (e.g. PET, nylon)	Blended or contaminated fabrics
Fiber quality	Degrades with each cycle	Can achieve virgin-like quality
Microplastic risk	High during shredding and handling	Moderate in pre-processing; lower in output
Energy use	Lower than chemical recycling	Higher, but potentially renewable powered
Economic viability	More mature and cost-effective	Emerging, high investment required
Scalability	Limited by feedstock and quality degradation	Growing with new technologies and investments

Additionally, technical textiles, sportswear, and outdoor gear commonly have multi-layered structures in which multiple materials are bonded together, which further complicates the process of recycling used materials. Additionally, several textiles feature micro-scale fiber or coatings, such as polyurethane membranes, which provide extra challenges in the manufacturing process. When one considers the complexities of the material composition of contemporary fabrics, it becomes clear that the process of isolating these components for recycling is both technologically difficult and energetically demanding. Recycling by mechanical means frequently leads to a decrease in the quality of the fiber, whereas recycling by chemical means is restricted by the heterogeneity of the components that are involved. It is also possible for the recycling process to be responsible for the accidental release of microplastics into the environment, particularly during the phases of shredding, melting, and washing. During these procedures, fibers that become dislodged have the potential to leak into water systems if they are not adequately filtered. Furthermore, the lack of effective sorting technologies that are able to reliably identify textiles that are loaded with microplastics is a barrier to the implementation of large-scale recycling initiatives. Examples of technologies that have difficulty distinguishing between fiber mixes include optical sensors and NIR technology. This is especially true in garments that are dark in color or polluted. Because of high expenses, a poor yield of usable fibers, and a restricted market demand for recycled materials with degraded performance attributes, recycling textiles that are loaded with microplastics continues to be commercially unviable in many instances. In spite of these obstacles, it is essential to make investments in technology and infrastructure, as well as to innovate, in order to guarantee the economic viability of textile recycling and to reduce the negative effects that textile waste has on the environment. The comparative analysis of fiber mechanical, thermomechanical, and chemical recycling emphasizes compatibility for blends, quality of outcomes, microplastic risk, and scalability, as detailed in Table 11.3.

Table 11.3 Comparative overview of recycling techniques.

Criteria	Mechanical	Thermo-mechanical	Chemical
Suitable for blends	High potential	Limited, expected for mono materials	Limited scope and need efficient sorting are required
Output quality	Degraded, mixed color, less interest by consumers	Medium, however, it depends on the quality of the feed materials	Virgin-equivalent
Microplastic risk	High	Moderate	Low (post-processing)
Scalability	High (with limitations)	Moderate, but difficult to process the mixed waste	Low (but growing)
Energy use	Low to moderate	High	High
Cost	Low	Medium to high	High, due to the initial and process running cost
Circular potential	Limited	Moderate	High

Gao [32] performed a study on four distinct varieties of polyester men's sportswear, comprising two variants of virgin polyester and one variant of recycled polyester derived from PET bottles. Martindale abrasion tests were performed to check their impacts on microplastic generation. The abrasion condition was established at 5000 rubs under 12 kPa. The quantity of microplastic fibers emitted after 5000 rubs differed across various textile groups, showing no distinction between virgin and recycled textiles. The predominant microplastic fibers were between 14 and 500 μm in length, while the fibrils were of lesser length. The total length of fibers (including microplastic fibers and fibrils) emitted from recycled polyester textiles was marginally greater than that from pristine polyester fabrics; however, this discrepancy lacked statistical significance. Notwithstanding slight discrepancies in the data, there is no evidence indicating that mechanically recycled PET textiles exhibit any substantial variations or inadequacies in microplastic fiber and fibril creation, and release capability compared to virgin polyesters.

11.3 Conclusion

The recycling of textiles embedded with microplastics is a multifaceted challenge encompassing environmental science, materials engineering, politics, and consumer behavior. Existing methods are insufficient for managing synthetic textile recycling; nevertheless, increasing awareness and legislative progress provide a route to a sustainable textile industry. Integrating reduction techniques with enhanced recycling infrastructure is essential for environmental preservation and the advancement of green manufacturing. Fiber mechanical and chemical recycling have unique benefits, however, encounter difficulties in addressing microplastic contamination. The future depends on a cohesive approach that integrates advanced recycling technology, sustainable textile design, and strong regulatory backing. This comprehensive strategy tackles the microplastic issue and facilitates the establishment of a circular textile sector. Significant advancement necessitates technical innovations and cooperative endeavors along the value chain, guaranteeing a sustainable, circular future for future generations.

References

1 European Environment Agency. Textiles in Europe's circular economy. https://www.eea.europa.eu/en/analysis/publications/textiles-in-europes-circular-economy (accessed on 29 January 2025).

2 European Environment Agency (2024) Management of used and waste textiles in Europe's circular economy. *European Environment Agency*.

3 European Environment Agency (2024) Microplastics from textiles: towards a circular economy for textiles in Europe. *European Environment Agency.*

4 Periyasamy, A.P. and Periyasami, S. (2023). *Critical Review on Sustainability in Denim: A Step toward Sustainable Production and Consumption of Denim.* ACS Omega.

5 Periyasamy, A.P. and Venkatesan, H. (2019). Eco-materials in textile finishing. In: *Handbook of Ecomaterials*, vol. 3, 1461–1482. Cham: Springer International Publishing.

6 Hedrich, S., Janmark, J., Langguth, N., Magnus, K.H., and Strand, M. (2022) Scaling textile recycling in Europe-turning waste into value. McKinsey.

7 Raiskio, S., Periyasamy, A., Hummel, M., and Heikkilä, P. (2025). Transforming mechanically recycled cotton and linen from post-consumer textiles into quality ring yarns and knitted fabrics. *Waste Management Bulletin* 3 (1): 76–86.

8 Periyasamy, A.P. and Tehrani-Bagha, A. (2022). A review on microplastic emission from textile materials and its reduction techniques. *Polymer Degradation and Stability* 199: 109901.

9 Periyasamy, A.P. (2024). Sustainable approaches in textile finishing to control microfiber releases. In: *Microfibre Pollution from Textiles.* London UK: Taylor & Francis Group.

10 Jahnz, A. (2022). *EU Strategy for Sustainable and Circular Textiles.* EU Publications.

11 Andritz (2021) Create Value with Mechanical Recycling. *Andritz Web page.*

12 Zhou, J., Yu, L., Ding, Q., and Wang, R. (2019). Textile fiber identification using near-infrared spectroscopy and pattern recognition. *Autex Research Journal* 19 (2): 201–209.

13 Quintero Balbas, D., Lanterna, G., Cirrincione, C. et al. (2022). Non-invasive identification of textile fibres using near-infrared fibre optics reflectance spectroscopy and multivariate classification techniques. *The European Physical Journal Plus* 137 (1): 85.

14 Yuksekkaya, M.E., Celep, G., Dogan, G. et al. (2016). A comparative study of physical properties of yarns and fabrics produced from virgin and recycled fibers. *Journal of Engineered Fibers and Fabrics* 11 (2): 155892501601100209.

15 Frost, H., Zambrano, M.C., Leonas, K. et al. (2020). Do recycled cotton or polyester fibers influence the shedding propensity of fabrics during laundering? *AATCC Journal of Research* 7 (1_suppl): 32–41.

16 Gliaudelytė, U., Persson, M., and Daukantienė, V. (2025). Impact of textile composition, structure, and treatment on microplastic release during washing: a review. *Textile Research Journal* 95 (1–2): 220–232.

17 O'Loughlin, C. (2018). The impact of activewear/swimwear laundering: investigating microplastic-fibre emissions from recycled and non-recycled

synthetic textiles. *Journal of the Home Economics Institute of Australia* 25 (1): 2–12.

18 Jönsson, C., Levenstam Arturin, O., Hanning, A.-C. et al. (2018). Microplastics shedding from textiles – developing analytical method for measurement of shed material representing release during domestic washing. *Sustainability* 10 (7): 2457.

19 Özkan, İ. and Gündoğdu, S. (2021). Investigation on the microfiber release under controlled washings from the knitted fabrics produced by recycled and virgin polyester yarns. *The Journal of The Textile Institute* 112 (2): 264–272.

20 Wu, H., Lv, S., He, Y., and Qu, J.-P. (2019). The study of the thermomechanical degradation and mechanical properties of PET recycled by industrial-scale elongational processing. *Polymer Testing* 77: 105882.

21 Björquist, S. (2017) Dissertation on Separation for regeneration: Chemical recycling of cotton and polyester textiles. University of Borås.

22 Mu, B. and Yang, Y. (2022). Complete separation of colorants from polymeric materials for cost-effective recycling of waste textiles. *Chemical Engineering Journal* 427: 131570.

23 Xia, G., Han, W., Xu, Z. et al. (2021). Complete recycling and valorization of waste textiles for value-added transparent films via an ionic liquid. *Journal of Environmental Chemical Engineering* 9 (5): 106182.

24 Fink, H.-P., Weigel, P., Purz, H.J., and Ganster, J. (2001). Structure formation of regenerated cellulose materials from NMMO-solutions. *Progress in Polymer Science* 26 (9): 1473–1524.

25 Jiang, Z., Yan, D., Xin, J. et al. (2022). Poly(ionic liquid)s as efficient and recyclable catalysts for methanolysis of PET. *Polymer Degradation and Stability* 199: 109905.

26 Li, Z. (2020). Separation of waste cotton-polyester blended fabric. *IOP Conference Series: Materials Science and Engineering* 768 (2): 022062.

27 Zhongkai, Y., Guoliang, Z., and Rihui, L. (2017) Method for separating components of waste polyester-cotton blended fabric. CN107308905B, issued 2017.

28 Seppälä, H. (2023) The hornification of cellulose in chemical textile recycling. Master's thesis, Aalto University School of Chemical Engineering, Aalto University, Espoo, Finland. http://urn.fi/URN:NBN:fi:aalto-202303262541 (accessed on 20 January 2025).

29 Andini, E., Bhalode, P., Gantert, E. et al. (2024). Chemical recycling of mixed textile waste. *Science Advances* 10 (27): eado6827.

30 Renewcell: We make fashion circular. *Renewcell.* https://solidynex.org/ (accessed on 20 January 2025).

31 Infinited Fiber (2022) Infinna fiber: it's time for the textile industry to lose its virginity. *Infinited Fiber*. https://infinitedfiber.com/ (accessed on 22 January 2025).

32 Gao, M., Yang, T., Som, C., and Nowack, B. (2024). Differences in the release of microplastic fibers and fibrils from virgin and recycled polyester textiles. *Resources, Conservation and Recycling* 207: 107659.

12

Wastewater Treatment Methods for Removal of Microplastics from Effluents

Jaromír Marek[1], Mohanapriya Venkataraman[1], Jiří Militký[1], and Aravin Prince Periyasamy[2]

[1] *Department of Material Engineering, Faculty of Textile Engineering, Technical University of Liberec, Czech Republic*
[2] *Textile and Nonwoven Materials, VTT Technical Research Center of Finland, Espoo, Finland*

12.1 Introduction

The separation of microplastics from waste, technical and drinking water, or air is still a big challenge for both academia and industry [1–3]. In principle, the separation of microplastics will be realized by membrane technologies and their combinations [4–6]. However, microplastics capture is only part of the whole environmental problem. The second important part will be to prevent further contamination of the environment by microplastics by suitable water treatment and specific microplastic disposal [7–9]. The aim of the common technologies and design/production of the special equipment is to achieve the maximum concentration of microplastics in the retentate from membrane technologies and treatment with suitable technologies [10]. The problem of microplastic disposal should also be generally solved. The plastic materials in waste waters are commonly removed by a set of operations such as: clarification, coagulation and settling, flotation, activated sludge processing, membrane separation, biological treatment, and filtration [11]. The modern ways of wastewater treatments are using of combined electro-membrane processes, filtration based on nanoporous membranes, and (bio) chemical and/or physical oxidation of microplastics [12–14]. Removal of microplastics from polluted water can also be realized by degradation (as photo oxidation and biological treatment) or preferably by separation. Many articles on effective separation and capturing microplastics are based in most cases on standard filtration, microfiltration, or ultrafiltration. However, none of these processes cannot guarantee a zero concentration of microplastics

Textile Microplastics in the Environment: Sources, Impacts, and Solutions,
First Edition. Aravin Prince Periyasamy.

and nanoplastics in the permeate. This chapter is devoted to general aspects of microplastics capturing by different membrane pressure filtration techniques, characterization of membranes, and solution of problems with possible fouling.

12.2 Microplastics and Nanoplastics Characterization for Waste Treatment

Microplastics are in fact chemical contaminants in the form of particles based on synthetic polymers contaminated often by other pollutants ([15]) in size of 1–5 mm or other [1, 16, 17]. *Chapter* 1 provides a comprehensive explanation of the categorization of plastic waste, and Figs 1.4 and 1.5 may be seen to illustrate the concept. All these groups of particles form total plastic debris and are often assumed for particles of similar size in all directions (nearly spherical). Really, microplastics have a broad range of sizes, colors, and morphologies; are composed of different polymers and mixtures; and contain a broad spectrum of chemical additives. They also accumulate organic chemicals and trace metals from the surrounding environment [15, 18]. According to the chemical nature of polymers, these particles are neutral or negatively charged (due to ionic groups as results of degradation). Nanoplastics have common behavior of nanoparticles, i.e. their collisions with water molecules prevent the sedimentation and they are moving by Brownian random motion throughout the solution (typical behavior of colloidal systems) [16, 19]. Relative portion of microplastics from various sources is given in Table 12.1.

Special categories are fibrous microplastics as one of the main sources of deterioration of our environment due to the use of fibrous structures [20]. Globally detected polymers in representative studies are in order: polyethylene ~ polypropylene > polystyrene > PVC > polyethylene terephthalate (PET) [21]. The major problem is the right definition of size of fibrous microplastics because their length is less than 15 mm [22] and thickness is in order of micrometers (Fig. 12.1). The source of fibrous microplastics is primarily textile materials, which make up 35% of all microplastics. Fibrous microplastics are usually

Table 12.1 Major sources of microplastics.

Source	Percentage
Textile branch	35
Tires	28
Urban dust	24
Road marks	7
Rest	6

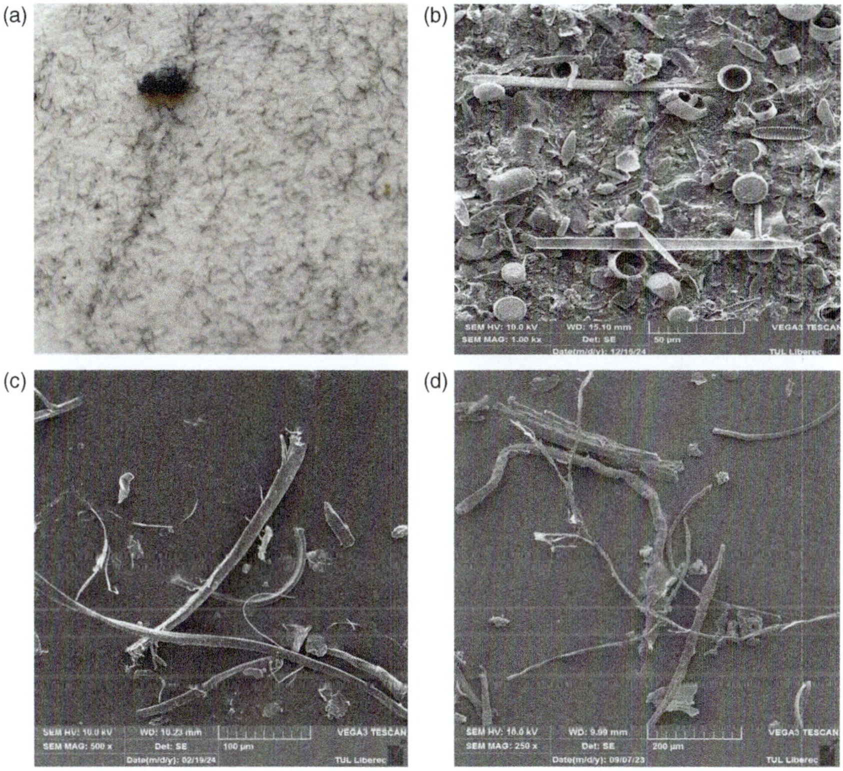

Figure 12.1 Different forms of microplastics: microplastics on filter after suction of the textile (a); microplastics in running water captured on the filter (b); and fibrous polyester microplastics after washing (c and d).

persistent and their removal is more complex due to the possible longitudinal penetration of fibers and fibrous segments into membrane pores (see Fig. 12.13). One solution to trap fibers passing the membrane longitudinally is to use combination of filtration and coagulation. Obviously, fiber fragments are also negatively charged, and therefore, the use of negatively charged membranes will prevent fiber passage. Fibrous micro/nanoplastics are usually characterized by their length (maximum dimension), but for selection of pores in separation layers, it is better to use their diameter (minimal dimension). Removal of micro and nanoplastics is closely connected with their shape (Fig. 12.1) and size distribution. This distribution is usually asymmetric, shifted to higher values, and common utilization of modus as microplastics size characterization is shifted down (more probability of particles occurrence will be above the median) [23]. Some example of the size distribution of PP microplastics is shown in Fig. 12.2.

In limited range of sizes (over median), the distribution of microplastics size can be approximated by simple power-law [25, 26]. Indication is straight line in

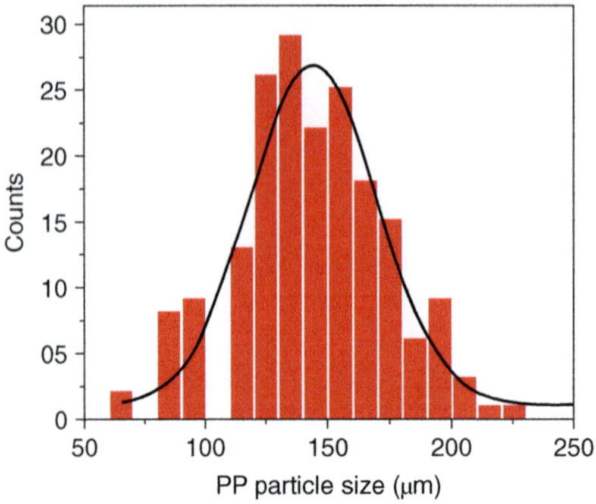

Figure 12.2 Size distribution of the polypropylene microplastic particles. (*Source:* Williams and Aravamudhan [24] / MDPI / CC BY 4.0.)

log–log plot of particle size histogram (constant size of bins). For characterization of microplastics size, the arithmetic mean and standard deviation s_d are often estimated as moments [27], where symmetric distribution is assumed. The α quantile of corresponding normal distribution is then simply approximated by the following relation:

$$Q(\alpha) = \bar{d} - 9,4\, s_d \,\ln(1/\alpha - 1)/[\mathrm{abs}(\ln(1/\alpha - 1)) + 14] \tag{12.1}$$

This α quantile is interpreted as value for which the probability of occurrence of particle size over this value is $1 - \alpha$. Value of then can be exceeded by 99% of microplastics size and it can be good for the selection of membranes pore size (99% of particles dimensions will be then bigger than the pore size). For data from Fig. 1.4 is $Q(0.01) = 89.7$ nm, which is significantly less than the mean nanoplastic size 92.5 nm. So, for membranes that should capture microplastics, it would be more suitable to use the 0.01% quantile of the microplastics size distribution for design of filtration membrane pore size. Limitation of this approach is assumption of symmetry (approximate normality). Particle size distribution is often asymmetric (shifted to the right) and can be approximated by a lognormal one [27]. Under the assumption of lognormal distribution validity, it is possible to calculate the corresponding α quantile from the well-known statement that lognormal distribution is obtained as distribution of logarithm of normally distributed variable [23]. A random variable d with a log-normal distribution is related to random quantity u, which has a standardized normal distribution, by the relation.

$$u = \left[\ln(d) - \bar{d}_{\ln}\right] / s_{d\ln} \tag{12.2}$$

Here is the arithmetic mean and is standard deviation of microplastic sizes in logarithms. Quantiles of variable u can therefore be approximated by Eq. 12.1.

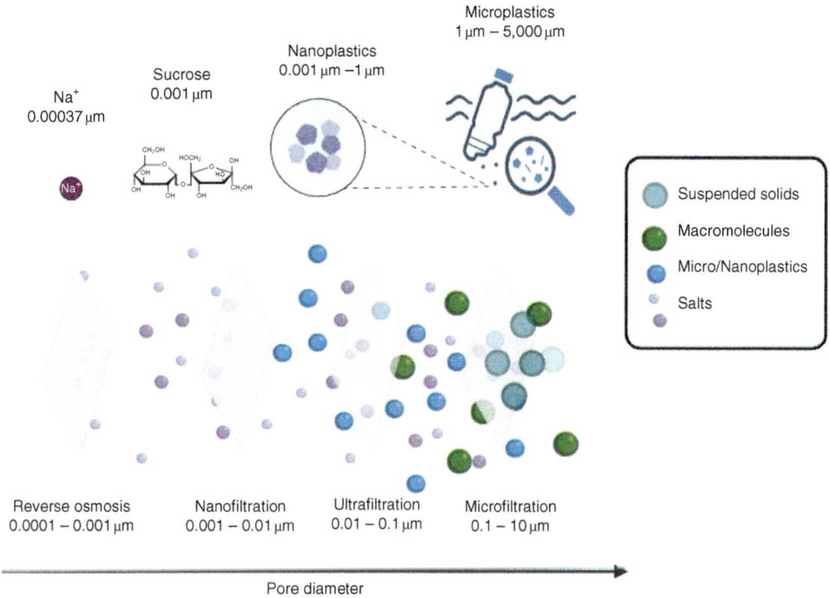

Figure 12.3 Pressure-driven membrane processes, range of membrane pore diameters, and some examples of contaminants capable of being removed. (*Source:* Pinto et al. [28] / MDPI / CC BY 4.0.)

12.2.1 Use of Membranes

Membranes are layers differing in the size of pores. If pores are smaller, molecules of penetrating substances cannot enter the membranes and are called non-porous or solution diffusion ones [4]. Pressure-driven membrane processes, a range of membrane pore diameters, and some examples of contaminants capable of being removed are shown in Fig. 12.3.

In the ideal case, membranes are a selective barrier allowing some components to pass through while others are not. Are membranes imperfect barriers allowing penetration of some components faster? Membranes are categorized according to their structure and type of material from which they are composed (Fig. 12.4) [29].

The membranes are created from organic or inorganic materials and often have some special filler enhancing their functionality. Inorganic membranes are based on metals, ceramics, zeolites, and metal oxides like Al_2O_3, TiO_2, ZrO_2, and other oxides as SiO_2 [29]. Polymeric membranes are more flexible with better formability and usually lower price. They are composed of a huge variety of polymers such as polysulfone, polyethersulfone, polyamides, cellulose acetate, polyvinylidene fluoride (PVDF), polyvinyl chloride, polyacrylonitrile (PAN),

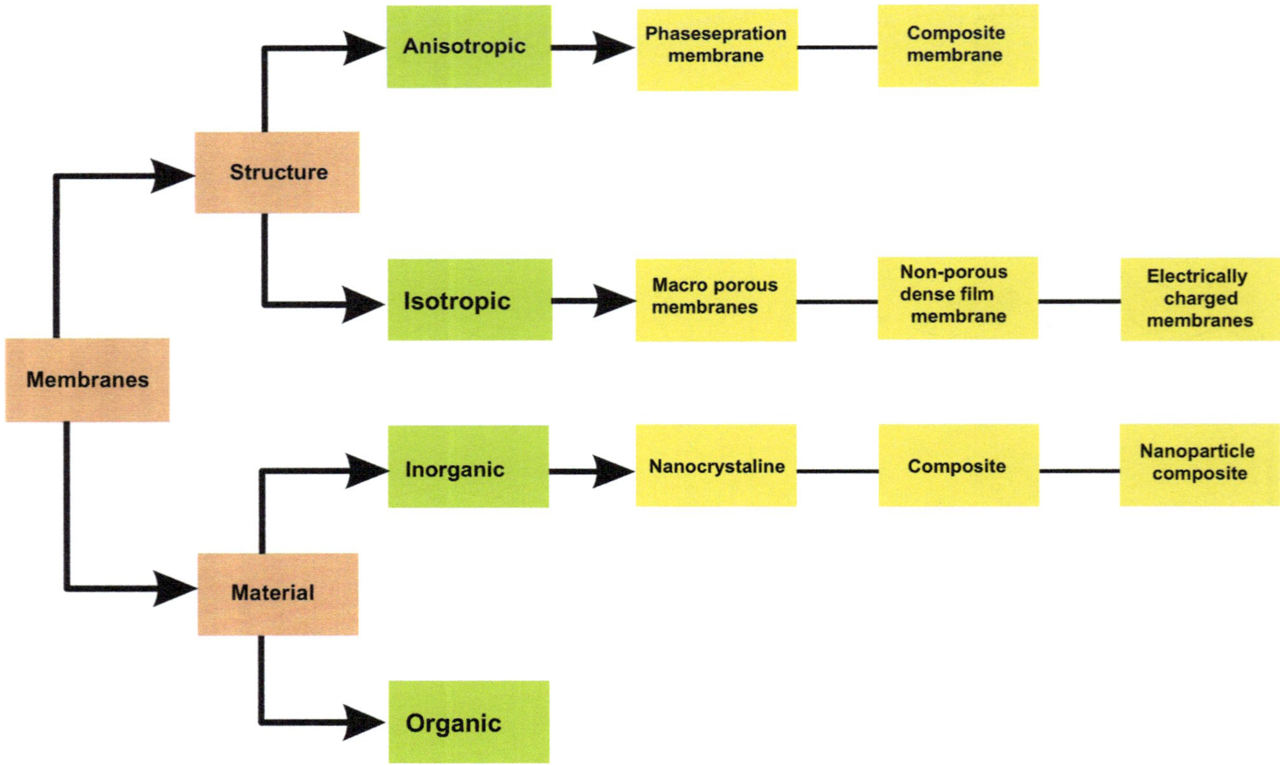

Figure 12.4 Classification of membranes. (*Source:* Kayvani Fard et al. [29] / MDPI / CC BY 4.0.)

and polymethacrylic acid [30]. Doping of membranes by proper particles can improve their hydrophobicity and tune pore size. One example is the use of polytetrafluoroethylene particles for enhancing the properties of a highly porous PVDF membrane [31]. Similar effects have inorganic additives such as SiO_2 or TiO_2 [30]. In the majority of membrane applications pressure difference is used for feeding of input mixture or polluted solvent to separate individual components or purified permeate. Mass transport of fluid through porous media is described by well-known Darcy's law [32].

$$J = \frac{B}{\mu}\frac{dp}{dx} \tag{12.3}$$

where J is the mass transfer rate (flux) [kg m^{-2} s^{-1}], B is constant characterizing membrane (permeability), dp/dx is the pressure gradient in the porous membrane, and μ is dynamic viscosity of the fluid. The *is* used for the case of straight pores with radius r. In the case of pore flow, mass flux J of a solution of density ρ and viscosity μ through porous membrane with a porosity P_o modeled as flow through a circular tubular pore of radius r and length l. Final form is known as Hagen-Poiseuille equation for pressure convective flow Eq. 12.4 [4].

$$J = \frac{\rho\, r^2 P_o}{8\,\mu\, l}(p_0 - p_L) \tag{12.4}$$

Porosity of the membrane can be calculated as, where N is the number of pores related to the membrane surface.

In the case of solution-diffusion, the penetrants diffuse (by Fickean mechanism) through the membranes and desorb at their opposite side. Mass flux is here a linear function of concentration gradient across the membrane with diffusion coefficient D as proportionality constant [32]. For the diffusion of very large spherical particles (molecules) in liquid (water) composed from small molecules (W), is diffusion coefficient D_{AW}, [m^2 s^{-1}] defined by the Stokes-Einstein formula [33], Eq. 12.5.

$$D_{AW} = \frac{k\,T}{3\,\pi\, d_s\, \eta} \tag{12.5}$$

where k is Boltzmann's constant (1.34 10^{-23} J/K), d_s is the diameter of particles [m], and η is the solute dynamic viscosity [cP].

In the pressure membranes, mass transport by diffusion is improved by the pressure difference between membrane sides, causing convection flow [34]. For the specification of fluid flow type through pores of known dimension (diameter), the Knudsen number K_n is used [35].

$$K_n = \lambda/\Lambda \tag{12.6}$$

where λ is mean free path of penetrant (gas) and Λ is length scale of system (pore size). For the case of transport of particles through of narrow pore, the Knudsen

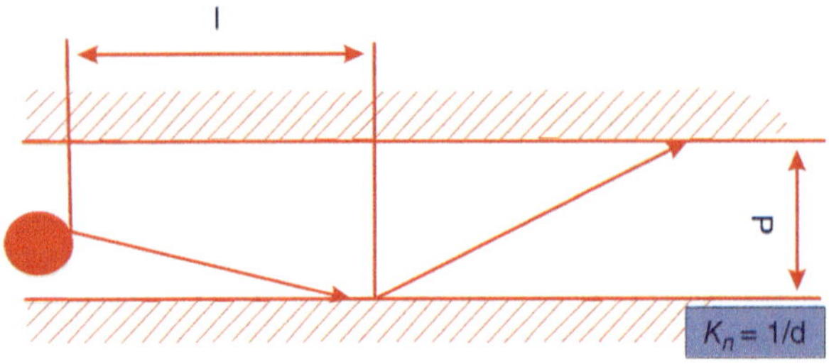

Figure 12.5 Knudsen number for transport of particles through narrow pore. (*Source:* Adapted from Wikipedia.)

number is defined as ratio between mean path *l* to collision with pore wall and pore diameter *d* (Fig. 12.5). Knudsen diffusion is a result of collisions of particles (gas molecules) with the pore walls, rather than intramolecular collisions (Brownian motion).

Transport through membranes covers the mechanisms from pure diffusion in solution, through Knudsen diffusional transport to molecular sieving transport. Various mechanisms for molecular transport in a pore/capillary are shown in Fig. 12.6.

If K_n increases, the rarefaction effects are more important, and the continuum assumption is not valid. For low K_n less than 10^{-3}, it is possible to use assumption

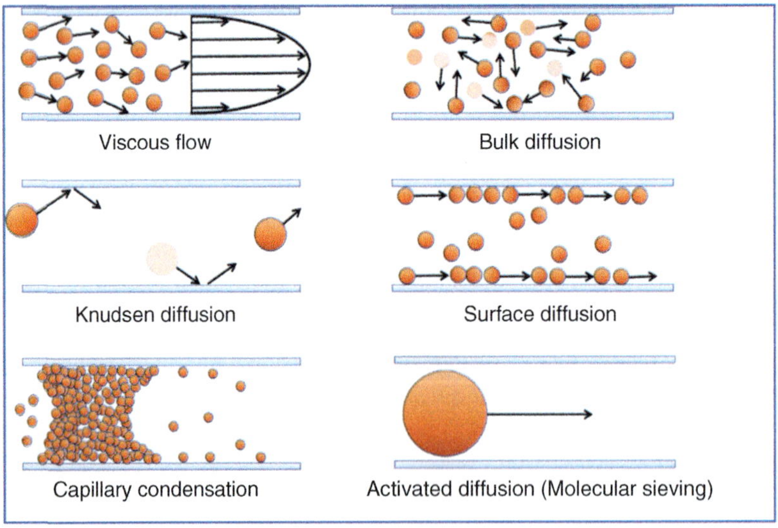

Figure 12.6 Various mechanisms for molecular transport in a pore/capillary. (*Source:* Ratnakar and Dindoruk [36] / MDPI / CC BY 4.0.)

of continuous flow. Knudsen diffusion coefficient (derived from kinetic gas theory) has the form [34].

$$D_K = \frac{2\,r\,P_o}{3}\sqrt{\frac{8\,R\,T}{\pi M}} \tag{12.7}$$

where R is the universal gas constant (8.314 J mol^{-1} K^{-1}), M is the molecular weight of particles [kg mol^{-1}], r [m] is pore radius, and P_0 is porosity of the membrane. Basic types of selectively permeable membranes with different mechanisms of action are:

- A macroporous membrane allowing a convective flow of air, aerosols, vapors, etc., through its large pores (see Eq. 12.5). No separation occurs.
- A microporous membrane following Knudsen diffusion through pores with diameters less than the mean free path of the molecules allowing lighter molecules to preferentially diffuse through its pores.
- An ultra-porous membrane has also been referred to as a molecular sieving membrane, where large molecules are excluded from the pores by virtue of their size.
- A non-porous membrane follows Fickian permeation through the non-porous membrane where penetrant dissolves into the membrane, diffuses across it, and desorbs on the other side based on concentration gradient, time, and membrane thickness.

In porous membranes, the hindered flow through pores is prevailing, and in non-porous (hydrophilic) membranes, diffusion through membranes is the transport mechanism of solute (Fig. 12.7), and practically all kinds of nano/microplastics are captured.

According to common definition non-porous material is one that has a volume porosity of less than 0.25 and a porous material usually has a porosity greater than 0.4 According to IUPAC classification, there are three groups of pores: micropores (dimensional less than 2 nm), mesopores (dimensions from 2 to 50 nm) and

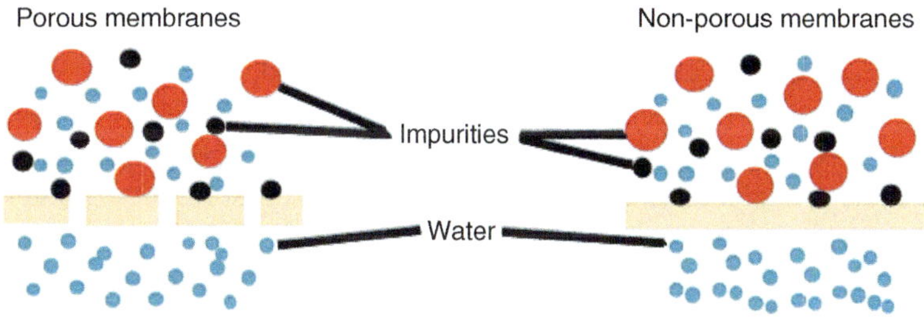

Figure 12.7 Porous and non-porous membranes flow. (*Source:* Covaliu-Mierlă et al. [37] / MDPI / CC BY 4.0.)

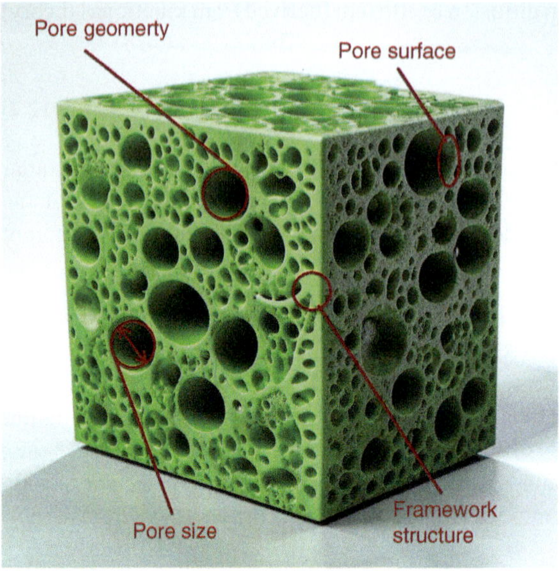

Figure 12.8 Pore geometry, pore surface, pore size, and framework structure of porous materials. (*Source:* Davletbaeva and Sazonov [40] / MDPI / CC BY 4.0.)

macropores (dimension over 50 nm) [38]. Micro and mesopores are typical for membranes and fibers (typical fiber diameter is over 10 μm) and macropores are typical for fibrous assemblies (structural units dimension in mm). There are structural characteristics of porous materials such as pore geometry, pore size, and pore surface functionality responsible for behavior of porous materials (Fig. 12.8) [39].

There are different technologies of membrane preparation [41].

Porous membranes

Porous membranes where the flow through pores hinders the penetration of particles (e.g. microplastics) prevail. Especially for the creation of microporous membranes and layers, there exist following basic possibilities [42–44]:

- *Wet coagulation:* Where microporosity is created by leaching out the salt from membrane with water. Incorporation of about 1% non-porous inorganic filler (silica or magnesium oxide of particle size <0.1 μm) is used.
- *Thermo coagulation:* The membrane polymer is applied from mixture of relatively volatile solvent and higher temperature boiling non-solvent. The polymer of membrane precipitates as a microporous layer when the solvent evaporates faster than the non-solvent.
- Non-solvent induced phase separation, where a polymeric solution is cast onto a support substrate and submerged in a non-solvent bath, causing a phase inversion that changes the polymer from a liquid to a solid phase. Throughout this inversion process, the solvent moves from the polymer solution to the

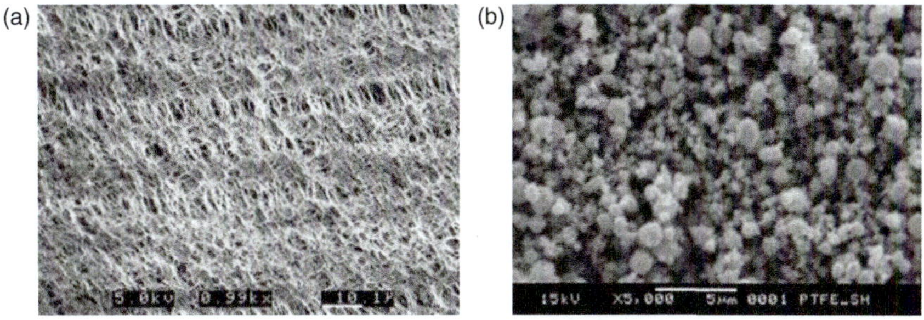

Figure 12.9 Two types of polytetrafluoroethylene membranes: Gore-Tex (pores 10–200 nm) (a), and nano membrane prepared by electro spraying (b).

coagulation bath (CB), and simultaneously, the non-solvent takes the opposite route, leading to the creation of porous membranes [41].

- Creation of micro and meso pores by biaxial stretching. An example is creation of microporous polytetrafluoroethylene (material GORE-TEX), see Fig. 12.9a.
- Microporous structures formed from aqueous dispersions as polyurethane UV/E Beam Polymerized Membranes (example is material Repel).
- Perforation in compact coated layer (high voltage).
- Extraction of soluble component from a polymer mixture.
- Electrospinning/electrospraying, see Fig. 12.9b.

Non-porous membranes

Non-porous membranes, where the basic mechanism is diffusion through membranes. Water first adsorbs on the surface of the layer and then diffuses to the other side. Basic variants of these membranes are:

- Hydrophilic polyurethanes (Comfortex),
- Hydrophilic polyester (Sympatex),
- Penta block copolymers-Nexar type characterized by hydrophilic and hydrophobic regions (Kraton),
- Cellulosic derivatives [45].

The physical attributes and chemical composition of the membrane materials are fundamental to determining the membrane characteristics and performance. The most common polymers for membranes are polyolefins (polyethylene, polypropylene), polyamides, polyimides, fluorinated polymers such as PVDF, polyvinylchloride, cellulose acetate, nitrocellulose, cellulose esters, polysulfones, polyether sulfones (PES), and polyacrylonitrile (PAN) [42]. Versatile materials for electrospinning and electrospraying are PUR and PVDF. Promising reveals the utilization of PTFE (Teflon) dispersion. Especially for the separation of small molecules (as water) from nanoplastics, the polymers with cavities can be effectively used. The most permeable polymer, poly(1-trimethylsilyl-1-propyne) (PTMSP), exhibits an approximately bimodal cavity size distribution centered at around $r = 0.3$ nm and $r = 0.6$–0.7 nm. Its ability to separate small molecules (less than 0.45 nm) is too low to be useful, and the large cavities collapse over time due to physical aging. An approach based on the preparation of intermediate-sized cavities through the spatial rearrangement of the rigid polymer chain segments in the glassy phase by thermal post-processing is described in [46]. Some glassy polymers, such as polyimides (or other polymers with rigid polymer chain segments in the glassy state, enabling phase-controlled free-volume element formation through spatial rearrangement), produce membranes with superior mechanical properties, thermal/chemical stability, and tunable free cavities [46]. The block copolymers are widely applicable for the same purposes. Sulfonated polystyrene is one of the basic blocks of Nexar membranes. This material has excellent dimensional stability and wet strength. Nexar phase separates into hydrophilic and hydrophobic domains. This phase separation gives the material its high wet strength at high degrees of sulfonation. The hydraulic water permeability and water vapor transport rate (at 50% relative humidity) of Nexar increase as sulfonation is increased (characterized by ion exchange capacity) due to increased hydrophilicity of the membrane [42]. Semi-porous nanofiltration membranes are compositionally like reverse osmosis membranes. For enhancement of filtration membrane properties, the surface activation, e.g. by plasma treatment [47] or adding different fillers, can be used. The addition of POSS particles greatly enhances hydrophobicity. Superhydrophobic/super oleophilic membrane was obtained as the POSS concentration increased to 40 wt.% [48]. Plenty of other nanoparticles can also be used for the functionalization of surface layers [49].

Membrane processes

There are three basic groups of water treatment. The first group consists of cleaning water for industrial purposes, the second group is the treatment of drinking water, and the third group is the treatment of wastewater. For water containing microplastics treatment and purification, the various physico-chemical and biochemical methods were traditionally used. Main of them were distillation, precipitation, and extraction methods, electrochemical methods, sludge formation methods, sedimentation and filtration methods, degradation methods, and methods based on the ion exchange [50]. Today, most of them are replaced by pressure membrane methods or their combination with other ones. Penetration of polluted fluid is here driven by pressure drop which represents the resistance to flow through the membranes. The pressure drop is expressed by the pressure difference

$$\Delta p = p_1 - p_2 \tag{12.8}$$

where p_1 is the initial pressure of the filtered medium on the membrane input and p_2 is the pressure behind the membrane.

Table 12.2 Range of pressure for membrane processes.

Membrane process	Pressure range [bar]
Microfiltration	0.5–3
Ultrafiltration	1–10
Nanofiltration	7–40
Reverse osmosis	25–100

The pressure drop is related to the flow rate. As the flow rate increases, the pressure drop increases mostly linearly. The pressure range required for various membrane filtration processes is shown in Table 12.2. Generally, smaller pores require greater pressure drops across the membrane.

Removal of nano/microplastics in pressure membranes is highly influenced by their shape (Fig. 12.15). Main membrane processes are adsorption, electrostatic repulsion (between charged solutes and charged membrane surface), and size exclusion or sieving [48] (Fig. 12.10).

Diffusion-based membranes (reverse osmosis) and porous membranes (microfiltration, ultrafiltration, nanofiltration) are used. Through the diffusion membrane pass only molecules of solvent and solutes, but not particles (nanoplastics). The pore size of porous membranes determines the particles that can pass through them. Typically, microfiltration will retain particles >1 μm, ultrafiltration >0.01 μm, and nanofiltration >0.001 μm. For all porous membranes, these limit values are below the size of microplastics, so no microplastics should pass through till the membranes are destroyed. There are two basic flow configurations in membranes. In standard dead-end mode, the fluid flows normally (perpendicular) to the membrane surface, and in cross-flow, the fluid flows parallel, or "tangential," to the membrane surface. One of the main reasons for the cross-flow mode is the partial removal of the filtration cake (highly concentrated suspension

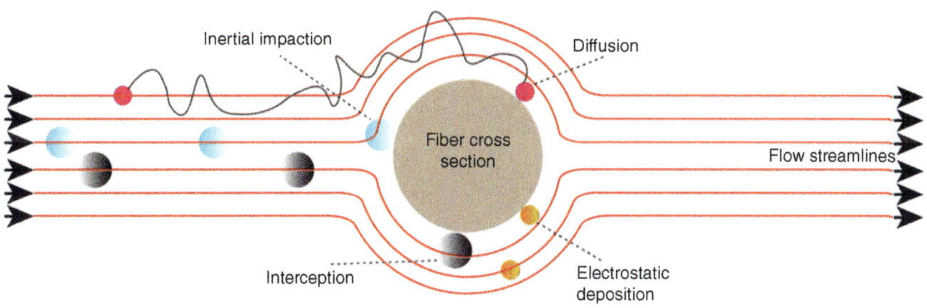

Figure 12.10 Mechanisms of particle captures. (*Source:* Zhou et al. [51] / MDPI / CC BY 4.0.)

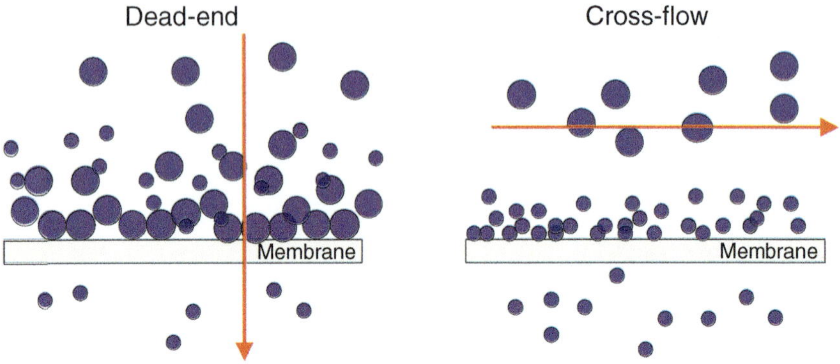

Figure 12.11 Dead-end and cross-flow modes. (*Source:* Ruiz-García et al. [53] / MDPI / CC BY 4.0.)

near the membrane surface) [5, 52]. Dead-end and cross-flow modes are shown in Fig. 12.11.

Cake formation is typical for normal filtration and microfiltration solely. The concentration polarization, i.e. the formation of a gradient in solute concentration (highest is near membrane and lowest in the bulk), is typical for ultrafiltration, nanofiltration, and reverse osmosis [5]. Pressure membrane technologies now represent the standard in the industrial treatment of drinking and wastewater. Only exceptionally are other methods used, mostly for a specific purpose separation (e.g. sorption of heavy metals using ionic systems), or if there is the possibility of using waste heat (distillation) or in cases when there exists a lack of investment funds for membrane technologies. Total operating costs of pressure membrane technologies are usually lower. A comparison of membrane processes with classical technologies is discussed in the article [2], where investment and operating calculations are given and data measured on membrane technologies versus classic technologies in real-life operation, i.e. during long-term feasibility tests on pilot membrane units are presented. A major portion of microplastics (approximately 60–98% [54]) in wastewater is commonly removed in municipal wastewater treatment plants by standard preliminary, primary, and secondary water treatment before membrane or other advanced processes [55, 56]. The preliminary treatment involves the removal of large objects and macroscopic scraps by fine screening [57]. The primary treatment is a combination of flocculation and sedimentation called clarification or activated sludge.

12.2.2 Clarification

Generally, for coagulation and flocculation, microplastics' properties such as surface charge, size, and shape can improve their agglomeration. Clarification is

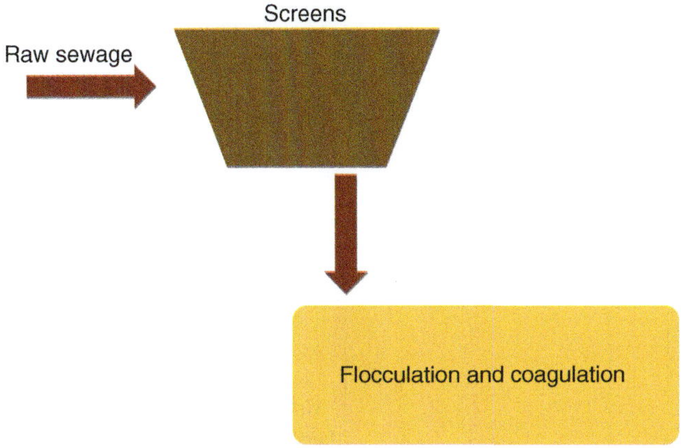

Figure 12.12 Processes of wastewater pre-treatment. (*Source:* Adapted from Teh et al. [58].)

realized by using flocculants and coagulants. Coagulants' (e.g. aluminum sulfate, sodium aluminate, ferric chloride, polyaluminum sulfate, diallyl dimethyl ammonium chloride, cationic polyelectrolytes) aim is to neutralize suspended particles' charge, and a flocculant's (e.g. modified starch, polyacrylamides, polyacrylic acid) role is floc formation from neutralized particles [58]. The processes of coagulation and flocculation are schematically shown in Fig. 12.12. Clarification efficiency decreases with high concentrations of polymeric debris in the effluent because negatively charged polymeric particles interact with some coagulants and require higher concentrations which are unpredictable. Some negatively charged polymeric particles are absorbed on metal hydroxides ($Fe(OH)_3$ and $Al(OH)_3$) produced during coagulation which leads to more stable aggregates that can precipitate [59]. The coagulation process was 25% more efficient with Al^{3+} than with Fe^{2+} for removing 0.1–5 mm microplastics and the addition of an anionic surfactant improved efficiency by more than two times [12]. The greener approach is to use bio-based flocculants as lignin or tannin derivatives [60]. The secondary treatment stage is removal of biodegradable organic matter, usually by aeration, oxidation, photooxidation [61], or biological treatment [13].

12.2.3 Filtration

Basic filtration processes for wastewater treatment differ in the porosity of active layer and filtration mechanism. The granular filtration (pores 10 μm–100 μm) mechanism is sieving

- Microfiltration (pores ≈ particles 10 mm–10 μm), mechanism is sieving
- Ultrafiltration (pores 2 nm up to 50 nm), mechanism is sieving

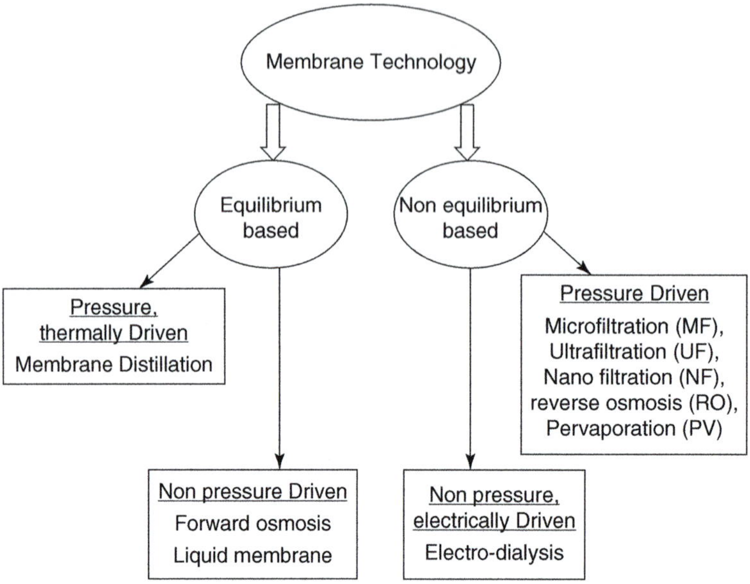

Figure 12.13 Pressure membrane filtration systems. (*Source:* Obotey Ezugbe and Rathilal [63] / MDPI / CC BY 4.0.)

- Nanofiltration (1–2 nm), mechanism is solution-diffusion mainly
- Reverse osmosis ("pores" <1 nm), mechanism is solution-diffusion
- Electrodialysis (electrodeionization, ion selective)

Granular filtration, microfiltration, and ultrafiltration functionality are based on the mechanism of pore-flow (sieving), but nanofiltration is based on a combination of solution-diffusion and pore-flow mechanisms, and reverse osmosis is based solely on the solution-diffusion mechanism. A summary of pressure membrane technologies for different kinds of materials is shown in Fig. 12.13 [62].

Granular filtration
For granular filtration, sand filters and activated porous carbon-based filters are commonly used. Sand filtration can separate microplastics from water by mechanical deposition in the silica-based bed. Typically, sand grains with a diameter around 500 μm are used. Microplastics between 80 and 100 μm cannot pass through the spaces between the sand grains and are removed from the water. Particles smaller than a micrometer (nanoplastics) can be removed when they attract the sand grains in the filter. The attachment mechanisms depend on the size, shape, and charge of the nanoplastics and on the hydraulic behavior of the filter. Plastic particles can be absorbed on filter sand grain surfaces via hydrogen bonds created between silica grain blocks and hydroxyl groups present in some

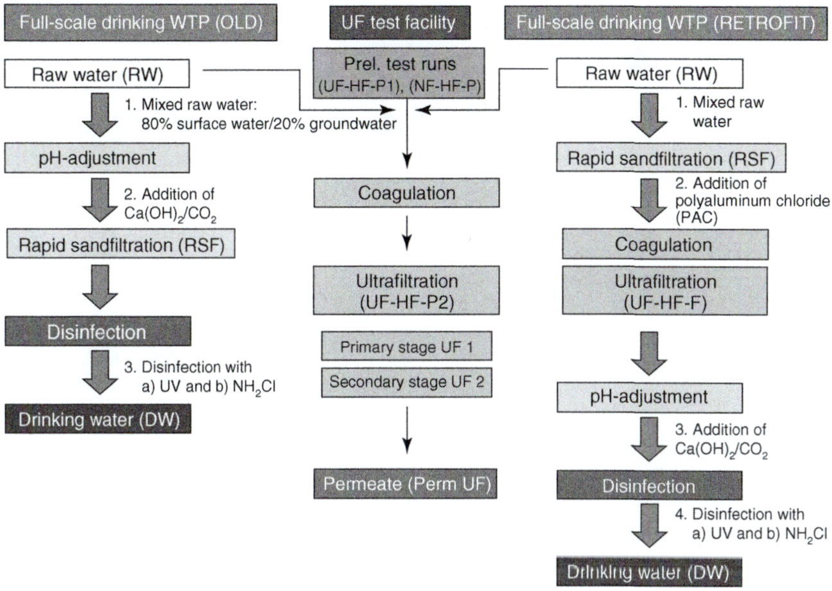

Figure 12.14 Coagulation and ultrafiltration process for MPs removal. (*Source:* Keucken et al. [67] / MDPI / CC BY 4.0.)

microplastics or arising as product of degradation processes. Their removal from filter layers will be more complicated and performance of sand filters will then be reduced [11, 64, 65]. Sand filtration can be effective for stopping plastic scraps, macroplastics, and some microplastics but its removal efficiency depends on plastic debris size and their chemical nature. Filtration and clarification cannot be replaced even technically, nor economically [66]. Microplastics removal is most frequently realized by a combination of clarification processes (coagulation, flocculation, sedimentation/flotation, and sand filtration) and membrane processes [6]. The scheme of removal and fouling of polyethylene nanoplastics (<0.5–5 mm) by a combination of $FeCl_3$-supported coagulation and an ultrafiltration PVDF membrane with a pore size of 30 nm was used. Coagulation and ultrafiltration process for MPs removal is shown in Fig. 12.14 [6].

Microfiltration

Microfiltration is usually applied as a pre-treatment for other separation processes such as ultrafiltration, and a post-treatment for standard granular media filtration [68]. When filtering with microfiltration membranes, the lowest pressure (<0.2 MPa) is required from all pressure processes. The pore size ranges from 500 nm to 10 μm which is below the range of granular microplastics size but the cross-section of fibrous microplastics is often below 10 μm (i.e. diameter of

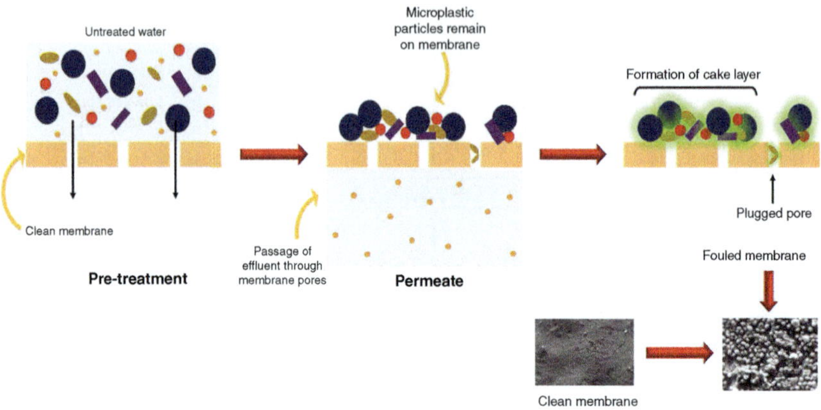

Figure 12.15 Schematic representation of microplastic removal by membrane filtration and membrane fouling. (*Source:* Pandey et al. [70] / MDPI / CC BY 4.0.)

fibers [69]). Schematic representation of microplastic removal by membrane filtration and membrane fouling is shown in Fig. 12.15.

Microfiltration can be used as pre-treatment of water for its further processing, e.g. by reverse osmosis or electrodialysis. Microfilters (ceramic units) will be sufficient for capturing the coarser fractions of microplastics. Emulsions and suspensions are most often processed by microfiltration. It finds its use in several branches of industry, e.g. in the food industry components in the processing of the whey or in the removal of bacteria and beer yeast, also during the cleaning and sterilization of fruit juices. It is also used for sterilization in the pharmaceutical industry or in metallurgy when processing colloidal solutions and dispersions of metals and their oxides [3].

Ultrafiltration

Ultrafiltration membranes have pore sizes from 3 to 50 nm and capture particles on the basis of the network effect. The required working pressures range from 0.1 to 0.5 MPa, which is more than in the case of microfiltration, but significantly less than in the case of reverse osmosis. Ultrafiltration is used for capturing the finer microplastic fractions and for the purification and thickening of high-molecular substances solutions, such as proteins and polysaccharides. The principle consists in introducing a dispersion of particles to the membrane through which dispersion containing part of the smaller particles passes (it is permeated), and part of the dispersion containing bigger particles (concentrate) is drained away. Ultrafiltration is used similarly to microfiltration in the food industry, further as pre-treatment of water for other processes or in engineering during maintenance of degreasers baths and separation of petroleum substances from wastewater. It was proven that ultrafiltration units reliably remove only microplastics and on the capture of nanotraps have essentially no effect, even if the modules were used with a guaranteed pore diameter of 50 nm [3].

Nanofiltration

The nanofiltration process uses membranes with pore sizes from 1 to 3 nm. Membranes can carry functional groups, most often with a negative charge. The mechanism of adsorption of nanoparticles on the membrane surface, diffusion through the membrane, and desorption on the other side of the membrane is used. Operating pressures range from 0.5 to 1.5 MPa. Nanofiltration membranes are used for water softening due to their ability to selectively trap polyvalent ions. It is also used for the removal of pesticides, pathogenic substances, and precursors of chlorinated derivatives of organic substances, which arise during water disinfection [3].

12.2.4 Reverse Osmosis

Pressure processes also include reverse osmosis, where porous membranes are no longer used. The diameter of their micropores is approx. 0.5 nm, which represents the size of some larger molecules (molecules with a molecular weight of up to approx. 100 Da) [3]. Reverse osmosis membranes' functioning is based on the solution-diffusion mechanism [4]. Reverse osmosis is a process that captures dissolved salts and low molecular weight components on the membrane based on different solution concentrations on their sides. Osmosis occurs when water flows through a semi-permeable membrane in the direction of a more concentrated solution for equalizing the pressures. The level of solution difference on both sides of the membrane corresponds to osmotic pressure. In reverse osmosis, pressure is applied, which, when equalized with the osmotic pressure π, causes the stop of flow through the membrane (osmotic equilibrium, see Fig. 12.16). At further increase of pressure, the water flows from a side of higher concentration to a side of lower concentration. In this way, the input solution can be separated into permeate and retentate, because only certain components pass through the semi-membrane and higher molecules are concentrated in front of the membrane. Reverse osmosis is used in the preparation of water with the lowest possible salinity. Work pressures are significantly greater than in the previous membrane processes [71].

Reverse osmosis membranes are non-porous and therefore only allow those components that are soluble in water to pass through. Microplastics and nanoplastics do not belong to them. The rejection of microplastics and nanoplastics is therefore 100%. Modern reverse osmosis membranes contain a wide variety of nanoparticles, including zeolites, carbon materials (including carbon nanotubes and graphene oxide), silica, metal oxides (titanium dioxide), metal nanoparticles, and organic–inorganic hybrid materials (*zeolitic imidazolate framework-8*) [73] in matrix mainly from aromatic polyamide films of composite membranes [74]. The proposed transport mechanism in thin film nanocomposite reverse osmosis

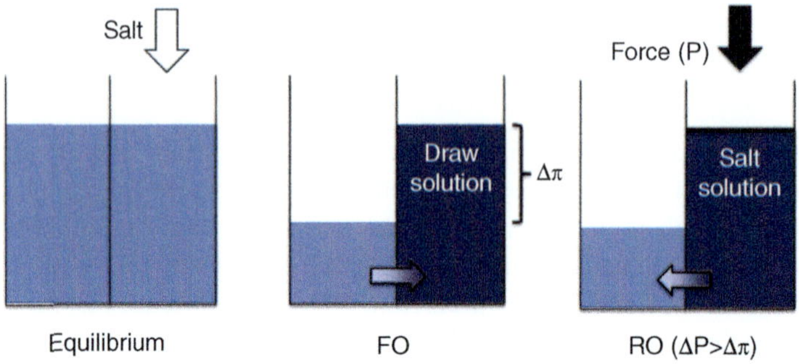

Figure 12.16 Schematic of forward osmosis (FO) and reverse osmosis (RO); $\Delta\pi$ is the osmotic pressure gradient. (*Source:* Mohammadifakhr et al. [72] / MDPI / CC BY 4.0.)

membranes with incorporated hydrotalcite (HT) into polyamide layer during interfacial polymerization process is described in [74].

12.2.5 Filtration Using Nanofiber Filters

According to the ISO definition, the fibers with diameters less than 100 nm are classified as nanofibers. Nanofibrous assemblies which are for better functionality often coated by special polymeric layers and are prepared mainly by electrostatic spinning [75] (see Fig. 12.17).

Nanofibrous membranes have very good filtering properties due to their high specific surface area and small pore size, and therefore are a prerequisite for wide use. Additional advantages include specific flow boundary conditions as well as the use of a small amount of material. The disadvantage, however, remains the low mechanical resistance that comes with it results in poor filter cake removal. Poor mechanical stability is solved, for example, by a suitable carrier, i.e. support layer. Basic methods on nanofibrous filtration layers creation are deeply discussed in books [68, 76]. One problem with mechanical stability is the generation of nanoplastics as a result of breaking some nanofibrous segments of membranes. For continuous preparation of nanofibrous assemblies, the needleless system Nanospider was developed (see Fig. 12.18).

12.2.6 Electrodialysis

Electrodialysis and reverse electrodialysis are using a set of alternating cationic and anionic membranes between anode and cathode [78]. Cationic membranes have negative fixed charge groups and anionic membranes have positive fixed charge groups. In these technologies, an alternating series of membranes creates

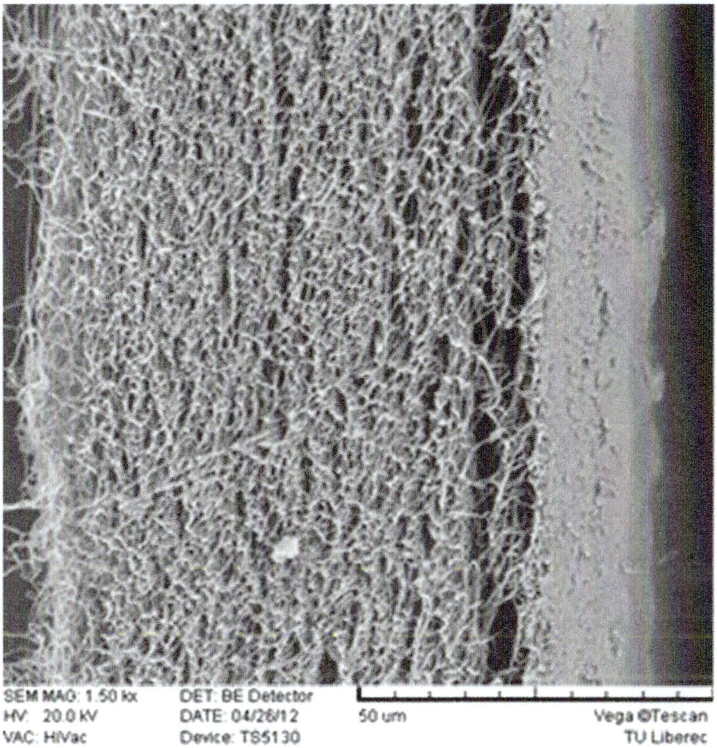

Figure 12.17 Hydrophobic membrane from PA 6 nanofibrous assembly and hydrophobic coating based on silicones.

flow channels [4]. One of the possibilities to increase the efficiency of the electrodialysis unit is the use of bipolar membranes. They are commonly used to form weak acids and bases from salt solutions, but they also increase considerably efficiency and economy of the process, as they ensure water splitting [79]. A relatively new technique is shock electrodialysis (SED), which enhances the electrodialysis process, reaching over-limiting conductance. SED is using a stack of one type of membranes (e.g. cation or anion exchange membrane). Space between them is filled with porous material. The non-homogeneous space charge region in the channels of porous materials is created by the normal component of the applied electrical field. The current is realized by the ions accumulated along the channel's walls [80]. A sharp concentration gradient propagating through the channels of the porous medium is obtained. SED is highly efficient for ion removal from water (desalination) [81]. Microplastics and nanoplastics would be trapped on the porous media when separated by SED, and this is the main reason for not directly using this technology for, they are capturing.

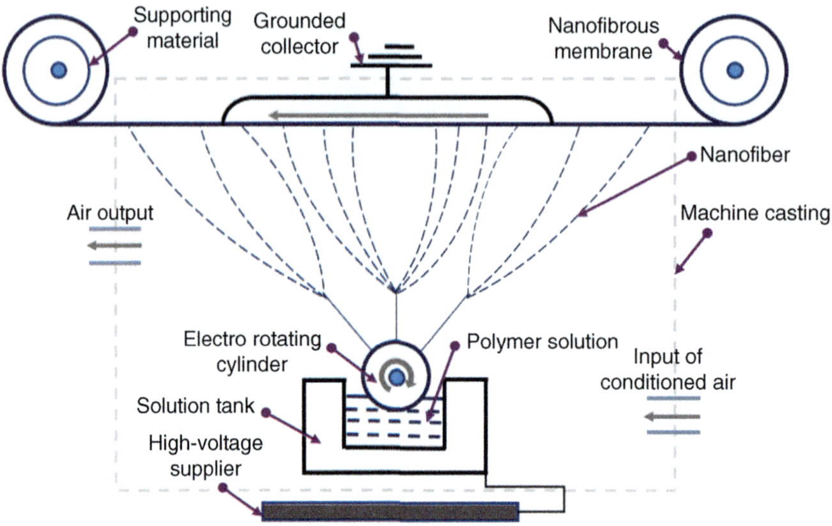

Figure 12.18 Continuous needleless electrospinning system (Nanospider). (*Source:* Venkataraman et al. [77] / MDPI / CC BY 4.0.)

The term microplastics and nanoplastics often gives the false impression that it is possible to filter these particles by pressure membrane units by microfiltration, and nanoplastics by nanofiltration or by ultrafiltration. This is not the case, and the removal of microplastics by microfiltration is far from complete [82]. However, even ultrafiltration is not capable of removing microplastics completely [83]. In membranes that have a given pore size, it is therefore possible to find pores that are orders of magnitude larger. This allows the filtered particles of larger dimensions to penetrate into the permeate. The real polydispersity of microplastic particles in marine and fresh water is in the range from 10 μm to over 500 μm [84, 85]. This wide size distribution is one of the reasons for lower efficiency of microplastics capturing by membrane processes. It should be noted that the distribution of pore diameters is wide, and the reported size value is usually the mode (better is the proper quantile $Q[\alpha]$, see Eq. 12.1 but here for $\alpha = 0.99$).

12.2.7 Membrane Fouling

Main problems of membrane use are concentration polarization and fouling (clogging) by microplastics and nanoplastics, which make them poorly working [86]. Microplastics with hydrophobic surfaces showed hydrophobic interaction with hydrophobic polymer membranes in water filtration processes. This interaction occurred through a noncovalent force formed a cluster of nonpolar species in

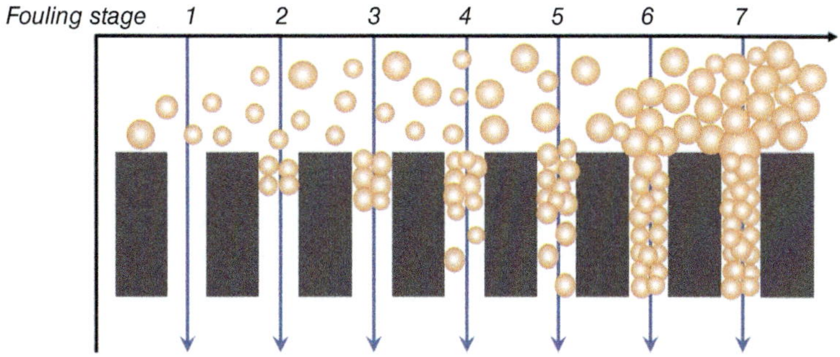

Figure 12.19 Stages of membrane fouling. (*Source:* Xu et al. [88] / MDPI / CC BY 4.0.)

water. Flux recovery ratios for hydrophobic membranes were found to be less than those for hydrophilic membranes at pH values of 5, 7, and 9. Concentration polarization appeared because of increasing of the concentration of waste dispersion near the membrane surface, which creates a boundary layer. Membrane fouling and cake creation depend on their porosity and size of penetrating particles. Reverse fouling and porous surface cake are caused by particles larger than pore size. If particles are smaller than membrane pores, the irreversible fouling is created [87]. The four main mechanisms of membrane fouling are: complete blocking, intermediate blocking, standard blocking, and cake layer formation were considered for modeling of time depending on fouling [76]. The stages of membrane fouling are shown in Fig. 12.19.

The simplest model of membrane fouling is resistance-in-series model based on the concept of resistance R [m^{-1}], which is related to permeate flux J [$l\ m^{-2}\ s^{-1}$] by relation [89]

$$R_c + R_m + R_f = \frac{\Delta p_T}{\eta J} \tag{12.9}$$

where Δp_T is transmembrane pressure loss, η is dynamic viscosity of permeate, and R_c, R_m, R_f are the cake, membrane, and fouling resistances, respectively. If the membrane resistance and the fouling resistance are negligible compared to the cake layer resistance, the flux can be written as [89]

$$J = \frac{\Delta p}{\eta R_{cs} \delta} \tag{12.10}$$

where R_{cs} [m^{-2}] is the specific cake resistance and δ [m] is the cake layer thickness.

Membrane fouling is a resistance-in-series model used for the simulation of time-dependent flux in conditions of time-dependent fouling and cake formation [89]. It was found that flux is directly proportional to the square root of the transmembrane pressure drop *for* incompressible cakes and it is not significant for compressible cakes. The mechanistic model is based on classical constant-pressure dead-end filtration proposed by Hermia [90, 91]. The adapted model to cross-flow ultrafiltration has the form [92–94].

$$\frac{dJ}{dt} = K\,(J - J_s)J^{2-n} \tag{12.11}$$

where J [L h^{-1} m^{-2}] is the permeate flux, J_s is permeate maximum flux to avoid fouling and K is rate constant, and n is kinetic order parameter specifying the mechanism of fouling. For complete blocking is $n = 2$. In this model, a particle cannot penetrate inside the pores and a single particle layer is created on the membrane surface. Under these conditions is a solution of Eq. 12.11 in the form [87].

$$J = J_s + (J_0 - J_s)\exp(-K_2\,t) \tag{12.12}$$

where J_0 is initial flux and K_2 is rate constant.

For internal (standard) blocking is $n = 1.5$. In this model, all membrane pores have the same length and diameter, and the particles are smaller than the membrane pore size and can penetrate inside the pores. Under these conditions is the solution of Eq. 12.11 in the form [87]:

$$J = \frac{1}{(1/\sqrt{J_0} + K_1\,t)^2} \tag{12.13}$$

where K_1 is rate constant.

For intermediate blocking is $n = 1$. In this model, similar to the complete blocking, the particles can create multiple layers. Under these conditions is the solution of Eq. 12.11 in the form [87]:

$$\frac{J}{J - J_s} = \frac{J_0}{J_0 - J_s}\exp(J_s K_3\,t) \tag{12.14}$$

where K_3 is reduced rate constant

For cake formation is $n = 0$. In this model, particles that are larger than the membrane pores are deposited on the membrane surface, creating a porous cake. Under these conditions is the solution of Eq. 12.11 in the form [87]:

$$K_4\,t = \frac{1}{J_s^2}\left[\ln\left(\frac{J(J_0 - J_s)}{J_0(J - J_s)}\right) - J_s(1/J - 1/J_0)\right] \tag{12.15}$$

where K_4 is rate constant.

These integral forms of fouling models were practically applied in the work [92]. Several other models describing fouling and cake formation kinetics and corresponding codes in MATLAB are described in the book [5]. Simulation of membranes fouling by substances including micro/nanoplastics which cannot be removed from surface water without the use of reverse osmosis or nanofiltration unit is described in [95]. Fouling of commercial ultrafiltration poly(sulfone) membrane by micro/nano plastics was described by intermediate, complete pore blockage, and cake formation mechanisms. The adsorption rate of the nano- and microplastics onto the membrane surface was caused by hydrophobic interactions and surface repulsion forces mainly [87]. Fouling can be suppressed by proper modification of membrane characteristics such as hydrophilicity, surface charge, and surface morphology. The creation of the thin water layer on the surface of hydrophilic membranes suppresses hydrophobic adhesion particles due to repulsion. Similar effects have repulsive forces between the charged coagulants (ionic salts supporting microplastics coagulation-see Fig. 12.8) and the charged membrane surface. Narrower pore size distribution, larger porosity, and a smoother surface of membranes can be reasons for the suppression of fouling [6, 86]. Some inorganic fillers such as TiO_2, SiO_2, Al_2O_3, zeolites, and carbon-based materials were proposed as modifiers of the membrane for fouling suppressing [86, 96]. Advanced oxidative processes (AOPs) such as ozonation, photocatalysis, and advanced oxidation with hydrogen peroxide can be used for partial degradation and elimination of microplastics and suppression of fouling [97].

12.2.8 Membrane Filtration Models

There were proposed many of proposed for description of filtration of fluids through membranes [98–105]. Eulerian methods are based on the replacement of filtered fluid with a continuous medium, and Lagrangian methods are based on the replacement of filtered fluid by system of interacted particles, used e.g. for the creation of surface cake [106]. The particle size distribution of the particles and their interaction with the membrane are here fundamental. For characterization of porous membranes, their pore structure and the pore size distribution are used [82]. One computational model for the description of the diffusion of a hard sphere containing fluid through a membrane is presented in [82]. The membrane matrix here is approximated by a series of parallel layers containing circular pores with some spatial distribution.

This model was used for the prediction of the filtration kinetics influenced by the distribution of particle size and membrane pores. A simulation study based on the Brownian dynamics of normally distributed particle diameters d characterized by probability density function of $f(d)$ through a membrane with normally distributed pore radii r characterized by probability density function of $f(r)$ was realized. Distribution of suspended solid spheres before filtration was characterized by the mean value μ_s and standard deviation σ_s and pores were characterized by the mean value μ_p and standard deviation σ_p. The simulation program generated an initial configuration where the particles were randomly distributed in the first half of the simulation box (see Fig. 12.17) and were then interacting with membrane pores due to their Brownian motion and a constant driving force. Simulation was realized for 500 particles diffusing through a membrane composed from 5 parallel planes, after 5000 time units. It was found that the particles in the filtrate (passed through membrane) size distribution are bounded from the left by the minimal size of particles and from the right by the maximum size of pores. Reduction of the mean pore size leads to a narrower distribution of particles in the filtrate and shifts their distribution to the left. Filtration with membranes having less variation in porosity (lower σ_p) leads to a decrease of the polydispersity and the mean size of the particles in filtration. The simulation results predicted the mean size as a function of membrane pore parameters μ_p and σ_p. For a constant mean $\mu_s = 1$ and a standard deviation of particle size $\sigma_s = 0.3$ as shown in ref. [82].

The proposed simulation is versatile and can be easily extended to any kind of distribution according to the needs of designers [82]. The model can be extended to any process in which the statistical behavior of particles and membranes is at their disposal.

12.3 Conclusion

Membranes are now very advanced materials with different structures and properties. For their construction and modifications, the advanced nanomaterials and nanotechnologies are beneficial [86]. Membrane technologies (microfiltration, ultrafiltration, nanofiltration, and reverse osmosis) can be very effective for capturing micro and nanoplastics including fibrous forms. Pressure membrane processes use both porous and non-porous membranes, and their filtration properties are dependent on pore size distribution. There are effective solutions to suppress fouling due to the surface deposition of microplastics. For practical purposes, it is advantageous to combine pre-treatment as clarification (coagulation and flocculation) with microfiltration and reverse osmosis for complete removal whole spectrum of plastic particles, including fibrous ones. There are also some post-processing strategies on how to remove nano and microplastics by destruction processes like photodegradation [107, 108] and electro-oxidation [109, 110]. The exotic processes of nano and microplastics capturing by metal organic frameworks [111] and magnetic extraction [112] were proposed. Some nano and microplastics can also be removed, especially from drinking water containing calcium carbonate by co-precipitating upon boiling [113], which is one simple way to suppress their human intake.

References

1 Abeynayaka, A., Werellagama, I., Ngoc-Bao, P. et al. (2022). Microplastics in Wastewater Treatment Plants. *Current Developments in Biotechnology and Bioengineering*, 311–337. Elsevier.

2 Marek, J. (2021). State-of-the-art water treatment in Czech power sector: industry-proven case studies showing economic and technical benefits of membrane and other novel Technologies for each Particular Water Cycle. *Membranes (Basel)* 11 (2): 98.

3 Jelínek, L. (2009) Desalination and separation methods for water treatment.

4 Geise, G.M., Lee, H., Miller, D.J. et al. (2010). Water purification by membranes: the role of polymer science. *Journal of Polymer Science Part B: Polymer Physics* 48 (15): 1685–1718.

5 Foley, G. (2013). *Membrane Filtration: A Problem Solving Approach with MATLAB*. Cambridge: Cambridge University Press.

6 Ma, J., Dai, R., Chen, M. et al. (2018). Applications of membrane bioreactors for water reclamation: micropollutant removal, mechanisms and perspectives. *Bioresource Technology* 269: 532–543.

7 Periyasamy, A.P. (2025). Functionalized textile microplastics: a closer look at the issues, strategy, and legislation on the microplastic reduction. *Kuwait Journal of Science* 100395.

8 Periyasamy, A.P. and Tehrani-Bagha, A. (2022). A review on microplastic emission from textile materials and its reduction techniques. *Polymer Degradation and Stability* 199: 109901.

9 Periyasamy, A.P. (2023). Environmentally friendly approach to the reduction of microplastics during domestic washing: prospects for machine vision in microplastics reduction. *Toxics* 11 (7): 575.

10 Pabby, A.K.S., Wickramasinghe, R., and Sastre, A.-M. (ed.) (2024). *Handbook of Membrane Separations Chemical, Pharmaceutical, Food, and Biotechnological Applications*. CRC Press.

11 Okoffo, E.D., O'Brien, S., O'Brien, J.W. et al. (2019). Wastewater treatment plants as a source of plastics in the environment: a review of occurrence, methods for identification, quantification and fate. *Environmental Science: Water Research & Technology* 5 (11): 1908–1931.

12 Ma, B., Xue, W., Hu, C. et al. (2019). Characteristics of microplastic removal via coagulation and ultrafiltration during drinking water treatment. *Chemical Engineering Journal* 359: 159–167.

13 Sun, J., Dai, X., Wang, Q. et al. (2019). Microplastics in wastewater treatment plants: detection, occurrence and removal. *Water Research* 152: 21–37.

14 Pabby, A.K.S., Wickramasinghe, R., and Sastre, A.-M. (ed.) (2024). *Handbook of Membrane Separations Chemical, Pharmaceutical, Food, and Biotechnological Applications*. CRC Press.

15 Ryberg, M.W., Hauschild, M.Z., Wang, F. et al. (2019). Global environmental losses of plastics across their value chains. *Resources, Conservation and Recycling* 151: 104459.

16 Tabraiz, S., Zeeshan, M., Asif, M.B. et al. (2023). Chapter 11 – hybrid membrane bioreactors for wastewater treatment. In: *Current Developments in Biotechnology and Bioengineering* (ed. X.-T. Bui, W. Guo, C. Chiemchaisri, and A. Pandey), 239–265. Elsevier.

17 Wu, P., Huang, J., Zheng, Y. et al. (2019). Environmental occurrences, fate, and impacts of microplastics. *Ecotoxicology and Environmental Safety* 184: 109612.

18 Rochman, C.M. et al. (2019). Rethinking microplastics as a diverse contaminant suite. *Environmental Toxicology and Chemistry* 38 (4): 703–711.

19 Gigault, J., ter Halle, A., Baudrimont, M. et al. (2018). Current opinion: what is a nanoplastic? *Environmental Pollution* 235: 1030–1034.

20 Zhang, Y.-Q., Lykaki, M., Alrajoula, M.T. et al. (2021). Microplastics from textile origin – emission and reduction measures. *Green Chemistry* 23 (15): 5247–5271.

21 Koelmans, A.A., Mohamed Nor, N.H., Hermsen, E. et al. (2019). Microplastics in freshwaters and drinking water: critical review and assessment of data quality. *Water Research* 155: 410–422.

22 Anonym: Background document to RAC and SEAC opinions on intentionally added microplastics. (2020). https://echa.europa.eu/hot-topics/microplastics.

23 Meloun, M. and Militký, J. (2011). 2 – the exploratory and confirmatory analysis of univariate data. In: *Statistical Data Analysis* (ed. M. Meloun and J. Militký), 25–71. Woodhead Publishing India.

24 Williams, W.A. and Aravamudhan, S. (2024). Micro-nanoparticle characterization: establishing underpinnings for proper identification and nanotechnology-enabled remediation. *Polymers (Basel)* 16 (19): 2837.

25 Newman, M. (2005). Power laws, Pareto distributions and Zipf's law. *Contemporary Physics* 46 (5): 323–351.

26 Park, H.-J., Oh, M.-J., Kim, P.-G. et al. (2020). National reconnaissance survey of microplastics in municipal wastewater treatment plants in Korea. *Environmental Science & Technology* 54 (3): 1503–1512.

27 Meloun, M. and Militký, J. (2011). 3 – statistical analysis of univariate data. In: *Statistical Data Analysis* (ed. M. Meloun and J. Militký), 73–150. Woodhead Publishing India.

28 Pinto, P.E., Giacobbo, A., de Almeida, G.M. et al. (2025). Pressure-driven membrane processes for removing microplastics. *Membranes (Basel)* 15 (3): 81.

29 Kayvani Fard, A., McKay, G., Buekenhoudt, A. et al. (2018). Inorganic membranes: preparation and application for water treatment and desalination. *Materials* 11 (1): 74.

30 Nadda, A.K., Banerjee, P., Sharma, S., and Nguyen-Tri, P. (ed.) (2023). *Membranes for Water Treatment and Remediation*. Singapore: Springer Nature Singapore.

31 Teoh, M.M. and Chung, T.-S. (2009). Membrane distillation with hydrophobic macrovoid-free PVDF–PTFE hollow fiber membranes. *Separation and Purification Technology* 66 (2): 229–236.

32 Nagy, E. (2012). *Basic Equations of the Mass Transport through a Membrane Layer*. Elsevier.

33 Bird, R.B., Lightfoot, E.N., and Stewart, W.E. (2006). *Transport Phenomena*. New York: Wiley.

34 Nagy, E. (2012). *Basic Equations of the Mass Transport through a Membrane Layer*. Elsevier.

35 Roy, S., Raju, R., Chuang, H.F. et al. (2003). Modeling gas flow through microchannels and nanopores. *Journal of Applied Physics* 93 (8): 4870–4879.

36 Ratnakar, R.R. and Dindoruk, B. (2022). The role of diffusivity in oil and gas industries: fundamentals, measurement, and correlative techniques. *PRO* 10 (6): 1194.

37 Covaliu-Mierlă, C.I., Matei, E., Stoian, O. et al. (2022). TiO_2–based nanofibrous membranes for environmental protection. *Membranes (Basel)* 12 (2): 236.

38 Everett, D.H. (2003) Definitions, terminology, and symbols in colloid and surface chemistry. *Chemistry International – Newsmagazine for IUPAC*, **25** (1).

39 Wu, D., Xu, F., Sun, B. et al. (2012). Design and preparation of porous polymers. *Chemical Reviews* 112 (7): 3959–4015.

40 Davletbaeva, I.M. and Sazonov, O.O. (2024). Macromolecular architecture in the synthesis of micro- and mesoporous polymers. *Polymers (Basel)* 16 (23): 3267.

41 Kahrs, C. and Schwellenbach, J. (2020). Membrane formation via non-solvent induced phase separation using sustainable solvents: a comparative study. *Polymer (Guildf)* 186: 122071.

42 Geise, G.M., Lee, H., Miller, D.J. et al. (2010). Water purification by membranes: the role of polymer science. *Journal of Polymer Science Part B: Polymer Physics* 48 (15): 1685–1718.

43 Sen, A.K. (2007). *Coated Textiles*. CRC Press.

44 Hu, J. (2016). *Active Coatings for Smart Textiles*. Elsevier.

45 Lukáš, D., Sarkar, A., Martinová, L. et al. (2009). Physical principles of electrospinning (electrospinning as a nano-scale technology of the twenty-first century). *Textile Progress* 41 (2): 59–140.

46 Park, H.B., Jung, C.H., Lee, Y.M. et al. (2007). Polymers with cavities tuned for fast selective transport of small molecules and ions. *Science (1979)* 318 (5848): 254–258.

47 Zou, X. and Zhu, G. (2019). *Microporous Materials for Separation Membranes*. Wiley.

48 Zhang, C., Li, P., and Cao, B. (2015). Electrospun microfibrous membranes based on PIM-1/POSS with high oil wettability for separation of oil–water mixtures and cleanup of oil soluble contaminants. *Industrial and Engineering Chemistry Research* 54 (35): 8772–8781.

49 Mishra, R., Militky, J., Baheti, V. et al. (2014). The production, characterization and applications of nanoparticles in the textile industry. *Textile Progress* 46 (2): 133–226.

50 Tyagi, R.D., Pandey, A., Drogui, P. et al. (ed.) (2022). *Current Developments in Biotechnology and Bioengineering Microplastics and Nanoplastics: Occurrence, Environmental Impacts and Treatment Processes.* Elsevier.

51 Zhou, G., Liu, R., Xu, Q. et al. (2022). Dual-structure PVDF/SDS nanofibrous membranes for highly efficient personal protection in mines. *Membranes (Basel)* 12 (5): 482.

52 Sanguanpak, S., Chiemchaisri, W., and Chiemchaisri, C. (2019). Membrane fouling and micro-pollutant removal of membrane bioreactor treating landfill leachate. *Reviews in Environmental Science and Biotechnology* 18 (4): 715–740.

53 Ruiz-García, A., Melián-Martel, N., and Nuez, I. (2017). Short review on predicting fouling in RO desalination. *Membranes (Basel)* 7 (4): 62.

54 Ngo, P.L., Pramanik, B.K., Shah, K., and Roychand, R. (2019). Pathway, classification and removal efficiency of microplastics in wastewater treatment plants. *Environmental Pollution* 255: 113326.

55 Murphy, F., Ewins, C., Carbonnier, F., and Quinn, B. (2016). Wastewater treatment works (WwTW) as a source of microplastics in the aquatic environment. *Environmental Science & Technology* 50 (11): 5800–5808.

56 Talvitie, J., Heinonen, M., Pääkkönen, J.-P. et al. (2015). Do wastewater treatment plants act as a potential point source of microplastics? Preliminary study in the coastal Gulf of Finland, Baltic Sea. *Water Science and Technology* 72 (9): 1495–1504.

57 Carr, S.A., Liu, J., and Tesoro, A.G. (2016). Transport and fate of microplastic particles in wastewater treatment plants. *Water Research* 91: 174–182.

58 Teh, C.Y., Budiman, P.M., Shak, K.P.Y., and Wu, T.Y. (2016). Recent advancement of coagulation–flocculation and its application in wastewater treatment. *Industrial and Engineering Chemistry Research* 55 (16): 4363–4389.

59 Enfrin, M., Dumée, L.F., and Lee, J. (2019). Nano/microplastics in water and wastewater treatment processes – origin, impact and potential solutions. *Water Research* 161: 621–638.

60 Magalhães, S., Alves, L., Medronho, B. et al. (2020). Microplastics in ecosystems: from current trends to bio-based removal strategies. *Molecules* 25 (17): 3954.

61 Allé, P.H., Garcia-Muñoz, P., Adouby, K. et al. (2021). Efficient photocatalytic mineralization of polymethylmethacrylate and polystyrene nanoplastics by TiO_2/β-SiC alveolar foams. *Environmental Chemistry Letters* 19 (2): 1803–1808.

62 Liao, Y., Loh, C.-H., Tian, M. et al. (2018). Progress in electrospun polymeric nanofibrous membranes for water treatment: fabrication, modification and applications. *Progress in Polymer Science* 77: 69–94.

63 Obotey Ezugbe, E. and Rathilal, S. (2020). Membrane technologies in wastewater treatment: a review. *Membranes (Basel)* 10 (5): 89.

64 Talvitie, J., Mikola, A., Koistinen, A., and Setälä, O. (2017). Solutions to microplastic pollution – removal of microplastics from wastewater effluent with advanced wastewater treatment technologies. *Water Research* 123: 401–407.

65 Pivokonsky, M., Cermakova, L., Novotna, K. et al. (2018). Occurrence of microplastics in raw and treated drinking water. *Science of the Total Environment* 643: 1644–1651.

66 Marek, J. (2023). Microplastics and nanoplastics released from textiles and their seperation from water. Habilitation Thesis, Technological University of Liberec, Liberec, Czech Republic, 2022.

67 Keucken, A., Heinicke, G., Persson, K., and Köhler, S. (2017). Combined coagulation and ultrafiltration process to counteract increasing NOM in brown surface water. *Water (Basel)* 9 (9): 697.

68 Khulbe, K.C. and Matsuura, T. (2021). *Nanotechnology in Membrane Processes*. Cham: Springer International Publishing.

69 Militký, J., Venkataraman, M., and Periyasamy, A.P. (ed.) (2023). *Fibrous Structures and their Impact on Textile Design*. Singapore: Springer Nature Singapore.

70 Pandey, B., Pathak, J., Singh, P. et al. (2022). Microplastics in the ecosystem: an overview on detection, removal, toxicity assessment, and control release. *Water (Basel)* 15 (1): 51.

71 Baker, R.W. (2012). *Membrane Technology and Applications*. Wiley.

72 Mohammadifakhr, M., de Grooth, J., Roesink, H.D.W., and Kemperman, A.J.B. (2020). Forward osmosis: a critical review. *Processes* 8 (4): 404.

73 Duan, J., Pan, Y., Pacheco, F. et al. (2015). High-performance polyamide thin-film-nanocomposite reverse osmosis membranes containing hydrophobic zeolitic imidazolate framework-8. *Journal of Membrane Science* 476: 303–310.

74 Tian, X., Wang, J., Zhang, H. et al. (2018). Establishment of transport channels with carriers for water in reverse osmosis membrane by incorporating hydrotalcite into the polyamide layer. *RSC Advances* 8 (22): 12439–12448.

75 Lukáš, D., Sarkar, A., Martinová, L. et al. (2009). Physical principles of electrospinning (electrospinning as a nano-scale technology of the twenty-first century). *Textile Progress* 41 (2): 59–140.

76 Ahmad, A. and Alshammari, M.B. (ed.) (2023). *Nanofiltration Membrane for Water Purification*. Singapore: Springer Nature Singapore.

77 Venkataraman, M., Yang, K., Xiong, X. et al. (2020). Preparation of electrosprayed, microporous particle filled layers. *Polymers (Basel)* 12 (6): 1352.

78 Strathmann, H. (2004). Chapter 3 – preparation and characterization of ion-exchange membranes. In: *Membrane Science and Technology*, vol. 9, 89–146. Elsevier.

79 Neděla, D., Křivčík, J., Válek, R. et al. (2014). Influence of water content on properties of a heterogeneous bipolar membrane. *Desalination and Water Treatment* 56 (12): 1–4.

80 Marek, J., Čížek, J., and Tvrzník, D. (2019). Optimizing porous material in shock electrodialysis unit. *Desalination and Water Treatment* 170: 38–45.

81 Schlumpberger, S. (2016) Dissertation on Shock electrodialysis for water purification and electrostatic correlations in electrolytes. Massachusetts Institute of Technology.

82 Marrufo-Hernández, N.A., Hernández-Guerrero, M., Nápoles-Duarte, J.M. et al. (2018). Prediction of the filtrate particle size distribution from the pore size distribution in membrane filtration: numerical correlations from computer simulations. *AIP Advances* 8 (3): 035308.

83 Tadsuwan, K. and Babel, S. (2022). Microplastic abundance and removal via an ultrafiltration system coupled to a conventional municipal wastewater treatment plant in Thailand. *Journal of Environmental Chemical Engineering* 10 (2): 107142.

84 Strungaru, S.-A., Jijie, R., Nicoara, M. et al. (2019). Micro- (nano) plastics in freshwater ecosystems: abundance, toxicological impact and quantification methodology. *TrAC Trends in Analytical Chemistry* 110: 116–128.

85 Chae, Y. and An, Y.-J. (2017). Effects of micro- and nanoplastics on aquatic ecosystems: current research trends and perspectives. *Marine Pollution Bulletin* 124 (2): 624–632.

86 Golgoli, M., Khiadani, M., Shafieian, A. et al. (2021). Microplastics fouling and interaction with polymeric membranes: a review. *Chemosphere* 283: 131185.

87 Enfrin, M., Lee, J., Le-Clech, P., and Dumée, L.F. (2020). Kinetic and mechanistic aspects of ultrafiltration membrane fouling by nano- and microplastics. *Journal of Membrane Science* 601: 117890.

88 Xu, H., Xiao, K., Yu, J. et al. (2020). A simple method to identify the dominant fouling mechanisms during membrane filtration based on piecewise multiple linear regression. *Membranes (Basel)* 10 (8): 171.

89 Silva, C.M., Reeve, D.W., Husain, H. et al. (2000). Model for flux prediction in high-shear microfiltration systems. *Journal of Membrane Science* 173 (1): 87–98.

90 Hermia, J. (1982). Constant pressure blocking filtration laws – application to power-law non-Newtonian fluids. *Trans. IChemE* 60: 183–187.

91 Cheng, Y.-L., Lee, D.-J., and Lai, J.-Y. (2011). Filtration blocking laws: revisited. *Journal of the Taiwan Institute of Chemical Engineers* 42 (3): 506–508.

92 Field, R.W., Wu, D., Howell, J.A., and Gupta, B.B. (1995). Critical flux concept for microfiltration fouling. *Journal of Membrane Science* 100 (3): 259–272.

93 Jarusutthirak, C., Mattaraj, S., and Jiraratananon, R. (2007). Influence of inorganic scalants and natural organic matter on nanofiltration membrane fouling. *Journal of Membrane Science* 287 (1): 138–145.

94 Corbatón-Báguena, M.-J., Álvarez-Blanco, S., and Vincent-Vela, M.-C. (2015). Fouling mechanisms of ultrafiltration membranes fouled with whey model solutions. *Desalination* 360: 87–96.

95 Marek, J., Šimunkova, M., and Parschova, H. (2014). Clogging of the electrodeionization chamber. *Desalination and Water Treatment* 56 (12): 1–5.

96 Enfrin, M., Wang, J., Merenda, A. et al. (2021). Mitigation of membrane fouling by nano/microplastics via surface chemistry control. *Journal of Membrane Science* 633: 119379.

97 Sacco, N.A., Zoppas, F.M., Devard, A. et al. (2023). Recent advances in microplastics removal from water with special attention given to photocatalytic degradation: review of scientific research. *Microplastics* 2 (3): 278–303.

98 Bowen, W.R. and Jenner, F. (1995). Theoretical descriptions of membrane filtration of colloids and fine particles: an assessment and review. *Advances in Colloid and Interface Science* 56: 141–200.

99 Ghidossi, R., Veyret, D., and Moulin, P. (2006). Computational fluid dynamics applied to membranes: state of the art and opportunities. *Chemical Engineering and Processing: Process Intensification* 45 (6): 437–454.

100 Kumar, V. and Upadhyay, S.N. (2000). Computer simulation of membrane processes: ultrafiltration and dialysis units. *Computers and Chemical Engineering* 23 (11–12): 1713–1724.

101 Saiz, L. and Klein, M.L. (2002). Computer simulation studies of model biological membranes. *Accounts of Chemical Research* 35 (6): 482–489.

102 Chen, J.C. and Kim, A.S. (2006). Monte Carlo simulation of colloidal membrane filtration: principal issues for modeling. *Advances in Colloid and Interface Science* 119 (1): 35–53.

103 Chen, J., Elimelech, M., and Kim, A. (2005). Monte Carlo simulation of colloidal membrane filtration: model development with application to characterization of colloid phase transition. *Journal of Membrane Science* 255 (1–2): 291–305.

104 Kim, A.S. and Hoek, E.M.V. (2002). Cake structure in dead-end membrane filtration: Monte Carlo simulations. *Environmental Engineering Science* 19 (6): 373–386.

105 Ariza, M.J. and Puertas, A.M. (2009). Colloidal permeability of liquid membranes consisting of hard particles by nonequilibrium simulations. *The Journal of Chemical Physics* 131 (16): 164903.

106 Wang, J., Dlamini, D.S., Mishra, A.K. et al. (2014). A critical review of transport through osmotic membranes. *Journal of Membrane Science* 454: 516–537.

107 Miao, F., Liu, Y., Gao, M. et al. (2020). Degradation of polyvinyl chloride microplastics via an electro-Fenton-like system with a TiO_2/graphite cathode. *Journal of Hazardous Materials* 399: 123023.

108 Luo, H., Zeng, Y., Zhao, Y. et al. (2021). Effects of advanced oxidation processes on leachates and properties of microplastics. *Journal of Hazardous Materials* 413: 125342.

109 Kiendrebeogo, M., Karimi Estahbanati, M.R., Khosravanipour Mostafazadeh, A. et al. (2021). Treatment of microplastics in water by anodic oxidation: a case study for polystyrene. *Environmental Pollution* 269: 116168.

110 Durán, F.E., de Araújo, D.M., do Nascimento Brito, C. et al. (2018). Electrochemical technology for the treatment of real washing machine effluent at pre-pilot plant scale by using active and non-active anodes. *Journal of Electroanalytical Chemistry* 818: 216–222.

111 Chen, Y.-J., Chen, Y., Miao, C. et al. (2020). Metal–organic framework-based foams for efficient microplastics removal. *Journal of Materials Chemistry A* 8 (29): 14644–14652.

112 Grbic, J., Nguyen, B., Guo, E. et al. (2019). Magnetic extraction of microplastics from environmental samples. *Environmental Science & Technology Letters* 6 (2): 68–72.

113 Yu, Z., Wang, J.-J., Liu, L.-Y. et al. (2024). Drinking boiled tap water reduces human intake of nanoplastics and microplastics. *Environmental Science & Technology Letters* 11 (3): 273–279.

Index

Textile Microplastics in the Environment: Sources, Impacts, and Solutions,
First Edition. Aravin Prince Periyasamy.

Printed and bound by CPI Group (UK) Ltd, Croydon, CR0 4YY

17/09/2026

14964330-0002